Elements of
REFRACTORY TECHNOLOGY

Elements of REFRACTORY TECHNOLOGY

Om Prakash Gupta
B. Tech. (Chemical Engg.)
Ex DGM, HRD
Bokaro Steel Plant
Steel Authority of India Limited (SAIL)

KHANNA BOOK PUBLISHING CO. (P) LTD.
Publisher of Engineering and Computer Books
4C/4344, Ansari Road, Darya Ganj, New Delhi-110002
Phone : 011-23244447-48 **Mobile:** +91-9910909320
E-mail : contact@khannabooks.com
Website : www.khannabooks.com

Price: ₹ 299.00

ELEMENTS OF REFRACTORY TECHNOLOGY

O.P. GUPTA

Copyright © Khanna Book Publishing Co. (P) Ltd.

This book is sold subject to the condition that it shall not, by way of trade or otherwise, be lent, resold, hired out, or otherwise circulated without the publisher's prior consent in any form of binding or cover other than that in which it is published and without a similar condition including this condition being imposed on the subsequent purchaser and without limiting the rights under copyright reserved above, no part of this publication may be reproduced, stored in or introduced into retrieval system, or transmitted any form or by any means (electronic, mechanical, photocopying, recording or otherwise), without the prior written permission of both the copyright owner and the above mentioned publisher of this book.

ISBN: 978-93-86173-64-5

First Edition: 2017
Reprint: 2019

Published by:

Khanna Book Publishing Co. (P) Ltd.
4C/4344, Ansari Road, Darya Ganj, New Delhi-110 002
Phone: 011-23244447-48 **Mobile:** +91-9910909320
E-mail: contact@khannabooks.com

Printed in India by:
S.P.S. Printers & Binders, Delhi

Preface

Refractory is an integral part of the degree & diploma curriculum of various branches of engineering. Besides, it is a compulsory paper for various associate membership examinations conducted by professional bodies like Institution of Engineers(AMIE), Indian Institute of Metals (AMIIM), Indian Institute of Chemical Engineers (AMIIChE) *etc.* This book has been prepared strictly as per the syllabus of these examinations.

This book contains detailed description of the following topics:

- Introduction to refractories
- Manufacture, Properties & Applications of Important Refractories
- Refractories used in various Furnaces/Industries
- Properties and Testing of Refractories
- Phase Equilibrium in Refractory materials
- Insulating refractories/Ceramic fibres
- Special Refractories
- Refractory Cements/Mortars, Castables and Ramming Masses

Besides short questions & answers and multiple choice questions & answers drawn from the examination papers of various engineering colleges and professional bodies examinations given at the end of the book enhances its utility for the students.

Subject matter has been presented in simple, lucid and easy to understand language.

I hope that the book in its present form will be more useful for the students.

Constructive suggestion for the improvement of the book are welcome.

Om Prakash Gupta

Preface

[illegible] is an integral part of the degree & diploma [illegible] [illegible]. [illegible] [illegible] Institution of Engineers [illegible] [illegible] Institute of [illegible] [illegible] this book [illegible] [illegible] of the [illegible].

The textbook has been divided [illegible] of the following [illegible]:

- Introduction [illegible]
- [illegible] Properties & Applications of Important [illegible]
- [illegible] used in Various [illegible]
- [illegible] and Techniques [illegible]
- [illegible]
- [illegible]
- [illegible]
- [illegible]

Besides [illegible] questions and [illegible] the end of the book [illegible].

[illegible]

[illegible]

[illegible] for the improvement of the book [illegible].

[illegible] Gupta

CONTENTS

Introduction to Refractories

Definition. Refractories are materials which withstand high temperature and do not fuse even at a very high temperature e.g. fire clay, silica, chromite, magnesite *etc.*

ASTM defines refractories as "non-metallic materials having those chemical & physical properties that make them applicable for structures or as components of systems that are exposed to environments above 538°C.

Refractories are inorganic nonmetallic material which can withstand high temperature without undergoing physical or chemical changes while remaining in contact with molten slag, metal and gases

Refractories perform four basic functions:

1. They act as a thermal barrier between a hot medium (e.g., flue gases, liquid metal, molten slags, and molten salts) and the wall of the containing vessel.
2. They insure a strong physical protection, preventing the erosion of walls by the circulating hot medium.
3. They represent a chemical protective barrier against corrosion.
4. They act as thermal insulation, insuring heat retention.

The principal raw materials used in the production of refractories are: the oxides of silicon, aluminum, magnesium, calcium & zirconium and some non-oxide refractories like carbides, nitrides, borides, silicates & graphite.

What are refractories used for?

- Refractories are used by the metallurgical industry in the internal linings of furnaces, kilns, reactors and other vessels for holding and transporting metal & slag.
- In non-metallurgical industries, the refractories are mostly installed on fired heaters, hydrogen reformers, ammonia primary & secondary reformers, cracking furnaces, utility boilers, catalytic cracking units, coke calciner, sulfur melting furnaces, air heaters, ducting, stacks, *etc.*
- Majority of these listed equipment operate under high pressure, and operating temperature can vary from very low to very high (approximately 540 to 1600°C).
- The refractory materials are therefore needed to withstand temperatures over and above these temperatures.

- Due to the extremely high melting point of common metals like iron, nickel and copper, metallurgists have to raise furnace temperatures to over 1600°C.

Refractories are used in furnace for the following purposes:

- To protect the steel structure.
- To reduce the heat losses.
- To improve the heating efficiency of the furnace.
- To improve the furnace life.

Classification of refractories

Refractories can be classified on the basis of chemical composition , refractoriness(fusion point) , the methods of manufacture and physical form.

Classification Based on Chemical Composition: Three types of refractory materials namely acidic,basic & neutral refractory exist depending upon their chemical nature as given below in the table.

Table 1.1: Classification of Refractory Based on Chemical Composition

Type of refractory material	Examples	Remarks
Acidic refractory (silica is chief constituent) which readily combines with bases.	Fire clay, silica, Quartz, ganister sand, semi-silica, Alumino- silicate *etc.*	It should not be allowed to come in contact with basic products to avoid the reaction. These are used in areas where slag and atmosphere are acidic.
Basic refractory (consists of basic oxide without free silica)	Bauxite, magnesite dolomite, (lime is not used as refractory due to its dehydrating tendency), Alumina, Zirconia *etc.*	Should not come in contact with acidic products. They are attacked by acid slags but stable to alkaline slags, dusts & fumes at elevated temperatures.
Neutral refractory which does not combine with acids or bases.	Chromite, graphite carbon refractory, carbide, mullite, Kyanite, pure alumina	Can be used in conjunction with acidic or basic products. They are chemically stable to both acids & bases and are used where slag & atmosphere are either acidic or basic.

Classification based on refractoriness (fusion points).

Refractories are also classified based on their refractoriness as given below:

Table 1.2: Classification of Refractories Based on Refractoriness

Type of Refractory	PCE Value (Segar Cones)	Refractoriness/ fusion point, °C
Low heat duty refractories (LHD)	19—28	1520—1630
Medium/Intermediate heat duty refractories (MHD)	28—30	1630—1670
High heat duty refractories (HHD)	30—33	1670—1730
Super duty refractories (SD)	> 33	> 1730

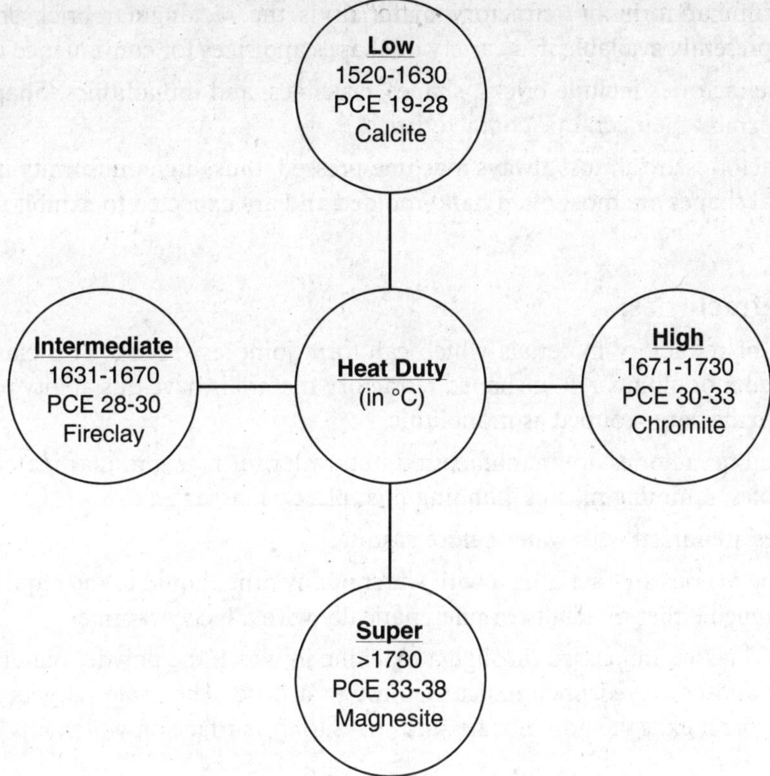

Fig.1.1: Classification of refractories based on refractoriness /fusion points

Classification based on method of refractory manufacture.

Based on the method of manufacture, refractories are classified as:

- Dry pressed
- Fusion cast or electro cast
- Hand moulded
- *Formed* (Normal, fired/burned or chemically bonded)
- *Unformed* (Monolithics-plastics, Ramming mass, Gunning, Castable, Spraying)

Classification Based on Physical Form of refractory

Refractories are classified according to their physical form. Thesc arc the shaped and unshaped refractories. The former is commonly known as refractory bricks and the latter as "monolithic" refractories.

Shaped Refractories

Shaped refractories are those which have fixed shaped when delivered to the user. These are what we call bricks. Brick shapes may be divided into two: standard shapes and special shapes. Standards shapes have dimension that are conformed to by most refractory manufacturers and are generally applicable to kilns and furnaces of the same type. Special shapes are specifically made for particular kilns and furnaces. This may not be applicable to another furnaces or kiln of the same type.

The most familiar form of refractory materials is the rectangular brick shape. However, refractories are presently available in a variety of shapes and sizes for convenience in construction.

Preshaped refractories include bricks, shapes, crucibles, and monolithics. Shaped refractories are pre-fired to exhibit their ceramic characteristics.

Shaped refractories are almost always machine-pressed, thus, high uniformity in properties are expected. Special shapes are most often hand-molded and are expected to exhibit slight variations in properties.

Unshaped Refractories:

There is a class of refractory materials which can form joint less lining. This class of refractory materials is called monolithic. All unshaped refractory materials have this ability to form jointless lining, and hence they are grouped as monolithic.

- Unshaped refractories are manufactured in powder form as granular material and known as castables, ramming masses, gunning mix, plastic masses, *etc.*
- Castables are mixed with water before casting.
- Ramming masses are first mixed with water or any other liquid to the required quality, and then rammed either manually or pneumatically with a heavy rammer.
- Gunning masses are passed through a machine in which the powder material is put under pressure and conveyed pneumatically through a hose. The material gets mixed up with water before it exits the hose nozzle, and sticks to the surface on which it is applied to form a lining.
- Plastic masses comprise ready-mix material that is applied manually in the furnace to form a lining.

Unshaped refractories are without definite form and are only given shape upon application. It forms joint less lining and are better known as monolithic refractories. These are categorized as Plastic refractories, ramming mixes, castables, gunning mixes, fettling mixes and mortars.

The manufacture of unshaped refractories differs slightly from shaped refractories. Unshaped refractories typically do not go through a firing process until they reach the final consumer. These unshaped refractories can be installed by spraying, casting, molding, or ramming.

Examples of shaped & unshaped refractories are tabulated below.

Table 1.3: Examples of shaped & unshaped refractories

SHAPED REFRACTORIES	UNSHAPED REFRACTORIES
1. Alumino Silicates • High Alumina • Low Alumina **2. Silica** **3. Basic** • Magnesite • Mag-Chrome • Mag-Carbon • Forsterite • Dolomite **4. Carbon** **5. Zirconia** • Zircon (Acidic) • Zirconia (Neutral) **6. Carbides/Nitrides** • Silicon Carbide • Boron Nitride **7. Pure Oxides** • Ferrites • Hofnium Oxides Boron	1. Mortars 2. Castables • LCC • ULCC • NCC • SFC 3. Gunning Mass 4. Ramming Mass 5. Filling Mass 6. Fettling Mass 7. Spraying Compound 8. Patching/Coating Mass

Table below lists each type of shaped & unshaped refractory and a description of its use.

Table 1.4: Types and Descriptions of Refractories uses

Types of refractories	Description of its use
Shaped Refractories	
Bricks	Refractories that have shapes and are used to line furnaces, kilns, glass tanks, incinerators, *etc.*
Insulating firebrick	Low thermal conductivity firebrick.
Unshaped Refractories (Monolithic)	
Mortar	Materials for bonding bricks in a lining. The three types of mortar—heat-setting, air-setting, and hydraulic-setting—have different setting mechanisms.
Castables	Refractories for which raw materials and hydraulic-setting cement are mixed. They are formed by casting and used to line furnaces, kilns, *etc.*
Plastics	Refractories in which raw materials and plastic materials are mixed with water. Plastic refractories are roughly formed, sometimes with chemical additives.

Gunning mixes	Refractories that are sprayed on the surface by a gun.
Ramming mixes	Granular refractories that are strengthened by gunning formulation of a ceramic bond after heating. Ramming mixes have less plasticity and are installed by an air rammer.
Slinger mixes	Refractories installed by a slinger machine.
Patching materials/ coating materials	Refractories with properties similar to refractory mortar. However, patching materials have controlled grain size for easy patching or coating.
Lightweight castables	Refractories in which porous lightweight materials and hydraulic cement are mixed. They are mixed with water and formed by casting. Lightweight castables are used to line furnaces, kilns, *etc.*
Fibrous Materials	
Ceramic fibre	Man-made fibrous refractory materials. There are several different types of ceramic fibre, including blanket, felt, module, vacuum form, rope, loo fiber, *etc.*

ACID REFRACTORIES

Those refractories which are attacked by basic slags (i.e. bases) are called acidic refractories and include:

(a) **Aluminium-silicate refractories** like fireclay bricks, high alumina bricks, kyanite, sillimanite, bauxite, diaspore, mullite, andalusite *etc.* These refractories vary widely in their physical, chemical and mineralogical characteristics depending on the nature and proportion of silica and alumina present in them. Free silica (SiO_2) should be as little as possible as it lowers the fusion point of these refractories. Addition of grog (ground pre-fired fireclay grains) improves the shrinkage properties of these refractories.

(b) **Silica refractories** consisting mainly of silica (and very low in alkalis and metallic oxide). Raw materials of various grades of silica refractories (containing silica upto 97%, alkalis $< 0.3\%$ and alumina $< 10\%$) include natural rock, prepared mass quartz sand (ganister), quartzite, electrically fused quartz *etc.*

BASIC REFRACTORIES

Those refractories which are attacked by acid slags (acid) are called basic refractories and include:

(a) **Alumina (Al_2O_3)** refractories manufactured from burnt bauxite, electrocast bauxite or electro fused corundum.

(b) **Dolomite refractories** produced by calcining natural dolomite which is mainly a mixture of lime and magnesia.

(c) **Magnesite refractories** produced from dead burnt magnesite or magnesia (MgO).

(d) **Chrome-magnesite and Magnesite-chrome refractories** which are mixture of calcined magnesite and chromium oxide in various proportion necessary to develop required properties in them.

NEUTRAL REFRACTORIES

Those refractories which are not attacked either by basic slag/bases or acid slag/ acids are called neutral refractories and include.

(a) chromite, carbon and graphite refractories

(b) artificially produced refractories like silicon carbide, zirconium carbide, titanium carbide *etc.*

(c) Metals used as refractory like thorium, tungsten, zirconium, platinum, nickel, iron, molybdenum, copper, vanadium, tantalum *etc.*

Alumino-silicate refractories are sometimes classified as neutral, refractories but when exposed to basic slag, they exhibit acidic reaction.

MONOLITHIC REFRACTORIES

Monolithic refractory, the name generally given to all unshaped refractory products, are materials installed as some form of suspension that ultimately harden to form a solid mass. Most monolithic formulations consist of large refractory particulates (an aggregate), fine filler materials (which fill the interparticle voids) and a binder phase (that gels the particulates together in the green state).

Monolithic refractories are replacing the conventional type fired refractories at a much faster rate in many applications including those of industrial furnaces.

The main advantages of monolithic refractory are:

- It eliminates joints which is an inherent weakness.
- Method of application is faster and skilled measures in large number are not required.
- Properties can be better than pressed bricks.
- Transportation and handling are simple.
- Offers considerable scope to reduce inventory and eliminate special shapes.
- Offers better scope to reduce downtime for repairs ,
- Has better spalling resistance and volume stability.
- It is a heat saver.
- Ability to install in "Hot Standby" mode.

Monolithic refractories (single piece cast in the shape of equipment such as one for a ladle shown in figure below) are replacing the conventional type fired refractories at a much faster rate in many applications including those of industrial furnaces.

Various means are employed in the placement of monolithics like ramming, casting, gunniting, spraying, sand slinging, *etc.* Ramming masses are used mostly in cold applications where proper consolidation of the material is important. The same practice can be adopted with both air setting and heat setting materials. Proper ramming tools need to be selected.

Castables by name implies a material of hydraulic setting in nature. Calcium aluminate cement being the binder, it will have to be stored properly to prevent moisture absorption. Further its strength starts deteriorating after a period of 6 to 12 months.

Various means are employed in the placement of monolithic refractories like ramming, casting, gunniting, spraying, sand slinging *etc.* Ramming masses are used mostly in cold applications where proper consolidation of the material is important. The same practice can be adopted with both air setting and heat setting materials. Proper ramming tools need to be selected.

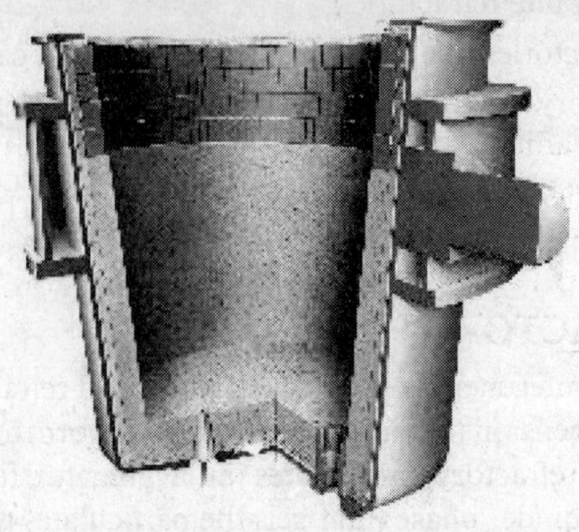

Fig.1.2: Monolithic lining for a ladle

Types of Monolithic Refractories

Castable Refractories: 'Castable' by name implies a material of hydraulic setting in nature. These are the materials that contain cement binder usually aluminate cement, which imparts hydraulic setting properties when mixed with water. Following the heat-up of the material the binder either transforms or volatilises facilitating the formation of a ceramic bond. The most common binder used in castables is HAC (high alumina cement). Other binders that are often used include hydratable aluminas and colloidal silica. These materials are installed by casting and are also known as refractory concretes.

Insulating Castables: Insulating castables are specialised monolithic refractories that are used on the cold face of applications. There are made from lightweight aggregate materials such as vermiculite, perlite, extend-o-spheres, bubble alumina and expanded clay. Their main function is to provide thermal insulation. They are typically of low density and low thermal conductivity. Insulating refractories have inferior mechanical strength to that of conventional castables.

Plastic Refractories: Plastic refractory is mixtures that is prepared in stiff plastic condition and are delivered in blocks wrapped in polyethylene. During application, the blocks are sliced into pieces and without further preparation, are rammed or poured into place with pneumatic rammer. Plastic are easily rammed to any shape or contour.

Ramming Mixes: Ramming refractory materials are very similar to plastic refractories but are much stiffer. The particle sizes are carefully graded and the final product is usually delivered dry and then mixed with a little amount of water just before application. Other ramming mixes are delivered in wet form and are ready for use immediately upon opening. Application is done with pneumatic rammer.

Patching Refractories: These materials are similar to plastic refractories though have a very soft plasticity allowing them to be pounded into place.

Coating Refractories: This type of product is used to protect refractory linings usually against chemical attack. Coating refractories are normally intended to cover just the working surface of a lining. They tend to be fairly thin layers.

Mortars: Mortars are generally neither classified under refractory brick nor monolithic refractories. These are finely ground refractory materials, which become plastic when mixed with water. These are used to bond the brickwork into solid unit, to provide cushion between the slightly irregular surfaces of the brick, to fill up spaces created by a deformed shell, and to make a wall gas-tight to prevent penetration of slag into the joints.

Gunning Mixes: Gunning mixes are granular refractory materials sprayed on application area using a variety of air placement guns. These are heat setting and are used for patching and maintenance works for kilns and furnaces.

Fettling Mixes: Fettling mixes are also granular refractory materials, similar to gunning mix function, but are applied by shoveling into the furnace needing patching.

INSULATING REFRACTORIES

A refractory suitable for minimising heat losses and thus achieving heat conservation in the furnaces is called insulating refractories. They have high porosity, low thermal conductivity & high thermal insulation properties and are produced from asbestos, fireclay, kieselguhr *etc.* At low temperature, slagwool, glasswool and vermiculite are also used as insulating materials. For high temperature thermal insulation applications foam ceramics, ceramic fibre and wool are used.

The These have lower densities in range of 65 to 1120 kg/m^3. The three basic types of insulating materials for industrial use are: 1) thin (less than 20 micrometers), low-density (less than 190 kg/m^3) fibers made from organic or inorganic materials; 2) cellular material in closed or open cell form made of organic or inorganic material; and 3) flaked or granular inorganic materials bonded in the desired form. In most cases, glass (silica), mineral wool, high alumina, mulite, or zirconia are the base materials and can be used to temperatures as high as 1600°C. This class of materials has a lower density that varies from 65 to 190 kg/m^3 and offers higher thermal resistance compared to firebricks. In all cases, thermal conductivity of the insulation increases significantly as temperature increases. Insulating materials greatly reduce the heat losses through walls. Insulation is effected by providing a layer of material having a low heat conductivity between the internal hot surface of a furnace and the external surface, thus causing the temperature of the external surface reduced.

The insulating materials may be classified into the following groups:

- Insulating bricks
- Insulating Castables
- Ceramic fibre
- Calcium silicate
- Ceramic coating

Insulating materials owe their low thermal conductivity to their pores while their heat capacity depends on the bulk density and specific heat. Structure of air insulating material consists of minute pores filled with air which have in themselves very low thermal conductivity. Excessive heat affects all insulation material adversely, but the temperatures to which the various materials can be heated before this adverse effect occurs differ widely. The choice of an insulating material depends upon its effectiveness to resist heat conductivity and upon the temperature that it will withstand.

One of the most widely used insulating materials is diatomite, also known as kieselguhr which is made up of a mass of skeletons of minute aquatic plants deposited thousands of years ago on the beds of seas and lakes. Chemically this consists of silica contaminated with clay and organic matter. A wide range of insulating refractories with wide combinations of properties are now available.

The important physical properties of some insulating refractories are shown in the table below:

Table 1.5: Physical properties of insulating refractories

Type	Thermal conductivity at 400°C	Maximum safe temperature ,°C	Cold Crushing Strength, Kg/cm^2	Porosity, %	Bulk density, Kg/m^3
Diatomite Solid Grade	0.025	1000	270	52	1090
Diatomite Porous Grade	0.014	800	110	77	10
Clay	0.030	1500	260	68	560
High Aluminia	0.028	1500-1600	300	66	910
Silica	0.040	1400	400	65	830

Castables and Concretes

Monolithic linings and furnace sections can be built up by casting refractory insulating concretes, and by stamping into place certain light weight aggregates suitably bonded. Other applications include the formation of the bases of tunnel kiln cars used in the ceramic industry. The ingredients are similar to those used for making piece refractories, except that concretes contain some kind of cement, either Portland or a high-alumina cement.

Ceramic fibre

Ceramic fibre is a low thermal mass insulation material, which has revolutionalised the furnace design lining systems.

Ceramic fibre is an alumino silicate material manufactured by blending and melting alumina & silica at temperature of 1800 – 2000°C and breaking the molten stream by blowing compressed air or dropping the melt on spinning disc to form loose or bulk ceramic fibre. The bulk fibre is converted to various products including blanket, strips, veneering and anchored modules, paper, vacuum formed boards and shapes, rope, wet felt, mastic cement *etc.* for insulation applications.

Fibres are usually produced in two temperature grades based on Al_2O_3 content. A recent addition is ZrO_2 added alumino silicate fibre, which helps to reduce shrinkage levels thereby rating the fibre for higher temperatures.

These fibres are generally produced in bulk wool form and needled into blanket mass of various densities ranging from 64 to 190 kg/m^3. Converted products and over 40 different forms are made from blankets to suit various requirements.

Properties of Ceramic Fibre

The characteristics of ceramic fibres are a remarkable combination of the properties of refractories and traditional insulation material.

- **Lower Thermal Conductivity**

  The low thermal conductivity – 0.1 kCal/m^3.°C at 600°C for 128 kg/m^3 density ceramic fibre blanket – allows construction of thinner linings with the same thermal efficiency as that of conventional refractories. Hence, for the same outer envelope dimension the furnace volume is much higher. It is 40 % more effective than good quality insulation brick and 2.5 times better than asbestos product. Insulating property of ceramic fibre is better than calcium silicate product.

- **Light Weight**

  Average density of ceramic fibre is 96 kg/m^3. It is one tenth of the weight of insulating brick and one third that of asbestos/calcium silicate boards. For new furnaces structural supports can be reduced by 40%.

- **Lower Heat Storage**

  Ceramic fibre linings absorb less heat because of lower density. Furnace can be heated and cooled at faster rates. Typically the heat stored in a ceramic fibre lining system is in the range of 2700 - 4050 kCal/m^2 as compared to 54200-493900 kCal/m^2 for conventionally lined system.

- **Thermal Shock Resistance**

  Ceramic fibre lining resist thermal shock due to their resilient matrix. Also faster heat up and cool down cycles are possible thereby improving furnace availability & productivity.

- **Chemical Resistance**

  Ceramic fibre resist most of the chemical attack and is unaffected by hydrocarbons, water and steam present in flue gases.

- **Mechanical Resilience**

  This property permits fibre lined furnaces to be shop fabricated and shipped to site in assembled form without fear of damage.

- **Low Installation Cost**

  No special skills are required as application practices are standardised. Fibre linings require no dry out or curing times and can be heated to the capacity of the burners after installation is completed without concern for cracking or spalling.

- **Simple Maintenance**

  In case of physical damage the defective section can be quickly removed and a replacement piece added. Whole panel sections can be pre fabricated for fast installation with minimal down time.

- **Ease of Handling**

  All product forms are easily handled and most can be quickly cut with a knife or scissors. Vacuum formed products may require cutting with a band saw.

- **Thermal Efficiency**

  The low thermal conductivity of ceramic fibre can be advantageously made use of by the lesser lining thickness and reduced furnace volume. The fast response of ceramic fibre lined furnace also allows for more accurate control and uniform temperature distribution within the furnace.

The other advantages offered by ceramic fibre are as follows:

- Light weight furnace
- Simple steel fabrication work
- Low down time
- Increased productivity
- Additional capacity
- Low maintenance cost
- Longer service life
- Higher thermal efficiency
- Faster response

High Emissivity ceramic Coatings

Emissivity, the measure of a material's ability to both absorb and radiate heat, is an inherent physical property which like density, specific heat & thermal conductivity, is not readily amenable to change. However, the development of high emissivity coatings now allows the surface emissivity of materials to be increased, with resultant benefits in heat transfer efficiency and in the service life of heat transfer components. High emissivity coatings are applied in the interior surface of furnaces. The Figure below shows emissivity of various insulating materials including high emissivity coatings. High emissivity coating shows a constant value over varying process temperatures.

The application of high-emissivity coatings in furnace chambers promotes rapid and efficient transfer of heat, uniform heating, and extended life of refractories & metallic components such as radiant tubes & heating elements. For intermittent furnaces or where rapid heating is required, use of such coatings was found to reduce fuel or power to tune of 25-45%. Other benefits are temperature uniformity and increased refractory life.

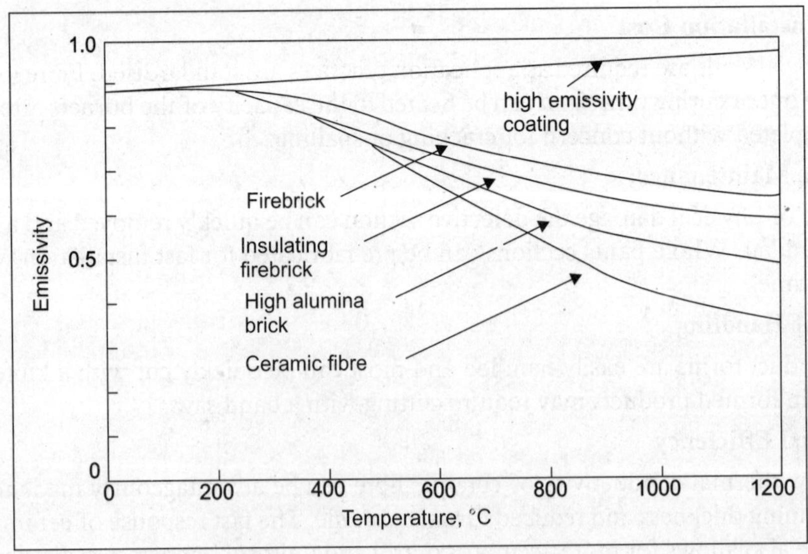

Fig.1.3: Emissivity of Refractory Materials at Different temperatures

Furnaces, which operate at high temperature, have emissivities of 0.3. By using high emissivity coatings this can go upto 0.8 thus effectively increasing the radiative heat transfer.

Special Refractories

These are very expensive refractory materials used for making crucibles and furnaces for special/experimental purposes where cost of refractory is no consideration. They are not very common due to their manufacturing limitation. Special refractories include pure oxides (e.g. magnesia, silica, alumina, beryllia, thoria *etc.*), borides, nitrides, silicides, carbides *etc.* Other special refractories are Sialons, zircon, carborundum, alundum (a mixture of fused alumina and clay), sillimanite, electrocast blocks of mullite, magnesia and mixtures of chromite, bauxite and magnesia.

Cermet Refractories

This refractory is a combination of ceramic materials (e.g. oxide, nitride, carbide, borides *etc.*) and metallic/metallic alloys materials. It has the combination of good properties of both metallic and ceramic materials e.g. high strength and resistance to high temperature. Cermets are used in nuclear reactors, missiles and space crafts *etc.*

Properties Of Refractories

Important properties of refractories are: chemical composition, bulk density, apparent porosity, apparent specific gravity and strength at atmospheric temperatures. These properties are among those which are used as 'control points' in the manufacturing and quality control process. The chemical composition serves as a basis for classification of refractories and the density, porosity and strength is influenced by many other factors. Among these are type & quality of the raw materials, the size & fit of the particles, moisture content at the time of pressing, pressure at mould, temperature, duration of firing and the rate of cooling.

Other important properties of refractories are resistance to high temperature, heavy load, fluxes, corrosive fluids like slag, thermal shocks due to rapid change in temperature (also called spalling resistance) *etc.* Other properties are strength, specific gravity , thermal & electrical conductivity *etc.* They are discussed below.

Porosity and Slag Permeability. The higher the porosity of the refractory, the greater is the susceptibility of refractory to chemical attack of molten slags, gases *etc.* By decreasing the porosity, strength & thermal conductivity of the refractory is increased ; but spalling resistance (thermal shock resistance) is decreased as the inner pore space for absorbing contraction & expansion is reduced.

Porosity is a measure of the effective open pore space in the refractory into which the molten metal, slag, fluxes, vapors etc can penetrate and thereby contribute to eventual degradation of the structure. The porosity of refractory is expressed as the average percentage of open pore space in the overall refractory volume.

High porosity materials tend to be highly insulating as a result of high volume of air they trap, because air is a very poor thermal conductor. As a result, low porosity materials are generally used in hotter zones, while the more porous materials are usually used for thermal backup.

Such materials, however, do not work with higher temperatures & direct flame impingement, and are likely to shrink when subjected to such conditions. Refractory materials with high porosity are usually NOT chosen when they will be in contact with molten slag because they can be penetrated easily.

Refractoriness or melting/ fusion point. It is the temperature at which the refractory fuses. It should be much higher than the operating temperature of the refractory /furnaces. Fusion point decreases when refractory is under load and this is called refractoriness under load (RUL).

Melting temperatures (melting points) specify the ability of materials to withstand high temperatures without chemical change and physical destruction. The melting point of few elements that constitute refractory composition in the pure state are indicated in the table below:

Table 1.6: Melting points of pure refractory compounds

Pure compound	Formula	Melting Temperature , ^0C
Alumina	Al_2O_3	2050
Lime	CaO	2570
Chromite	$FeO.Cr_2O_3$	2180
Chromium Oxide	Cr_2O_3	2275
Magnesia	MgO	2800
Silica	SiO_2	1715
Titania	TiO_2	1850

The melting point serves as a sufficient basis for considering the thermal stability of refractory mixtures and is an important characteristic indicating the maximum temperature of use.

Pyrometric Cone Equivalent (PCE)

Refractories due to their chemical complexity melt progressively over a range of temperature. Hence refractoriness or fusion point is ideally assessed by the cone fusion method. The equivalent standard cone which melts to the same extent as the test cone is known as the pyrometric cone equivalent (PCE). According to ASTM C24 - 01, PCE is measured by making a cone of the material and firing it until it bends to 3 O'clock.

Representative PCE values for selected refractories include cones 33-34 for super duty fireclay, cones 29-31 for medium duty fire clay and cones 36-37 for a 60% alumina product. The cone values reported for refractories are based on a defined standard time– temperature relationship, so different heating rates will result in different PCE values. PCE can be useful for quality control purposes to detect variations in batch chemistry that result from changes or errors in the raw material formulation.

"Fusibility temperature range" is a more appropriate term for refractory materials as mixture of refractory materials do not have a sharp fusion/melting point. The heat resistance of refractory material is expressed in terms of Pyrometric Cone equivalents (PCE) which is determined by comparing the softening of a test pyramid cone with that of standard pyramid cones of known fusion temperature by heating them under similar conditions.

Temperature at which a refractory will deform under its own weight is known as its softening temperature which is indicated by PCE. Refractories due to their chemical complexity melt progressively over a range of temperature. Hence refractoriness or fusion point is ideally assessed by the cone fusion method. The equivalent standard cone which melts to the same extent as the test cone is known as the pyrometric cone equivalent (PCE).

The pyrometric cone equivalent indicates only the softening temperature. But, in service the refractory is subjected to loads which would deform the refractory at a much lower temperature than that indicated by PCE. With change in the environmental conditions, such as reducing atmosphere, the PCE value changes drastically.

The standard cones are called 'Segar cones' (as Germany standard) and

'Orton cones' (as British standard) and they are numbered according to their refractoriness/ fusion points.

Refractoriness under load (RUL)

Refractoriness points to the resistance of extreme conditions of heat and corrosion when hot & molten materials are contained while being transported and/or processed.

The ability to withstand exposure to elevated temperatures without undergoing appreciable deformation is measured in terms of refractoriness. The refractoriness under load test (RUL test) gives an indication of the temperature at which the bricks will collapse, in service conditions with similar load. However, under actual service where the bricks are heated only on one face, most of the load is carried by the relatively cooler rigid portion of the bricks. Hence the RUL, test gives only an index of the refractory quality, rather than a figure which can be used in a refractory design. Under service conditions, where the refractory used is heating from all sides such as checkers, partition walls, *etc.* the RUL test data is quite significant.

The refractoriness under load test (RUL test) gives an indication of the temperature at which the bricks will collapse, in service conditions with similar load.

Refractoriness under load (RUL) of a refractory material is a measure of its failure resistance to the combined action of heat and load. RUL is the softening temperature of a refractory brick (indicated by breaking of the test specimen) when a load of 2 kgf/cm^2 is applied as per the standard test procedure.

Refractory bricks generally having more than two constituent compounds fail under load conditions at much lower temperature (i.e., at RUL) than the fusion point (i.e., refractoriness) of the highest melting and predominant constituent. That signifies that the temperature corresponding to the refractoriness under load of a refractory brick is much lower than its refractoriness (fusion/ melting point). A refractory material composed of a single pure compound will have a closer value of its refractoriness and refractoriness under load.

Refractory failure is also influenced by its exposure time to load and heat. Longer exposure time results in failure of refractory under relatively low load & temperature conditions. Similarly refractories will fail at a lower temperature with increased load.

Creep at high temperature:

Creep is a time dependent property which determines the deformation in a given time and at a given temperature by a material under stress. Refractory materials must maintain dimensional stability under extreme temperatures (including repeated thermal cycling) and constant corrosion from very hot liquids and gases. The criterion of acceptance usually adopted is; that compressive creep (deformation at a given time and temperature under stress) for normal working conditions of load and temperature shall not exceed 0.3% in the first 50 hours of the test. This figure has been found to indicate that the creep rate falls by a negligible amount at the end of the stipulated period, and therefore the refractory can be considered safe to use for a much longer time.

Volume stability, expansion and shrinkage at high temperature:

The contraction or expansion of the refractories can take place during service. Such permanent changes in dimensions may be due to:

- The changes in the allotropic forms which cause a change in specific gravity.
- A chemical reaction which produces a new material of altered specific gravity.
- The formation of liquid phase.
- Sintering reactions.
- It may also happen on account of fluxing with dust & slag or by the action of alkalies on fireclay refractories, to form alkali-alumina silicates, causing expansion and disruption. This is an example which is generally observed in blast furnaces.

While it is desirable that all these changes are effected during manufacturing, it is not possible due to economic reasons, as the process is time dependent. Permanent Linear Change (PLC) on reheating and cooling of the bricks give an indication on the volume stability of the product as well as the adequacy of the processing parameters during manufacture. It is particularly significant as a measure of the degree of conversion achieved in the manufacture of silica bricks.

Reversible Thermal Expansion

Any material when heated expands, and contracts on cooling. The reversible thermal expansion is a reflection on the phase transformations that occur during heating and cooling. The PLC and the reversible thermal expansion are followed in the design of refractory linings for provision of expansion joints. As a general rule, those with a lower thermal expansion co-efficient are less susceptible to thermal spalling.

Strength of the refractory

It is the resistance of the refractory to compressive loads, tension and shear stresses. Refractoriness under load (RUL) is important as some refractories fail at a lower temperature when subjected to load. In taller furnaces, the refractory has to support a heavy load, hence strength under the combined effect of temperature and load is important.

The physical strengths, in both cold and hot conditions, are often characterized as measures of the use of a refractory. Cold strengths indicate the handling and installation of the refractory, whereas hot strengths indicate how the refractory will perform when used at elevated temperatures. Initial strength develops in refractory materials during the forming process.

For shaped refractories, the strengths often develop during the physical processing of the products and sometimes followed by higher temperatures where the refractory goes through a firing process.

For monolithic refractories, the initial strength develops during the installation or forming process (for precast shapes), and the final strength develops while in application. In recent years, more importance has been given to high-temperature strengths of refractories rather than cold strengths since refractories are used at elevated temperatures and not at room temperatures. Strengths of refractories are measured as cold compressive strength (CCS) , cold modulus of rupture, or hot modulus of rupture. Hot modulus of rupture provides the best indication of the performance of a refractory material in use.

Strengths of refractories are usually reported in terms of the three-point bend strength or the flexural strength, frequently called the modulus of rupture (MOR). In case of refractories, there is a standard test for this simple strength measurement.

The flexural strength will depend on sample size, since with increasing size there is an increase in the probability of the existence of a crack-producing flaw, along with a corresponding decrease in the flexural strength.

Characteristic flexural strength values for several ceramic materials are given in table below.

Table 1.7: Flexural strength (Modulus of Rupture) of ceramics.

Material	Flexural strength	
	MPa	psi
Silicon nitride (Si_3N_4)	250-1000	35-145
Zirconia[a] (ZrO_2)	800-1500	115-215
Silicon carbide (SiC)	100-820	15-120
Aluminium oxide (Al_2O_3)	275-700	40-100
Glass-ceramic (Pyroceram)	247	36
Mullite ($3Al_2O_3$-$2SiO_2$)	185	27
Spinel ($MgAl_2O_4$)	110-245	16-35.5
Magnesium oxide (MgO)	105[b]	15[b]
Fused silica (SiO_2)	110	16
Soda-lime glass	69	10

[a] Partially stabilized with 3 mol% Y_2O_3.

[b] Sintered and containing approximately 5% porosity.

Cold Crushing strength or cold compressive strength (CCS)

The cold crushing strength, which is considered to be doubtful relevancc as a uscful property, other than it reveals little more than the ability to withstand the rigorous of transport. It can be seen as a useful indicator to the adequacy of firing and abrasion resistance in consonance with other properties such as bulk density & porosity.

The cold compressive strength or the cold crushing strength is the capacity of a refractory to provide resistance to a compressive load at room temperature. It is the load, in kg/cm^2, at which the refractory breaks.

The cold compressive strength of a refractory material is an indication of its suitability for use of refractories in construction. It is a combined measure of the refractory for the strength of the grains and also of the bonding system.

Cold Modulus of Rupture (CMOR): (ASTM C-133)

The cold modulus of rupture of a refractory material indicates the flexural strength and its suitability for use in construction. It is indicative of the strength of the bonding system of the refractory product. Since the test is done at room temperature, it can only show its suitability and its use in construction. It provides no indication of how the refractory will behave at elevated temperatures.

Hot Modulus of Rupture (HMOR): (ASTM C-583)

The hot modulus of rupture provides the indication of a refractory material about its flexural strength at elevated temperatures. Since refractories are used at elevated temperatures, the hot modulus of rupture is the true indicator of the suitability and performance of a refractory at high temperatures. Hence, in recent years, the hot modulus of rupture has been prescribed and required by users as the most important test criterion for selection & use of refractories.

Specific gravity. It should be low to produce more number of bricks per unit weight. Also lighter bricks in the lower part of a tall structure will not be subjected to a heavy load. Cost of bricks of higher specific gravity is more than that of lower specific gravity (assuming the price of raw material being same per unit). Increase in bulk density of refractory increases its volume stability, heat capacity and resistance to stag penetration.

Bulk Density

The bulk density is generally considered in conjunction with apparent porosity. It is a measure of the weight of a given volume of the refractory. For many refractories, the bulk density provides a general indication of the product quality; it is considered that the refractory with higher bulk density (low porosity) will be better in quality.

An increase in bulk density increases the volume stability, the heat capacity, as well as the resistance to abrasion and slag penetration.

Thermal Properties of refractories

(a) Thermal Conductivity

It is defined as the quantity of heat that will flow through a unit area in direction normal to the surface area in a defined time with a known temperature gradient under steady state conditions.

Thermal conductivity depends upon the chemical & mineralogical compositions as well as the glassy phase contained in the refractory and the application temperature. The conductivity usually changes with rise in temperature. In cases where heat transfer is required though the brick work, for example in regenerators, muffles, *etc.* the refractory should have high conductivity. Low thermal conductivity is desirable for conservation of heat by providing adequate insulation.

Effect of the structure and the microstructure of refractories on the thermal conductivity:

- The highest conductivity are achieved in the cluttered structure, that in structures consisting of single element, structure made up elements of similar atomic weight, and structure with no extraneous atoms in solid solution. Graphite is good example of single element refractory structure having high thermal conductivity.
- Refractories composed of elements of similar atomic weight and size, have high thermal conductivity. Good examples of this type of refractory are BeO, SiC and B_4C. The refractories

having large difference in atomic weight and size have low thermal conductivity as for example UO_2 & ThO_2.

- Impurities, porosity and microcracks exert the main influence on the effective thermal conductivity. Presence of these defects can reduced the thermal conductivity of refractories.

A more porous brick has, in general, lower thermal conductivity. However, the size and number of the pores is also important as heat transfer can take place across the pores by radiation and is prominent at high temperature.

Refractories used in melting furnace should have low thermal conductivity to ensure least heat losses and maximum heat efficiencies whereas recuperates require refractory with high thermal conductivity to ensure maximum heat transfer.

High thermal conductivity refractories are required for some applications where good heat transfer is essential such as coke oven walls, regenerators, muffles and water cooled furnace walls. However, refractories with lower thermal conductivity are preferred in industrial applications, as they help in conserving heat energy.

Porosity is a significant factor in heat flow through refractories. The thermal conductivity of a refractory decreases on increasing its porosity.

Although it is one of the least important properties as far as service performance is concerned, it evidently determines the thickness of brick work.

Ordinary furnaces block should have low thermal conductivity ; but refractory used in coke oven, regenerators, muffle furnace should have high heat conductivity. Thermal conductivity is increased when the porosity of refractory bricks decreases i.e. when the air entrapped in refractory decreases. Insulating bricks (of low thermal conductivity) are maintained at high temperature and hence their life is short.

(b) Thermal Diffusivity

Thermal diffusivity is the thermal conductivity divided by density and specific heat capacity at constant pressure. It measures the ability of a material to conduct thermal energy relative to its ability to store thermal energy. It has the SI unit of m^2/s or m^2/hr.

(c) Thermal Expansion

This is a measure of the refractory about its linear stability when it is exposed to different ranges of high temperatures and then cooled to room temperatures. It is defined as a permanent linear change (PLC) and is measured by the changes in thc longest linear dimensions. Most refractory materials expand when heated. Hence, when refractories are installed at room temperatures, the whole structure tightens up when heated.

Thermal expansion of ceramic materials depends on bonding strength between atoms. Ceramic with high degree of ionic bond have high thermal expansion (behavior similar to metals). As the bond strength increases or as the percent of covalent bond increases, the thermal expansion decreases. Both melting temperature & thermal expansion are controlled primarily by bond strength and the magnitude of thermal vibration. As the bond strength increases, the melting temperature increases and thermal expansion coefficient decreases.

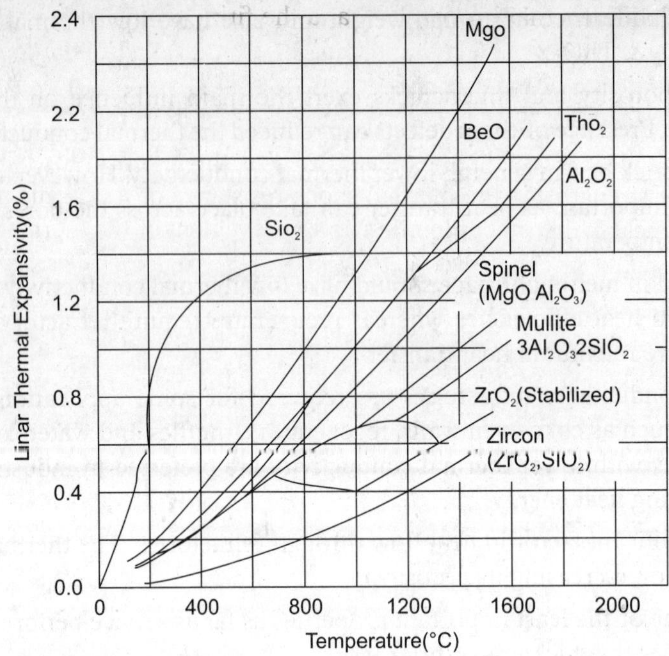

Fig.1.4: Thermal Expansion of Refractory Oxides as Function of Temperature

The expansion characteristic correlates with the nature of the structure, and the atomic bonding. Al_2O_3, TiO_2 and mullite have moderately close packed structures and atomic bond strength. As result, they have moderate thermal expansion. Within the graphite layers, the atomic bonding is very strong (C-C covalent bond) and the coefficient of expansion is low ($1\times10^{-6}/°C$). The atomic bonding between layers is very week (Van der waal's bonds) and the thermal expansion coefficient is high ($27\times10^{-6}/°C$).

Importance of thermal expansion: Many ceramic applications are exposed to a range of varying temperature. Mismatch in thermal expansion behavior between two adjacent materials, can result high enough stresses to fracture. Moderate close thermal expansion match is necessary to minimize thermal stress. Some applications require very low thermal expansion. A well-known domestic application is the use of (LAS) based polycrystalline ceramic ($LiAlSi_2O_6$) for heat-resistant cooking ware.

(d) Thermal Shock

Thermal shock is the fracture of a body resulting from thermal stresses induced by rapid temperature changes. Because ceramic materials are brittle, they are especially susceptible to this type of failure. The thermal shock resistance of many materials is proportional to the fracture strength & thermal conductivity, and inversely proportional to both the modulus of elasticity & the coefficient of thermal expansion.

Thermal shock is an important property for a refractory material. Most high-temperature processes experience heating and cooling. Thermal shock can be indicated by the number of cycles to withstand such temperature fluctuations. Thermal shock resistance can be defined as partial or complete fracture of material due to gradient temperature. The factor that affects the thermal shock resistance includes the shaping methods and the firing temperatures. Thermal shock refers to the thermal stresses that occur in a component as a result of exposure to temperature difference between surface and interior of the component.

As ceramic materials subjected to thermal shock generally fail in tension rather than in shear or compression, the tensile strength is generally used as the criterion for failure rather than the compressive or shear strength. As the thermal shock resistance of a ceramic material is governed mainly by its mechanical & thermal properties, an improvement of the thermal shock resistance of the high Young's modulus of ceramics.

(e) Thermal Spalling

Spalling is defined as the fracture of the refractory brick or block due to any of the following causes:

- A temperature gradient in the brick due to uneven heating or cooling that sets up stresses causing failure (Thermal Spalling). Thermal spalling is the fracture of refractory due to its uneven expansion on account of heat when they are subjected to rapid heating and cooling. For reducing spalling, a refractory should be well fired and its porosity should be more.
- Compression in a structure of refractories due to expansion of the whole from a rise of temperature causing shear failure.
- Variation in coefficient of thermal expansion between the surface layer and the body of the brick, due to the surface slag penetration or due to structural changes in service resulting in shearing off the surface layer.

As a general rule, those with a lower thermal expansion co-efficient are less susceptible to thermal spalling.

Thermal Spalling can be reduced by:

- Using materials with low coefficient of expansion and avoiding sudden temperature fluctuation for example addition of alumina in small quantities decreases expansion to a large extent.
- Using high porosity bricks.
- By designing the furnace such that stress is alienated.

Electrical Conductivity. For electrical furnaces, the electrical conductivity of refractory material should be low. Except, graphite, all other refractories are bad conductors of electricity. The electrical resistivity of the refractory drops rapidly with rise in temperature.

Chemical Composition. The chemical composition of the refractory should be such that the surroundings do not chemically attack the refractory and corrode it. Acidic refractory should not be used in a furnace heating a basic material, otherwise the brick will react with the furnace stock and get eroded and hence its life will be reduced.

Slag Resistance. Slag attacks the refractory particularly at high temperature by forming mixture or new compounds with the refractory substance in contact with. Slag dissolves some of the refractory material and forms a viscous slag film whose melting temperature may be higher or lower. Resistance of refractory to slag action depends on the nature of slag and the refractory. Relative slag resistance of various refractories depend upon temperature, change in slag composition (on combining with refractories) and the nature of contact with the slag i.e. whether the slag is moving, stationary or is carried as fine particles by the gases/flame. A basic slag attacks an acidic refractory while an acid slag attacks a basic refractory very rapidly.

Rate of slag attack is accelerated by the following factors:

- defective joints and cracks in the refractory structures.
- presence of fusible glass of refractories.
- rough and porous surfaces of refractories.
- moving slag as compared to stationary slag.

Rate of slag attack is reduced, if the refractory is dense and well vitrified (i.e. offering a smoother surface) and the slag is stationary.

Abrasion Resistance. A refractory is subjected to wear, when it comes in contact with a moving charge as in case of blast furnace, descending charge materials result in abrasion/ wearing away of the refractory lining material. Abrasion resistance of refractory is increased by using a dense (low porosity), fine grained and wear resistant refractory lining.

Erosion Resistance. Erosion (damage due to mechanical action) occurs when metal or gas carrying dust and slag particles hit against the refractory lining resulting in 'chipping off' of particles from the refractory. Splashing of slag erodes the refractory lining whereas slag particles entrained by flame/gases erode particularly the arches and bends of the structure.

Size , Shape and dimensional stability. It is a part of the design of furnace, since it affects the stability of the furnace structure. Accurate refractory size is important to properly fit the refractory shape inside the furnace and to minimise space between construction joints.

Dimensional accuracy and size is extremely important to enable proper fitting of the refractory shape and to minimize the thickness and joints in construction.

GENERAL METHOD OF MANUFACTURE OF REFRACTORY

Manufacture of refractory consists of the following processes:

Grinding. Raw materials are ground so that processes sizes of material are obtained. Ratio of coarse to fine particles should be even. This ratio is normally maintained at 55:45. Equipments used for this purpose are various types of crushers, pulverisers, hammer mills, ball mills, shaking tables and screens *etc.*

Pre-Treatment. The main pre-treatment given to refractory raw material is its firing/ calcination at high temperature for considerable periods of time which bring about the complete mineral conversion and hence stabilises the material. The constituents of the finished refractories produced by reaction between the solid particles of raw materials are rarely completely in equilibrium because it is seldom feasible to actually melt and cast these refractory materials. Natural silica mainly contains quartz which is not stable at high temperatures but undergoes transformation to its allotropic forms tridymite and cristobalite involving high volume changes. Quartz (a stable form of silica up to 870°C is transformed into tridymite on very prolonged heating at this temperature i.e. 870°C. A shorter duration firing between 870 and 1470°C results in transformation of quartz into tridymite. On heating above 1470°C, this is further transformed into cristobalite. Conversion of quartz into tridymite involves 16% increase in volume whereas conversion of tridymite into cristobalite involves a little contraction in volume. Thus raw silica expands considerably when heated to high temperatures and pre stabilization of silica becomes necessary.

Similarly, fireclay which consists of numerous hydrous alumino-silicate minerals, break down to several constituents on heating. And these constituents tend to form stable mullite ($3Al_2O_3.2SiO_2$). Mullite formation begins at 1000°C and is accelerated at increasing temperatures. High temperature strength, resistance to slag action and other useful properties of fireclay depend upon the formation of mullite and properties of glassy bonds. Even weathering of mixed fireclay increases its plasticity.

Magnesite is dead burnt to eliminate all the CO_2 from it and to convert it into stable crystalline magnesia (MgO) known as 'Periclase' and involves least shrinkage in refractory shapes on firing.

Dolomite ($CaCO_3$ + $MgCO_3$ mainly) is dead burnt to calcium and periclase. Dolomite as such has the tendency to slaken and is not suitable for refractories and is hence made inactive by converting it into a silicate or some other compound. The slakening tendency of burnt dolomite is eliminated by conversion of calcium to dicalcium or tri-calcium silicate in presence of Fe_2O_3 and surpentine respectively. Presence of Fe_2O_3 and for conversion into dicalcium silicate prevents the formation of y-dicalcium silicate which otherwise results in explosions in service because of about 10% increase in its volume at the working temperature. Calcination of mixture of dolomite and serpentine at 1600°C results in tricalcium silicate and periclase (MgO) which forms a very good refractory material.

Mixing. Ground refractory material is mixed with the binding material in such a way that plastic materials are equally distributed throughout the mass to facilitate easy moulding.

Mixing is usually carried out in pug mills for even distribution of fine and coarse particles in the whole mass. Pre calculated amount of water, additives, binding materials and mineralisers (i.e. grog, stabilisers *etc.*) are added and the mass is mixed thoroughly to ensure a product of uniform composition and uniform distribution of fine and coarse grains plastic and bond materials *etc.* for easy moulding and development of useful properties in the fired refractories. To ensure increased plasticity for easy moulding, the mixture is allowed to remain as such for a day or more.

Mixing can be dry, semi-plastic or wet type. 14-20% water is used in wet mixing which are usually used for hand moulding. Semi-plastic mixtures have lower percentages of water to achieve semi-state of plasticity whereas the dry mixture include non-plastic basic mixes and clay mixes containing < 5% water. Since mixing with so little water is difficult, water as a fine spray or mist is used to achieve proper mixing. Due to requirement of high moulding pressures, semi-plastic and dry mixture are machine pressed. Clay mixtures are also obtained in the form of colloidal solutions suitable for moulding by slip casting.

Moulding. It is done mechanically by applying high pressure or by hand to increase density (by deairing) and strength of the refractory. Moulding which follows mixing of refractory materials is done either by hand or machine (by pressing or extrusion). Hand moulding (successful with wet mixtures) is carried out in wooden boxes and is cheaper than machine on a jobbing basis.

However, machine moulding has more practical applications and is cheaper for mass production of standard refractory shapes. Machine moulded refractories have higher strength and density than hand moulded refractories. Machine moulding can be used for semi- plastic mixture using moderate moulding pressures. Extrusion is usually used to get the rough shapes of approximate dimensions which are subsequently pressed to exact shapes. Machine moulding of dry mixture requires a pressure of about 1000 kg/cm^2 or more.

At high moulding pressures, air is trapped in the refractories and cause laminations. Therefore de-airing of refractories is done in following ways by:

- pressing the material and then allowed to crack, finally it is repressed to close the voids.
- applying vacuum to the system through the vents in the mould
- passing a gas (such as butane) through the mould to displace air (the gas is absorbed by the clay on the application of pressure)

Density and strength of refractories increase on de-airing due to decrease in laminations and crack. Density increase is also due to elimination of air of a lower density than the refractory material and/or closing of voids.

The method used for moulding refractories of hollow and other special shapes is slip casting in.which a colloidal suspension is poured into a plaster of paris moulds which absorbs the water and causes deposition of uniform layer of clay on the walls of the mould. On deposition of desired thickness of clay (which depends on the time allowed for deposition), the extra mixture in suspension is poured out of the mould resulting in hollow refractories with irregular interval contours.-

Powder pressing is used for moulding refractories involving dry compacting or slightly damp refractory powder mixture in metallic dies using sufficiently high pressure to produce strong and dense refractory shapes. Recent moulding processes used for some typical refractories include reactive hot pressing and isostatic pressing.

Drying. It is done at slow rate to avoid voids and high shrinkage. Drying of moulded refractories increases its green strength by removing moisture and thus making them safe for subsequent handling. Drying is usually carried out under shade (not in direct sunlight) on large drying floors. Floors heated by waste heat from kilns are also used where refractories are laid out in open arrays. Drying is carried out in predetermined set of conditions (of temperature and humidity) and the drying rate is kept at minimum to avoid higher shrinkage and crack formations.

For faster drying of refractory of constant shape and sizes, tunnel kilns are used, where raw refractories are stacked in bogies or placed on belts moving through a tunnel against a stream of hot air. Drying conditions in this case need not be altered (as the shape and sizes of refractories are constant) to ensure drying without damaging the refractories due to high shrinkages and crack formations involved at high rates of drying.

For drying, heavy refractories of intricate shapes, controlled humidity driers are specially used. Drying rate of fireclay bricks should be very low to avoid crack formations as they are liable to maximum shrinkage on drying.

Firing/Burning. Bricks are burnt in Kiln (down draught kiln or tunnel kiln) to remove water of hydration, vitrification and development of stable mineral forms. Shrinkage in volume upto

30% occurs during burning. Burning or firing of refractories which follows their drying facilitates developments of stable mineral forms in them and high crushing strength of finished products.

Degree of firing of silica brick is indicated by its specific gravity. High specific gravity of 2.5 indicates the presence of considerable amount of unconverted quartz in the refractory whereas a specific gravity of 2.32 to 2.37 indicates adequate firing and transformation of quartz to cristobalite and/or tridymite. Rapid heating of silica refractories above transformation temperatures may result in their fracture or spalling due to internal stresses set up owing to large volume changes within the refractories. It can be prevented by following a careful warning of schedule even when fired refractories are finally used in furnaces as some transformation back to quartz takes place on cooling them. Silica bricks are loosely packed in the kiln to meet its expansion requirement during firing whereas fireclay bricks are closely packed (as they shrink on firing) to get maximum production.

Brick firing is done either in down draught, batch type or tunnel kilns. Batch type kiln firing is time consuming due to involvement of three stage operation namely brick charging, firing and cooling. Difficulty in achieving uniform temperature throughout the batch type kiln (as it is heated by coal/coke) results in well or overfired product from the hottest zone and under fired brick from coldest zone. However in a modern kiln (which is divided into different compartments), heating is carried out in one compartment and to a uniform temperature and only after that the fire is advanced to the next compartment. Predetermined quantity of combustion air is preheated through some cooling compartment of a kiln which also accelerates the cooling rate of the fired refractories and thereby reduces the production time. Similarly, hot flue gas from compartments on fire are passed through freshly charged compartments of the kiln to preheat the dried refractories therein which also leads to fuel conservation.

For mass production of refractory bricks however continuous tunnel kilns are more suitable in which dried refractory bricks are charged from one end and fired product at near room temperature is discharged from the other end. Firing is completed in the hottest central zone of the kiln in which the secondary air enters through the discharge end and gets preheated as it cools the fired bricks on their way out of the furnace. Preheated combustion air is used for burning fuel oil (some kilns still use coal/coke as fuel) in the central hottest zone. Air preheating, apart from increasing the temperature in the firing zone also reduces fuel consumption. Hot flue gases while moving towards charging end of the tunnel kiln give away their sensible heat to cold refractories and preheat them as they move towards the firing zone which is the hottest zone.

Firing temperature of some typical common bricks is tabulated in Table 1.8.

Table 1.8: Typical Firing Temperature of Refractories

Refractory Type	Firing temperature, °C
Fireclay bricks	1250 — 1400
High alumina bricks	1450 — 1550
Silica bricks	1450 — 1510
Chrome bricks	1450 — 1650
Magnesite bricks	1450 — 1650
Direct bonded basic bricks	1650 — 1760
Silicon carbide bricks	1370— 1510

Various steps involved in the manufacture of refractories are shown in figure 1.5.

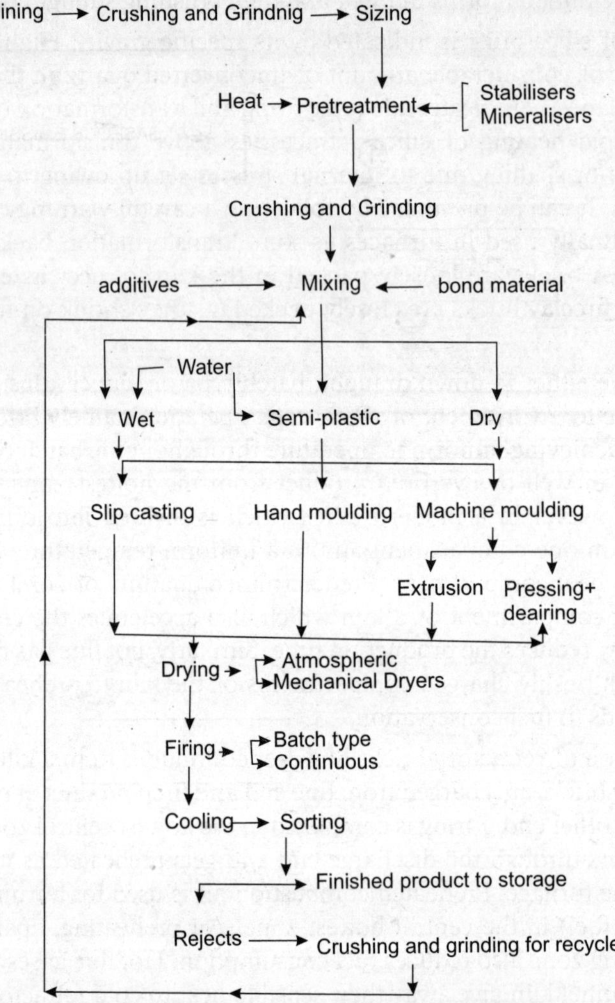

Fig.1.5: Various steps in the manufacture of refractories.

Refractory Manufacturing Process Flow Diagram is shown in Figure 1.6.

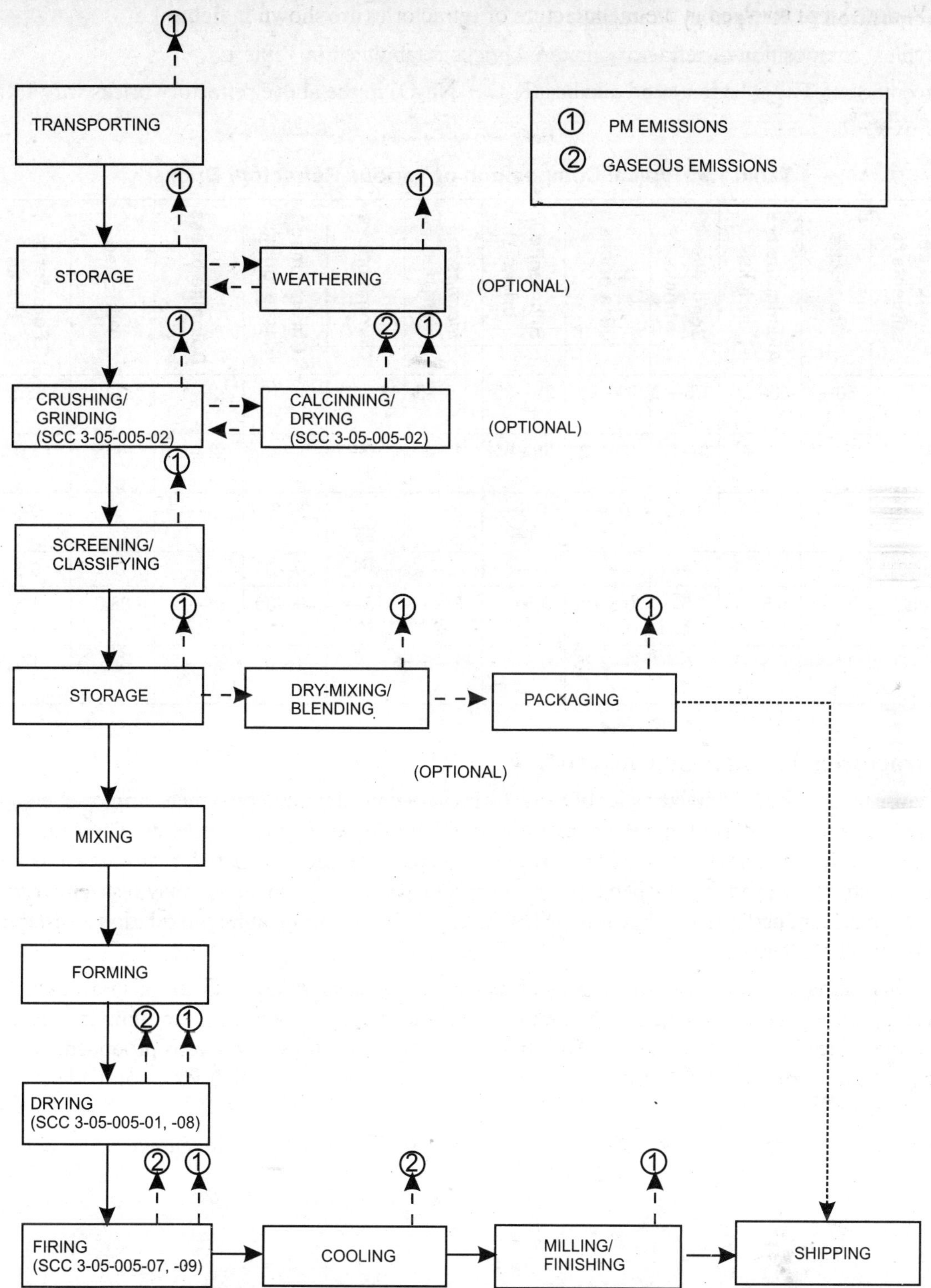

Fig.1.6: Refractory Manufacturing Process Flow Diagram

Composition of Refractory Materials:

A typical composition of refractory material bricks is tabulated in Table 1.7.

Percentage of TiO_2 (Titania) and alkalies (K_2O + Na_2O) in the above refractory bricks vary from traces to 2%.

Table 1.9: Typical Composition of Various Refractory Bricks

Principal constituents %	Fire bricks	Aluminious fire bricks	Semi silica bricks	Silica bricks	Sillimanite bricks	*magnesite. Bricks*	Chrome bricks	Chrome magnesite. bricks	Dolomite bricks	Periclase	Foresterite bricks
SiO_2	50—70	50—55	75—90	94—95	25—35	1.5—5	3—10	3—10	12—15	3.0	33.5
Al_2O_3	25—3	38—45	8—15	0.5—2	55—65	1—4	10—25	5—15	2—3	1.0	1.0
Fe_2O_3	2—5	2—4	1—2	0.5—1.5	0.5—1.5	2—8	12—25	8—20	2—4	3.0	0.1
Cr_2O_3	—	—	—	—	—	— 135—45		20—35	—	0.5	0.5
MgO	0.5—1	0.5—1	0.5—1	0.5—1	0.5—1	84—92	15—35	40—60	38—12	98.0	1.5
CaO	0.5—1	0.5—1	0.5—1	2—2.25	0.5—1	2—5	2	1—4	38—42	2.5	1.0

Refractories Product Characteristics

Because the quality & characteristics of refractories vary considerably, consumers employ chemical and physical tests to ensure that the refractories purchased meet their requirements. The American Society for Testing and Materials (ASTM) provides specifications and tests for various kinds and uses of refractory products. Depending on the intended end use, consumers may test refractories for thermal conductivity, resistance to abrasion & corrosion, permeability, oxidation resistance, and other characteristics.

Most refractory products are sold as preformed shapes. However, they are also available in special purpose clays; bonding mortars; and monolithic, plastic refractories; ramming mixes; and gunning mixes. A variety of processed refractory grains and powders are also produced. From the physical form, refractory products can be further classified into oxide bricks, nonoxide bricks, and composites.

Table below lists types of oxide, nonoxide, and composite refractories; their characteristics & their applications.

Table 1.10: General Characteristics & Applications of Various Types of Refractories

Refractory Type	General Characteristics	Application
Oxide Bricks: Silica	High strength at high temperatures, residual expansion, low specific gravity, high expansion coefficient at low temperatures, low expansion coefficient at high temperatures	Glass tank crown, copper refining furnace, electric arc furnace roof
Fused silica	Low thermal expansion coefficient, high thermal shock resistance, low thermal conductivity, low specific gravity, low specific heat	Coke oven, hot stove, soaking pit, glass tank crown
Chamotte (fireclay)	Low thermal expansion coefficient, low thermal conductivity, low specific gravity, low specific heat, low strength at high temperatures, less slag penetration	Ladle, runner, sleeve, coke oven, annealing furnace, blast furnace hot stove, reheating furnace, soaking pit
Alumina	High refractoriness, high mechanical strength, high slag resistance, high specific gravity, relatively high thermal conductivity	Hot stove, stopper head, sleeve, soaking pit cover, reheating furnace, glass tank, high-temperature kiln
High alumina	High refractoriness, high mechanical strength, high slag resistance, high specific gravity, relatively high thermal conductivity	Slide gate, aluminum melting furnace, skid rail, ladle, incinerator, reheating furnace hearth, skid rail, ladle, incinerator
Roseki	Low thermal expansion coefficient, high thermal shock resistance, low thermal conductivity, low specific gravity, low specific heat	Ladle, runner, sleeve, coke oven, annealing furnace, blast furnace hot stove, reheating furnace, soaking pit
Zircon	High thermal shock resistance, high slag resistance, high specific gravity	Ladle, nozzle, stopper head, sleeve
Zirconia	High melting point, low wettability against molten metal, low thermal conductivity, high corrosion resistance, high specific gravity	Nozzle for continuous casting, glass tank, high-temperature furnace, crucible
Alumina zirconia silica	High slag resistance, high corrosion resistance against molten glass	Glass tank, incinerator, ladle, nozzle for continuous casting
Lime	High slag resistance, low hydration resistance	Special refining surface
Magnesia	High refractoriness, relatively low strength at high temperature, high basic slag resistance, low thermal shock resistance, low durability at high humidity	Hot-metal mixer, secondary refining vessel, rotary kiln, checker chamber of glass tank, electric arc furnace
Magnesia- chrome	High refractoriness, high refractoriness under load, high basic slag resistance, relatively good thermal shock resistance (low MgO bricks), high strength at high temperature (direct bonded and fusion cast)	Hot-metal mixer, electric arc furnace, secondary refining vessel, nonferrous refining furnace, rotary cement kiln, lime and dolomite kiln, copper furnace, ladle, checker chamber for glass tank, slag line of electric arc furnace, degasser for copper, nonferrous smelter
Chrome	High refractoriness, low strength at high temperature, low thermal resistance	Buffer brick between acid and basic brick

Dolomite	High refractoriness, high refractoriness under load, high basic slag resistance, low durability in high humidity, high thermal expansion coefficient	Basic oxygen furnace, electric arc furnace, secondary refining vessel, rotary cement kiln
Spinel	High thermal shock resistance, high strength at high temperatures, high slag resistance	Rotary cement kiln, ladle
Non oxide Bricks:		
Carbon	High refractoriness, high slag resistance, low oxidation resistance	Blast furnace hearth, electric arc furnace
Silicon carbide	High refractoriness, high strength at high temperature, high thermal conductivity, high thermal shock resistance, reduced oxidation resistance at high temperature, high slag resistance	Kiln furniture, incinerator, blast furnace
Silicon carbide-graphite	High refractoriness, high strength at high temperature, high thermal conductivity, high thermal shock resistance	Incinerator
Silicon nitride	High strength, high thermal shock resistance, relatively high oxidation resistance	Kiln furniture, blast furnace
Composite bricks:		
Silicon carbide Containing bricks	High corrosion resistance against low iron oxide, high strength at high temperatures, high thermal shock resistance	Ladle, blast furnace, electric arc, torpedo ladle, iron ladle
Magnesia-carbon	High slag resistance, high thermal shock resistance	Basic oxygen furnace, electric arc furnace, ladle
Alumina-carbon	High refractoriness, high thermal shock resistance, high corrosion resistance	Submerged entry nozzle, slide gate

Types & characteristics of raw materials used in various clay & non-clay refractory manufacture is tabulated below.

Table 1.11: Types and Characteristics of Raw Materials used in Refractory Manufacture

Type	Characteristics
Clay Refractories	
Fireclay	Consists of kaolinite ($Al_2O_32SiO_22H_2O$) and minor amount of other clay materials. Fireclay refractories can be low, medium, high, or super-duty based on their resistance to high temperature or refractoriness. Fireclay refractories are used to produce bricks, insulating refractories, and two types of ladle brick.
High-alumina	Composed of bauxite or other raw materials that contain 50 to 87.5 percent alumina. High-alumina refractories are generally multipurpose, offering resistance to chipping and higher volume stability. High-alumina refractories are used to produce brick and insulating refractories.
Nonclay Refractories	
Basic	Produced from a composition of dead-burned magnesite, dolomite, chrome ore, and small amount of other minerals. Basic refractories can be further subdivided into magnesia, dolomite, chrome, and combination bricks. Basic refractories are typically used to make bricks.

Extra-high alumina	Made predominately from bauxite or alumina (Al_2O_3), extra-high alumina refractories contain from 87.5 to 100 percent alumina and offer good volume stability. They are typically poured into special shapes using a fused casting process.
Mullite	Made from kyanite, sillimanite, andalusite, bauxite, or mixtures of alumina silicate materials; mullite refractories are about 70% alumina. They maintain a low level of impurities and high resistance to loading in high temperatures.
Silica	Containing silica, silica refractories are characterized by a high coefficient of thermal expansion between room temperature and 500°C (930°F). Silica brick is available in three grades: super-duty (low alumina and alkali), regular, and coke oven quality. Silica compositions can be used for hot patching, shrouds, and bricks.
Silicon carbide	Produced by the reaction of sand and coke in an electric furnace, silicon carbide refractories are used to make special shapes, such as kiln furniture, to support ceramicware as it is fired in kilns. It has high thermal conductivity, good load bearing characteristics at high temperatures, and good resistance to changes in temperatures.
Zircon	Containing zirconium silicate (ZrO_2SiO_2), zircon refractories maintain good volume stability for extended periods or exposure to high temperatures. Zircon refractories are widely used for glass tank construction.

Clays

Clay is composed mainly of fine particles of hydrous aluminum silicates and other minerals that is plastic when moist but hard when fired. Fireclay is the predominate clay used in firebrick; bentonite, in foundry sand; common clay, in refractory mortar & cement; and kaolin, in calcine, grog, high alumina brick, kiln furniture, and plug, tap, & wad.

Non clays

Nonclay refractories are composed of alumina, mullite, chromite, magnesite, silica, silicon carbide, zircon, and other nonclays.

Typical Refractories used in Industries

The following are the common types of refractories used in various industries:

- **Silica bricks:** Contains at least 92% of silica; bricks of this type are normally bonded with a small amount of lime.
- **Siliceous brick:** Contains from 85-92% of silica, the remainder being essentially alumina.
- Semi-siliceous brick: Contains 77 % of silica, the remainder being essentially alumina.
- **Fire brick:** A brick that in the fired state, consists essentially of alumina-silicates and silica, with <78% of silica and <38% of alumina.
- **Aluminous brick:** Contains from 38-45% of alumina, the remainder being essentially silica.

The names of other types of refractory give a sufficient indication of their general composition. e.g. magnesite brick, chrome brick, chrome magnesite brick, silicon carbide brick, sillimanite brick, *etc.*

Depending on the area of application such as boilers, furnaces, kilns, ovens etc, temperatures and atmospheres encountered different types of refractories are used.

Fireclay refractories

Fireclay refractories, such as firebricks, siliceous fireclays and aluminous clay refractories consist of aluminium silicates with various amounts of silica ranging from SiO_2 content of less than 78% and containing less than 44% of Al_2O_3.

Table below shows that as the quantity of impurities increases and the amount of Al_2O_3 decreases, the melting point of fireclay brick decreases. Owing to its relative cheapness and widespread location of the raw materials used to manufacture firebricks, this material finds use in most furnaces, kilns, stoves, *etc.*

Table 1.12: Properties of typical fireclay bricks

Fireclay Brick	Percent SiO_2	Percent Al_2O_3	Other Constituents,%	PCE , °C
Super Duty	49-53	40-44	5-7	1745-1760
High Duty	50-80	35-40	5-9	1690-1745
Intermediate	60-70	26-36	5-9	1640-1680
High Duty (Siliceous)	65-80	18-30	3-8	1620-1680
Low Duty	60-70	23-33	6-10	1520-1595

Firebrick is the most common form of refractory material. It is used extensively in the iron & steel industry, nonferrous metallurgy, glass industry, pottery kilns, cement industry, and by many others.

Fireclays are hydrated aluminum silicates that occur naturally. They are sufficiently pure to serve as raw materials for refractories. The principal mineral in fireclays is kaolinite.

While other clay minerals may be present, the formula $Al_2O_3.2SiO_2.2H_2O$ can usually represent the clay fraction.

Plastic and semi-plastic fireclays, as their names indicate, develop varying degrees of plasticity when they are mixed with water. This is an important factor in the manufacture of fireclay bricks, because the plastic fireclays facilitate the forming process and act as a bonding phase for the raw & calcined flint clays, and they have greater variation in their impurity content.

High alumina refractories

Alumino silicate refractories containing more than 45% alumina are generally termed as high alumina materials. The alumina concentration ranges from 45 to 100%. The refractoriness of high alumina refractories increases with increase in alumina percentage. The applications of high alumina refractories includes the hearth & shaft of blast furnaces, ceramic kilns, cement kilns, glass tanks and crucibles for melting a wide range of metals.

High-alumina bricks serve as a multi-purpose refractory material for severe environments. They are used extensively in the steel industry for such applications as hot metal cars, electric furnace roofs, piers & muffles for a variety of furnaces, and numerous applications where strength at high temperature is an essential requirement. The aluminium and glass industries use high-alumina refractories to keep the melt in the molten state.

Most high-alumina refractories are classified according to their alumina content, which could range from 50 –99%. They are designated as 50%, 60%, 70%, 80%, 85% and 90% alumina.

Two classes of high-alumina refractories are distinguished by a microstructure that is essentially a single, crystalline phase.

These are: (1) mullite refractories and (2) corundum refractories.

Mullite refractories

Mullite is about 72% alumina with 28% silica. The manufacturing procedures are designed to maximize the formation of the compound mullite ($6Al_2O_6.2SiO_2$). A refractory with 71.8% alumina and 28.2% silica will be composed of only mullite ($6Al_2O_6.2SiO_2$) if fired at equilibrium conditions.

However, the extent to which well-developed mullite crystalline form occurs in a refractory depends on the purity of the raw materials used and the manufacturing processes, particularly firing. Therefore, all high-alumina bricks with around 70% alumina may have a well-developed mullite phase.

Mullite refractories have excellent volume stability and strength at high temperatures. They are highly suitable material for electric furnace roofs, blast furnaces and blast furnace stoves, hot metal cars, and the superstructure of glass tank furnaces.

Corundum refractories

The 99% alumina class of refractories is called corundum. These refractories comprise single-phase, polycrystalline, alpha- alumina.

Silica brick

Silica brick is a refractory material containing at least 93% SiO_2. The raw material is quality rocks. Various grades of silica brick have found extensive use in the iron and steel melting furnaces. In addition to high fusion point multi-type refractories, the other important properties are their high resistance to thermal shock (spalling) and their high refractoriness. It finds typical use in glass making and steel industry.

The outstanding property of silica brick is that it does not begin to soften under high loads until its fusion point is approached. This behaviour contrasts with that of many other refractories, for example alumino silicate materials, which begin to fuse and creep at temperatures considerably lower than their fusion points. Other advantages are flux & slag resistance, volume stability and high spalling resistance.

Silica bricks are made from raw materials that are essentially quartz. During the initial firing, alpha-quartz is first converted to beta-quartz, accompanied by an abrupt expansion at 576 °C. Slow rates of heating are required through this temperature range to prevent cracking, as the volume change is about 0.9%. Since the final firing temperature is somewhat over 1026 °C, the brick as it is put into service consists of cristobalite particles (properties of cristobalite are not affected by temperature fluctuation, provided the temperature does not drop below 600°C), with possibly some having residual unconverted quartz cores.

Magnesite Refractories

Magnesite refractories are chemically basic materials, containing at least 85% magnesium oxide. They are made from naturally occurring magnesite ($MgCO_3$). The properties of magnesite refractories depend on the concentration of silicate bond at the operating temperatures. Good quality magnesite usually results from a CaO-SiO_2 ratio of less than 2 with a minimum ferrite concentration, particularly if the furnaces lined with the refractory operate in oxidizing and reducing conditions. The slag resistance is very high particularly to lime and iron rich slags.

Chromite refractories

Here, a distinction must be made between chrome-magnesite refractories and magnesite- chromite-refractories. Chrome magnesite material usually contain 15-35% Cr_2O_3 and 42-50% MgO whereas magnesite-chromite refractories contain at least 60% MgO and 8-18% Cr_2O_3.

Chrome-magnesite refractories are made in a wide range of qualities and are used for building the critical parts of high temperature furnaces. These materials can withstand corrosive slags & gases and have high refractoriness. The magnesite-chromite products are suitable for service at the highest temperatures and in contact with the most basic slags used in steel melting. Magnesite-chromite usually has a better spalling resistance than chrome- magnesite.

Zirconia refractories

Zirconium dioxide (ZrO_2) is a polymorphic, material. There are certain difficulties in its usage and fabrication as a refractory material. It is essential to stabilize it before application as a refractory. This is achieved by incorporating small quantities of calcium, magnesium and cerium oxide, *etc.* Its properties depend mainly on the degree of stabilization and quantity of stabilizer as well as the quality of the original raw material. Zirconia refractories have a very high strength at room temperature which is maintained upto temperatures as high as 1500°C. They are, therefore, useful as high temperature constructional materials for furnaces and kilns. The thermal conductivity, of zirconium dioxide is found to be much lower than that of most other refractories and the material is therefore used as a high temperature insulating refractory.

Since Zirconia exhibits very low thermal losses and does not react readily with liquid metals, it is particularly useful for making refractory crucibles and other vessels for metallurgical purposes. Zirconia is a useful refractory material for glass furnaces primarily since it is not easily wetted by molten glasses and because of its low reaction with them.

Oxide refractories (Alumina)

Alumina refractory materials which consist of aluminium oxide with little traces of impurities are known as pure alumina. Alumina is one of the most chemically stable oxides known. It is mechanically very strong, insoluble in water & super heated steam, and in most inorganic acids & alkalies. Its properties make it suitable for the shaping of crucibles for fusing sodium carbonate, sodium hydroxide and sodium peroxide. It has a high resistance in oxidizing and reducing atmosphere. Alumina is extensively used in heat processing industries. Highly porous alumina is used for lining furnaces operating up to 1850°C.

Typical refractory lining pattern in different zones of a steel reheating furnace is given below:

Indicative lining pattern of a Reheating Furnace Roof:

- **Soaking Zone**: 250 mm thick special shaped bricks made of 60% Alumina Refractories backed by 50 mm thick Insulating castable.
- **Heating Zone**: 250 mm thick special shaped Refractories 60% Alumina quality backed by 50 mm thick Insulating castable.
- **Pre-Heating Zone**: 250 mm thick special shaped Refractories 40% Alumina quality backed by 50 mm thick Insulating castable.

Indicative lining pattern of Reheating Furnace side walls and end walls:

- **Discharge end wall**: 230 mm thick 60% Alumina quality refractories backed by 115 mm hot face insulation bricks, 115 mm thick Mica Insulation bricks (IS-2042) and 75 mm thick Calcium silicate blocks insulation and 5 mm thick asbestos sheet.
- **Side walls (Heating & Soaking zones):** 230 mm thick 60% Alumina quality refractory bricks backed by 115 mm thick hot face insulation bricks, 115 mm thick Mica Insulation bricks / Cold face insulation bricks, 75 mm thick Calcium silicate block insulation and 5 mm thick asbestos sheet.
- **Side walls (Preheating Zone):** 230 mm thick 40% Alumina quality (IS-8) refractory bricks backed by 115 mm thick Hot face insulation bricks, 115 mm thick Mica Insulation bricks, 75 mm thick calcium silicate block insulation and 5 mm thick asbestos sheet.
- **End wall (Charging side):** 230 mm thick 40% Alumina quality(IS-8) refractory bricks backed by 115 mm thick Hot face insulation bricks, 115 mm thick Mica Insulation bricks, 75 mm thick calcium silicate block insulation and 5 mm thick asbestos sheet..
- **Flue Off take/down corners**: 150 mm thick ceramic fibre blanket (RT128) to be held in position by heat resisting (Stainless steel) studs and washers in case of flue port at the roof. If the flue line is below the hearth, lining pattern will be same as that of charging side end wall.
- **Soaking Zone of the Hearth**: 150mm thick High Alumina fire bricks (60% Alumina) backed by 75 mm thick 40% Alumina quality firebricks, 230 mm thick cold face insulation bricks and 5 mm thick asbestos sheet.
- **Heating Zone of the Hearth:** 150 mm thick High Alumina quality fire bricks (60% Alumina) backed by 75 mm thick 40% Alumina (IS-8) Firebrick, 115 mm thick Hot face insulation bricks, 115 mm thick cold face insulation bricks and 5 mm thick asbestos sheet.
- **Pre-Heating Zone of the Hearth:** 150 mm thick High Alumina fire bricks (45% Alumina) backed by 75 mm thick 40% Alumina quality firebricks, 115 mm thick Hot face insulation bricks, 115 mm thick cold face insulation bricks and 5 mm thick asbestos sheet.
- **Charging end Doors:** 175 mm thick RT128 grade ceramic fibre module backed by 25 mm thick mineral wool insulation blanket.
- **Discharge end Doors:** 225 mm thick RT128 grade ceramic fibre module backed by 25 mm thick mineral wool insulation blanket.
- **Recuperator area:** 115 mm thick IS-8 quality firebricks backed by 115 mm thick Mica insulating bricks and 50 mm thick calcium silicate block insulation.

- **Flue duct from Recuperator up to chimney:** 115 mm thick IS-6 quality firebricks backed by 115 mm thick Mica insulating bricks.

REQUISITES OF A GOOD REFRACTORY FOR A PARTICULAR JOB

- It should be able to withstand the temperature to which it is liable to be exposed.
- It should not be corroded by the slag produced in the furnace and to which the refractory is to be exposed.
- It should be resistant to the cutting action of flue gases and flames.
- Contraction and expansion of the refractory should not be sudden but should be uniform.
- It should be able to withstand the pressure which is likely to be put on it.
- It should be able to conserve heat and prevent heat loss from the furnace.
- It should be able to withstand sudden changes of temperature.
- It should have the ability to resist the rubbing action of the material that comes in its contact (*i.e.* high abrasion resistant).
- It should be impermeable to liquids and gases {*i.e.* low permeability and porosity)
- It should have ability to promote heat transfer in recuperative type of furnaces (*i.e.* high thermal conductivity)
- It should be able to resist disintegration by undergoing least possible structural changes on cooling and heating (*i.e.* low co-efficient of thermal contraction and expansion)

The general requirements of a refractory material can be summed up as:

- Its ability to withstand high temperatures and trap heat within a limited area like a furnace.
- Its ability to withstand action of molten metal , hot gases and slag erosion *etc.*
- Its ability to withstand load at service conditions.
- Its ability to resist contamination of the material with which it comes into contact.
- Its ability to maintain sufficient dimensional stability at high temperatures and after/during repeated thermal cycling.
- Its ability to conserve heat.

SELECTION OF REFRACTORIES

The selection of refractories for any particular application is made with a view to achieving the best performance of the equipment, furnace, kiln or boiler and depends on certain properties of the refractories. Further, the choice of a refractory material for a given application will be determined by the type of furnace or heating unit and the prevailing conditions *e.g.* the gaseous atmosphere, the presence of slags, the type of metal charge *etc.* Therefore, temperature is not the only criterion for selection of refractories.

Any furnace designer or industry should have a clear idea about the service conditions which the refractory is required to face. The furnace manufacturers or users have to consider the following points, before selecting a refractory.

- Area of application.
- Working temperatures.
- Extent of abrasion and impact.
- Structural load of the furnace.
- Stress due to temperature gradient in the structures and temperatures fluctuations.
- Chemical compatibility to the furnace environment.
- Heat transfer and fuel conservation.
- Cost considerations.

It is essential to have an evaluation of the above conditions. The assessment of the desired properties would provide guidelines for selection of the proper refractory materials.

The furnace manufacturer or a user is also concerned with the conservation of energy. Fuel can be saved in two ways, either by insulation or by faster working. Both these methods give low energy cost per ton of product.

There is practically no application where one particular refractory quality can withstand all operating conditions. Usually one would like to have a refractory with low thermal conductivity so that heat may be more effectively contained within a furnace. Sometimes, however refractories and materials having high thermal conductivity are desired; for example, a protective muffle in certain ceramic kilns is designed to prevent combustion gases from reaching the ceramic ware. It must transfer as much heat to the ware as possible, so conductive ceramic materials like *silicon carbide* are often used for muffles.

We must take into account the specific metal we will be melting, the temperatures we will be reaching, the length of melt, how long we will be holding metal in the furnace, how much inductive stirring will take place, what additives or alloying agents we will be using, and furnace relining practices. Therefore refractory selection is based on the most critical factor vis-à-vis the wear out pattern and operating parameters in order to achieve the best possible techno economics. A right selection will not only gives a longer life but helps in reducing down hours ; thus a better productivity.

The five most important factors governing refractory service are:

- Area of application
- Type of refractory
- Operating conditions
- Quality of refractory
- Workmanship

It is essential to have an objective evaluation of the above considerations. These factors are not necessarily in order of importance. The user needs to adjust them in the order to suit the conditions existing in any particular furnace.

AREA OF APPLICATION

Selecting insulation materials and designing insulation systems requires evaluation of four factors:

1) Thermal performance; 2) Physical properties; 3) Chemical characteristics and 4) Costs

1. Thermal performance- involves temperature limit; melting or fusion temperature; thermal conductivity (or insulation capability); heat capacity or storage; thermal expansion; and thermal shock (spalling) resistance. Operating temperature is an important criterion for selection of refractories. Temperatures exceeding the refractory's rating can soften its surface and cause rapid erosion & failure. If the operating conditions heat or cool the refractory lining beyond its specified range, the resulting thermal shock can damage the integrity of the lining.

- **Design of furnace/vessels** -The performance and stability of the refractory lining depends on the structural design of the furnace and its configuration. Sometimes adequate attention is not given to the refractory lining and its engineering related issues during early plant equipment design. This may lead to several compromises as there is limited scope for change in vessel design, operating conditions and the process to reduce the impact of these factors on lining performance during later stages of detailed engineering.
- The refractory lining failure may occur due to insufficient combustion volume. The heat released within the system is generally more than absorbed by the process and is dissipated through walls or exhausted with flue gases. In such cases, there are possibilities of the lining approaching the flame temperature and causing severe problems. The burner type, its design, location, flame shape, possibilities of flame impingement, flowing pattern of flue gases, *etc.*, may affect the lining. In many cases, limited vessel dimensions, inaccessibility and complicated configuration restricts the best lining practices during initial construction as well as subsequent maintenance & repair.

2. Physical properties- includes density or porosity; abrasion; wear & erosion resistance; electrical resistivity for use in electric heating; mechanical strength; and other structural properties at high temperatures. High mechanical strength is required to resist load, impact, abrasion and erosion in refractory materials.

- **Physical Shock and Mechanical Stress -** Most refractory materials tend to be brittle and weak in tension. Bulky charge material dropped into an empty furnace can easily cause the lining to crack upon impact. If a crack goes unnoticed, molten metal may penetrate, leading to a run-out with the possibility of a water/molten metal explosion.

  The sudden or cumulative effects of physical shocks and mechanical stress can lead to failure of refractory lining.
- **Thermal Stress -** The entire refractory lining is subjected to thermal stress which also plays a dominant role in selection. The refractory lining system or design must take care of thermal resistance, expansion relief; prevent cracking and eliminating built-in barriers. Thermo-mechanical FEM analysis is a reliable tool to investigate thermal stresses and develop ways on improving the lining behavior.
- **Refractory Wear -** Refractory linings are subject to regular wear from the scraping of metal on the furnace walls. In theory, refractory wear should be uniform, but in practice this never occurs. The most intense wear occurs at the slag/metal interface; where sidewalls join the floor; and at thin spots caused by poor lining installation.

3. Chemical considerations- It includes uniformity of composition; reactions between base materials & operating environment; and issues such as volatilization of the constituents or binding agents, corrosion, chemical attack, or diffusion and reactions with the product. Chemical attack may occur from gases such as steam, hydrogen, carbon monoxide, acid & alkali slag, sulfurous gases, *etc.*; these acid gases can initiate various problems in the lining. All of the important operating factors and any other criteria specific to the process under consideration should be verified for their possible effect on the performance of the lining.

- **Chemical v/s Physical Characteristics.** Chemical characteristics of the furnace process usually determine the type of refractory required, although this is not universally true. We have the familiar example of silica brick, the most acidic of refractories, being successfully used in the roofs of basic open hearth steel furnaces. In this case the physical properties of the brick, particularly load bearing strength, are more important than the chemical characteristics. For successful refractory practice the physical & chemical properties of the refractory must be compatible with the furnace operation and nature of the process.

4. Cost / Economics – it means initial installation labor; maintenance, repair; & replacement costs. Selecting materials solely based on price and ease of installation should be avoided. Very often, monolithic lining system is selected in the lining design of critical equipment where brick lining or some other design may be more suitable. The selection may be due to cheap & easy availability of monolithics and easier installation than brick lining and to avoid preparation of too many engineering drawings for the complicated brick shapes & laying details. In many cases, recommendations of refractory manufacturers are biased and based on their available product ranges, which may not be appropriate for the required conditions.

TYPE OF REFRACTORY

Refractory materials are typically quasi brittle at low temperature and have a viscous behavior at high temperature. They are subject to considerable variability in strength, resulting from local variations in their microstructure and their lack of ductility. They are characterized by high-temperature creep or plastic deformation. Their brittleness and their high elastic modulus make them sensible to failure under thermal stresses and shocks.

In order to judge the type of refractory material most likely to give satisfactory service in a certain furnace, one must have knowledge of the limitations as well as the strong points of the principal types of refractories. Very often it is desirable to use different types of refractories in various parts of the same furnace to obtain uniformly good or balanced refractory service.

OPERATING CONDITIONS

Selecting materials should always be based on properties & specifications suitable for the specific application and operating conditions. We frequently hear from users that "the refractories fail prematurely and our operating practice has not been changed in years." This may be true, but very often there can be unintentional changes in furnace conditions long enough in duration to destroy a refractory lining but not long enough to be considered as a change in practice. For example, a burner getting out of adjustment so that the furnace atmosphere becomes highly reducing for a short period of time; possibly the operating temperature of the furnace may become abnormally high and damage the grid-work before it is corrected; or perhaps the chemical composition of raw material charge or the fuel may have changed. Such variations in practice are usually extremely difficult to discover because they leave no record except for their effect on the lining. Another complicating factor is that the failure of the refractory may not become apparent until weeks or even months after the actual damage was done.

Transportation, handling, timely arrival of refractory bricks, mortars, skill of masonry work, proper equipment for application e.g. mixer machine for castable, vibrator for installation, forms *etc.* are very important.

QUALITY OF REFRACTORY

Quality of refractories can be the controlling factor in the success or failure of thermal equipment as well as the safe and profitable operation of the plant. The performance of any refractory is considered reasonable when similar service lives are achieved on a regular basis. Premature lining failure may be defined as one that does not achieve normal or average performance and service life.

The most common refractory problems are:

- Hot spots (higher casing temperature);
- Excessive cracking;
- Spalling of lining (thermal, mechanical, structural);
- Erosion and thinning of lining;
- Chemical attack/corrosion from process gases (such as hydrogen, carbon monoxide, sulfur dioxide, alkalies), flue gases (sulfur, sodium, vanadium), steam, etc;
- Acid-gas dew-point corrosion of refractory and metallic parts;
- Partial melting and degradation of lining;
- Excessive shrinkage and development of gaps;
- Anchor failure and detachment of lining from wall;
- Failure of metal liner over refractory;
- Explosive spalling during dry out;
- Mechanical damages.

Critical applications, purchasing of materials based on a comparison of product datasheet or catalogue specifications or equivalent principles should be avoided. The actual performance references and records for specific products or brands should be verified against similar applications. Reviews of manufacturing facilities & quality control program, and random inspection & testing of important properties are essential.

WORKMANSHIP

Proper installation of the refractory is as important as selection of the right material. If the material is inadequately compacted during installation, voids or areas of low density may form, creating a weak spot easily attacked by molten metal. Maintaining proper expansion gaps, correct dimension of bricks and monolithics, fixing anchors *etc.* all are very important to achieve better life of refractories.

Some of the important aspects of refractory construction are:

Bond: Bond in brick work is an arrangement by which the joints between the bricks are staggered in all directions in a pre-determined manner. This ensures stability in construction as well as air tightness. An all header construction is recommended in the hot face of a furnace where heavy refractory wear is anticipated either due to action of molten metal slag or dust laden gas moving at high velocities. Where intermittent repairs to the walls are expected, the working layer is built separately from the main wall.

Wall Thickness: Structural stability should be taken into consideration while deciding the wall thickness. As a good engineering practice, furnace walls lined with 115mm thickness bricks should not be constructed beyond a height of one meter without proper anchorage. From the stability point of view minimum thickness should be around 230mm, which may be suitably increased as the height goes up. Use of metallic sheets with anchoring arrangements, improves the stability of the wall where basic refractories are used. Such anchoring of walls with the furnace structure should have sliding arrangement to take care of the movement of refractories during heating up and cooling down.

Joints: The joints form a small part of the lining of the vessels but they have two important functions.

- To assemble the pieces (bricks or blocks) together.
- To absorb thermal expansion and to limit the stresses generated in the masonry.

The joints may be either mortared joints (i.e. granular material with matrix and voids) or dry (without mortar) joints. In either case, the stress/strain behavior of the joint interface differs considerably. At room temperature, compressibility of the joints is around 20%. At 1200°C, the mortar becomes more and more plastic and its compressibility approaches 50%.

Mortar Joints: Mortar joints improve the stability of construction. It also takes care of slight variation in sizes as well as warpages by offering a uniform bedding surface. Chemical and physical properties of mortar should be identical to that of the bricks which they are joining. Workability of the mortar is important, in the absence of which, good construction is not possible. The excess mortar applied to the brick surface should come out from the edges when one brick is tapped against the other, to ensure that the joints are full and that there is not open gap.

Expansion Joints: Since refractories expand on heating up , suitable provisions should be made at the time of construction for this expansion to take place without creating undue stresses which may result in failure of the refractories. Usually, expansion joints are built in as the construction proceeds. The necessary provision is made depending on the expansion characteristic of the particular refractory at the operating temperature. Care should be taken to check the expansion allowance – in practice just one half of the theoretical expansion is sufficient. The remaining one half will be absorbed by the mortal joints, or even dry joints between the bricks.

Structural Support: Refractories are normally held in position with steel shell or a structural framework by tie rods as failure of structural by other reasons will cause partial failure or limit the total collapse of refractories. Their alignment, shape etc should also be equal concern. Also important is that the structural should not get overheated, and should be adequately insulated wherever required.

Good brickwork is extremely important but there are other factors involved like the proper size & location of expansion joints, the rise of an arch, or the thickness of a wall as properly coming under the head of masonry workmanship. If the refractory lining is improperly cantered or has somehow been distorted during storage or shipment, lining thickness will be uneven and it may fail prematurely before the end of its predicted service life. The refractory manufacturer's procedures for installation, drying and sintering must be followed.

The five most important factors governing refractory service are tabulated below.

Table 1.13: Factors governing refractory service

Area Of Application	Type Of Refractory	Operating Conditions	Quality of Refractory	Workmanship
Selecting insulation materials and designing insulation systems requires evaluation of four factors:	Refractory materials are typically quasi brittle at low temperature and have a viscous behavior at high temperature.	Selecting materials should always be based on properties and specifications suitable for the specific application and operating conditions.	The performance of any refractory is considered reasonable when similar service lives are achieved on a regular basis.	Proper installation of the refractory is as important as selection of the right material.
1) Thermal performance; involves temperature limit; melting or fusion temperature; thermal conductivity , heat capacity or storage; thermal expansion; and spalling resistance	They are subject to considerable variability in strength, resulting from local variations in their microstructure and their lack of ductility.	We frequently hear statements from users that "the refractories fail prematurely and our operating practice has not been changed in years."	Premature lining failure may be defined as one that does not achieve normal or average performance and service life.	If the material is inadequately compacted during installation, voids or areas of low density may form, creating a weak spot easily attacked by molten metal.
2) Physical properties; includes density or porosity; abrasion; wear and erosion resistance; electrical resistivity; mechanical strength; and other structural properties at high temperatures	In order to judge the type of refractory material most likely to give satisfactory service in a certain furnace, one must have knowledge of the limitations as well as the strong points of the principal types of refractories.	This may be true, but very often there can be unintentional changes in furnace conditions long enough in duration to destroy a refractory lining but not long enough to be considered as a change in practice.	The most common refractory problems are: • Hot spots (higher casing temperature); • Excessive cracking; • Spalling of lining (thermal, mechanical, structural);	Maintaining proper expansion gaps, correct dimension of bricks and monolithics, fixing anchors *etc.* all are very important to achieve better life of refractories. Some of the important aspects of refractory construction are:

3) Chemical characteristics; includes uniformity of composition; reactions between base materials and operating environment; and issues such as volatilization of binding agents, corrosion, chemical attack, or diffusion and reactions with the product.	Very often it is desirable to use different types of refractories in various parts of the same furnace to obtain uniformly good or balanced refractory service.	For example, a burner getting out of adjustment so that possibly the operating temperature of the furnace may become abnormally high and damage the grid-work before it is corrected.	• Erosion and thinning of lining; • Chemical attack/corrosion from process gases (such as hydrogen, carbon monoxide, sulfur dioxide, alkalies), flue gases (sulfur, sodium, vanadium), steam, etc; • Partial melting and degradation of lining; • Excessive shrinkage and development of gaps; • Detachment of lining from wall; • Mechanical damages.	Bond: Bond in brick work is an arrangement by which the joints between the bricks are staggered in all directions in a pre- determined manner.
4) Costs; mean initial installation labor; maintenance, repair; and replacement costs.		Such variations in practice are usually extremely difficult to discover because they leave no record except for their effect on the lining.		Wall Thicknes: Joints: Mortar Joints: Mortar joints improve the stability of construction.
		Transportation, handling, timely, skill of masonry work etc is very important.		Expansion Joints: Structural Support: Refractories are normally held in position with steel shell or a structural framework by tie rods

Key refractory Selection Facts

- Reaction/wear increases exponentially with increasing temperature.
- Temperature cycling will cause cracks, loss of strength,& spalling of material.Stronger is not always better.
- Lining life is controlled by the "weakest link".
- Shorter service life can be more cost effective.
- Refractory manufacture should use high purity, ultrafine materials.
- Microstructure engineering – control of aggregate shape is important.

Since, no single refractory material meets the above requirements/properties hence refractories with a maximum given set of properties are chosen for a specific purpose to meet the service conditions/requirements. Refractories are expensive and any failure in the refractories results in great loss of production, time, equipment and some time the product itself. The type of refractories also will influence energy consumption and product quality. Therefore, the problem of obtaining refractories best suited to each application is of supreme importance.

FACTORS AFFECTING SELECTION OF REFRACTORIES

Optimum use of refractories is achieved by careful study of furnace design and evaluation of operating conditions prior to selection of refractory products which meet the design and operating requirements.

From the multiple factors listed below, it may appear that the choice of most suitable material would be exceedingly difficult. Sometimes this is true.

Moreover, the best refractory selection often depends on a few requirements so important that other factors play a minor role. In some cases, refractoriness, i.e., maximum service temperature, alone will be the deciding factor, and in others high refractoriness will have to be coupled with resistance to thermal shock. Under other circumstances, resistance to metals, slags, or disintegration by reducing gases may be the governing factors. Sometimes high insulating value is desirable, but in other situations high thermal conductivity may be needed. When selecting refractories, the following major factors must be identified.

OPERATING FACTORS

- Function of the furnace
- Nature of material being processed
- Rate and continuity of operation
- Range and rapidity of temperature changes
- Chemical attack by metals, slags, ash, *etc.*
- Fluidity of molten metal or slag
- Velocity of furnace gases
- Abrasion from contained solids or gases
- Impact from charging
- Erosion by molten furnace contents
- Impinging flames or hot spots

FURNACE DESIGN & CONSTRUCTION RELATED FACTORS

- Type of furnace
- Design and dimensions of walls and arches
- Loads imposed on the lining
- Conditions of heating (one or more sides)
- Amount of insulation
- Air- or water-cooling
- Type of refractory construction (brick or monoliths)
- Methods of bonding or support
- Provision for thermal expansion
- Mechanics of any moving furnace parts

REFRACTORY RELATED FACTORS

- **Properties at room temperature–**
    - Workmanship and physical strength

  - Density, porosity, permeability
  - Chemical and mineral composition
  - Uniformity
  - Size and design
- **Properties at high temperatures**–
  - Refractoriness or maximum service temperature
  - Reversible thermal expansion
  - Resistance to thermal shock
  - Resistance to chemical attack
  - Resistance to mechanical impact or stress
  - Resistance to abrasion or erosion
  - Permeability to gases or liquids
  - Volume stability (bloating or shrinkage)
  - Resistance to gases and fumes
  - Thermal conductivity
  - Heat capacity
  - Electrical resistance
- **Economic factors** -
  - Delivered cost
  - Cost of installation (brick vs. monoliths)
  - Special shapes or forming required
  - Service life

TESTING OF REFRACTORIES

The durability of refractories is intimately connected with economics and efficiency of the process in they are employed. Refractories are characteristically anisotropic in nature which makes it all the more difficult to judge exactly the durability while it is in use. The furnace designer has to choose the refractories according to the data from different tests available to him. That means in order to fulfill the users or various application requirements and to achieve long service lives, refractory products must be assessed by testing.

The main functions of refractories testing may be classified into three types:

1. Evaluation of new materials before use.
2. Quality control by the manufacturer or user.
3. Post-mortem examination of refractory bricks that behaved differently than the normal.

The testing of refractories can be categorized as:

- Non-destructive type.
- Destructive type.

A list of refractory testing including both these types of testing is given below:

- Bulk density
- Apparent porosity
- Apparent specific gravity
- True specific gravity and true density
- Particle size analysis
- Permeability
- Pore size distribution
- Refractoriness or Pyrometric Cone Equivalent (P.C.E)
- Refractoriness Under Load (R.U.L)
- Cold Crushing Stress (C.C.S)
- Modulus of Rupture (M.O.R)
- Modulus of Elasticity (M.O.E)
- Permanent Linear Change (P.L.C)
- Reversible Thermal Expansion (R.T.E)
- Thermal Conductivity
- Differential Thermal Analysis (DTA)
- Thermo-gravimetric Analysis (TGA)
- Spalling Resistance (water quenching and air spalling)
- Hydration Resistance Test
- Creep in Compression Test
- Abrasion Resistance Test
- Resistance to Carbon Monoxide
- Slag Corrosion Test
- Microstructural analysis under Optical Microscope
- Mineral Phase identification by X-ray Diffractometer (XRD pattern)
- Chemical Analysis

The influence of refractories on fuel economy arises from many causes, among which are the following:

- The stability of the furnace structure is obviously an essential feature in the efficient performance of industrial operation.

- Maintenance of a sound construction free from cracks within the bricks of the joints in essential. Cracks in the brick work may cause short circuiting of flue gases or leakage of live gas into the air or air into the gas, with disastrous effects on temperature uniformity in the furnace, on the amount of fuel used and on the brickwork; this is particularly important in coke ovens and gas retorts.
- The action of slag on furnace brickwork may be such as to destroy it. In extreme circumstances it may give rise to holes through furnace walls, for example. into the heating flues.
- Correct thermal gradient is also important; in many firing operations bricks tend to wear down to a limiting thickness, no matter how thick they may be originally.
- In continuously operated furnaces the longer the periods between cooling down for repairs , the less fuel is needed for heating up the furnace from the cold. On the other hand, planned maintenance may be more economical even though it may mean doing repairs sooner than they are, strictly speaking required.
- In many industrial operational heat has to be transmitted through a wall of refractory material and in all furnaces a great deal of heat is stored within the bricks. Their thermal properties, therefore, affect the quantity of fuel used and are of interest to the furnace operator.

Machineries & equipments used in the refractory production process

Several types of machineries & equipments are used to produce refractories eg: mixing/kneading machines, presses, and kilns.

Mixing/Kneading Machines

Figure below illustrates different machines used to mix or knead refractory products.

There are two types of mixing and kneading machines: fixed vessel and driven vessel. Mixing homogenizes more than two types of bulk materials, and kneading machines make a uniform coating layer. Mixing and kneading machines are equipped with mixing blades or muller wheels. Heating, cooling, or de-airing equipment may also be applied to the vessel. Mixing and kneading machines are used for manufacturing shaped and unshaped refractories. Unshaped refractories, however, are not processed any further.

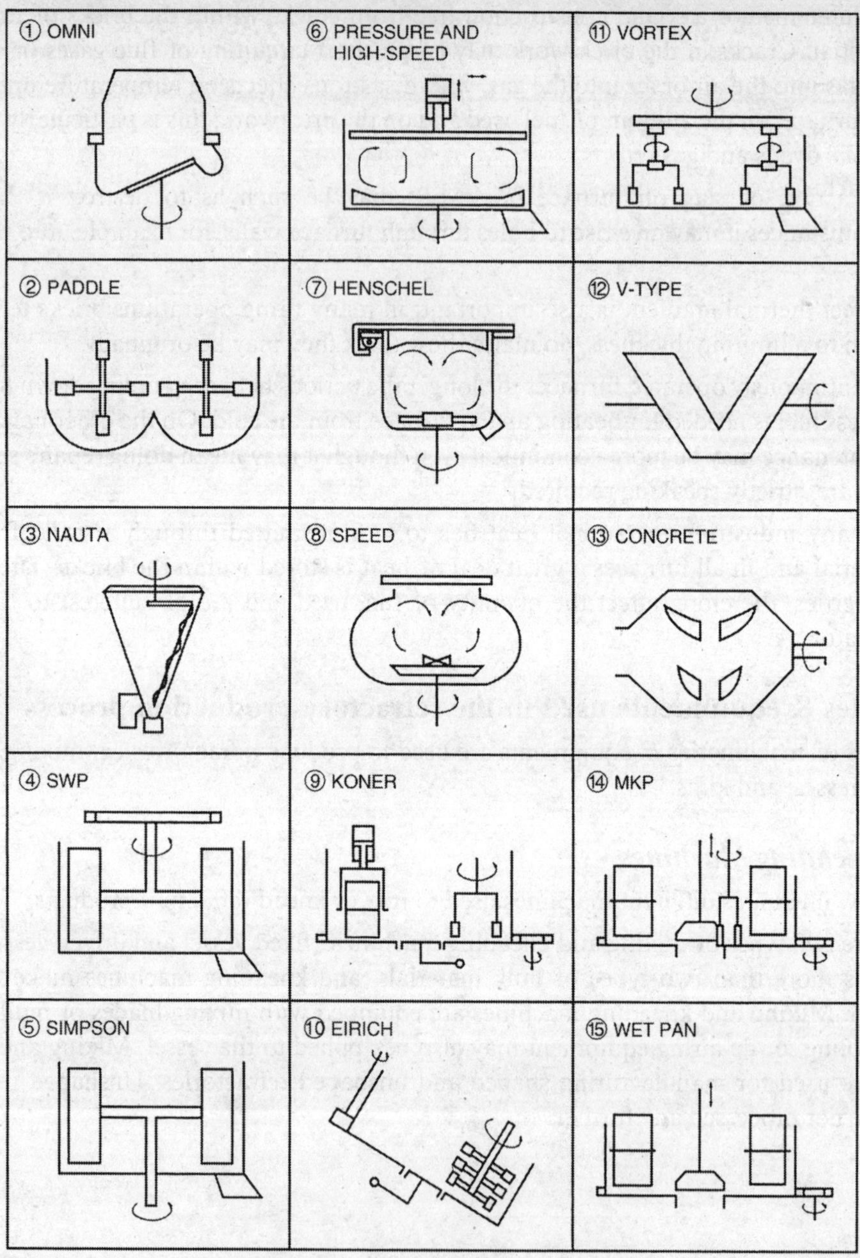

Fig.1.7: Mixing and Kneading Machines

Refractory Presses

Refractory pressing machines are broadly categorized into three groups: impact & static, vibrating, and cold isostatic press. Choosing between the three groups of presses largely depends on the type of raw materials used.

- **Impact & Static Presses**: Figure below illustrates a friction and a hydraulic screw press, two types of impact presses. Another Figure is a diagram of a hydraulic screw press, a type of static press. Impact and static presses are typically equipped with a vacuum deaerator. Impact presses have a higher allowable maximum compacting force than static presses. However, static presses find increasing application in the production of sophisticated refractories such as submerged nozzles & shrouds and in the production of industrial ceramics. Bricks formed with static presses are flat, uniform and, compact.
- **Vibrating Press**: Vibrating presses, shown in Figure below, are classified into two types: air cylinder type and hydraulic cylinder. The vibrator in the air cylinder type is attached to the table, and the air cylinder compacts the material. The hydraulic vibrating press is constructed with the hydraulic pulse generator attached to the pressure block, and the hydraulic cylinder compacts the material. Vibrating presses are typically used for the compaction of complexly shaped refractories.

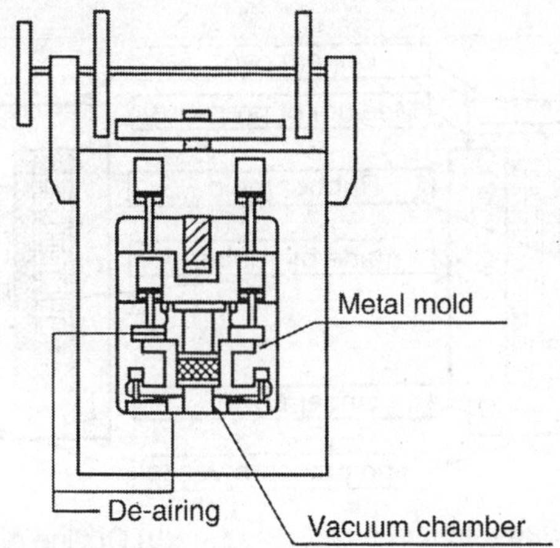

Fig.1.8: Vacuum Press (Friction, Hydraulic Press)

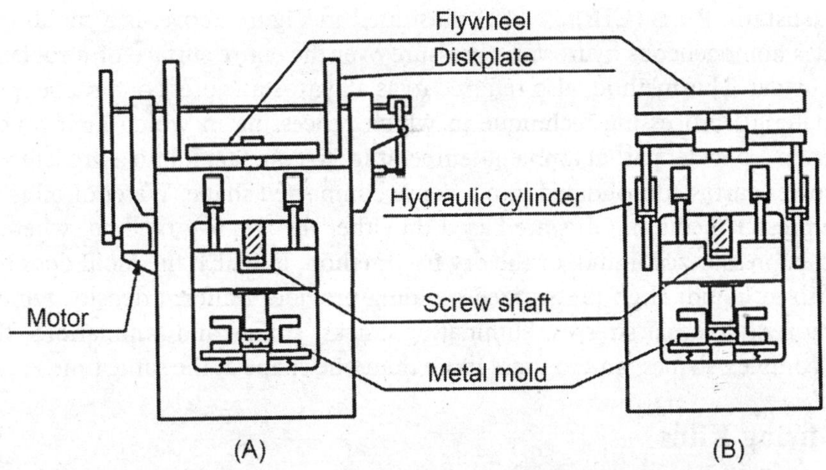

Fig.1.9: Friction Press (A), and Hydraulic Screw Press (B)

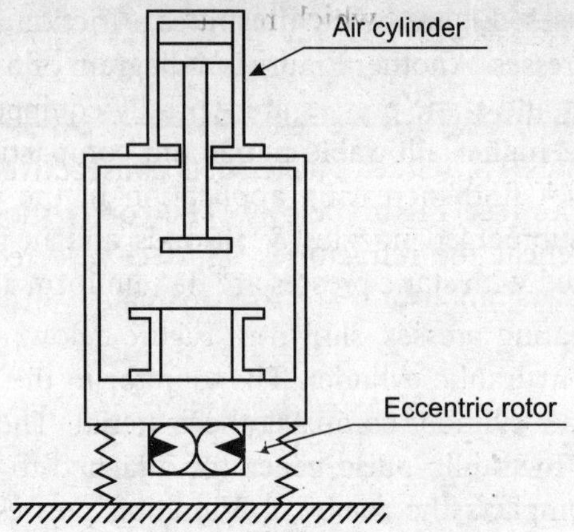

Fig.1.10: Vibrating Press

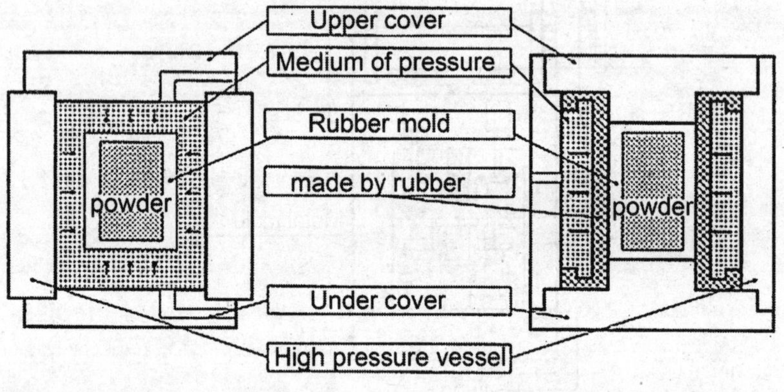

Fig.1.11: Cross Section of CIP (Cold Isostatic Press)

- **Cold Isostatic Press (CIP):** A CIP, illustrated in Figure above, is a molding device that provides homogeneous hydrostatic pressure over the entire surface of a rubber mold filled with powder. This method, also referred to as a hydrostatic press or a rubber press method, is a materials processing technique in which processing in which high fluid pressure is applied to a powder part at ambient temperature to compact it into a predetermined shape. The power part is consolidated into a dense compacted shape. Water or oil is usually used as the presser medium. CIPs are based on either the wet bag method, where the mold is placed in pressurized liquid, or the dry bag method, in which the mold does not touch the pressurized liquid. High pressurized molding provides uniform density, which leads to a reduction of internal stresses, eliminating cracks, strains and laminations, the ability to make complex shapes, and to press more than one shape at the same time.

Refractory firing Kilns

Refractories are fired to develop the materials refractory properties. The unfired ("green") refractories pass through a heat treatment, which results in a thermally stable refractory and/or crystallization.

The refractory industry uses three types of kilns:

- **Tunnel Kiln**: In a tunnel kiln, refractory products consecutively pass through preheating, firing, and cooling zones (see Figure below). The combustion gas from the firing zone is typically used to preheat the refractories. Heat can be recovered from cooling fired refractories and reused as combustion air. Approximately 80 % of shaped refractories are fired in tunnel kilns.

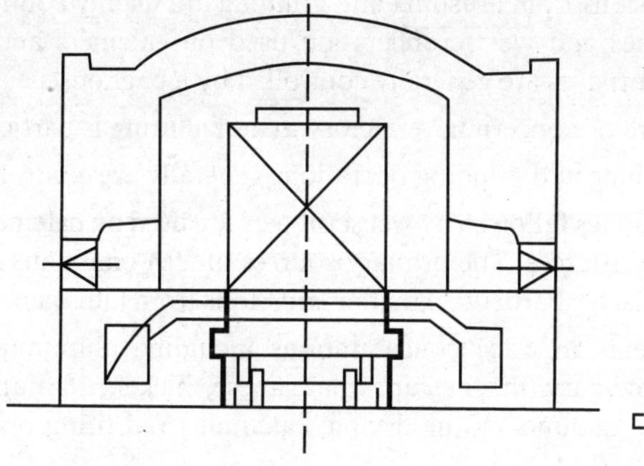

Fig.1.12: Tunnel Kiln

- **Round Periodic Kilns**: Round periodic kilns are typically used to fire silica bricks. Figure below is a diagram of a round periodic kiln. These kilns can be used to fire large refractory products that cannot be fired in a tunnel kiln and can easily accommodate changes in production.
- **Shuttle Kilns**: As illustrated in Figure below, the design of a shuttle kiln resembles the firing zone of a tunnel kiln. Shuttle kilns effectively store heat and are used to fire fireclay and specialty bricks.

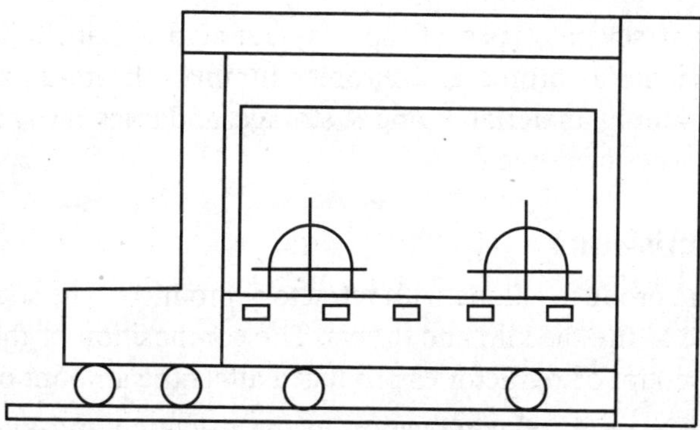

Fig.1.13: Shuttle Kiln

Pollutants Emissions & Controls in Refractory Manufacturing

Refractory production leads to emissions of organic particulate matter (PM); metals; and gaseous pollutants, including sulfur dioxide (SO_2), nitrogen oxides (NO_x), carbon monoxide (CO), carbon dioxide (CO_2), fluorides, & volatile organic compounds (VOCs).

PM & Metals Emissions

PM and metal emissions occur during the crushing, grinding, calcining, & drying of the raw materials; the drying & firing of unfired refractory bricks and tar & pitch operations; and finishing of the refractories. Emissions from crushing and grinding are readily controlled with fabric filters. Product recovery cyclones and wet scrubbers are used on calciners and dryers to reduce PM emissions. Electric arc furnaces are generally controlled by a baghouse.

The primary pollutant of concern in refractory manufacturing is particulate matter (PM).

Emissions from crushing and grinding operations generally are controlled with fabric filters.

Product recovery cyclones followed by wet scrubbers are used on calciners and dryers to control PM emissions from these sources. The primary sources of PM emissions are the refractory firing kilns and electric arc furnaces. Particulate matter emissions from kilns generally are not controlled.

A number of elements in trace concentrations including aluminum, beryllium, calcium, chromium, iron, lead, mercury, magnesium, manganese, nickel, titanium, vanadium, and zinc also are emitted in trace amounts by the drying, calcining, and firing operations of all types of refractory materials.

Emissions of PM from electric arc furnaces producing fused cast refractory material are controlled with baghouses. The efficiency of the fabric filters often exceeds 99.5 percent. Emissions of PM from the ceramic fiber process also are controlled with fabric filters, at an efficiency similar to that found in the fused cast refractory process. To control blow chamber emissions, a fabric filter is used to remove small pieces of fine threads formed in the fiberization stage. The efficiency of fabric filters in similar control devices exceeds 99 percent. Small particles of ceramic fiber are broken off or separated during the handling and forming of the fiber blankets in the curing oven. An oil is used in this process, and higher molecular weight organics may be emitted. However, these emissions generally are controlled with a fabric filter followed by incineration, at an expected overall efficiency in excess of 95 percent.

Chromium is used in several types of nonclay refractories, including chrome-magnesite, (chromite-magnesite), magnesia-chrome, and chrome-alumina. Chromium compounds are emitted from the ore crushing, grinding, material drying & storage, and brick firing and finishing processes used in producing these types of refractories.

Gaseous & VOCs Emissions

SO_2, NO_x, CO & CO_2 are produced along with refractory products. The source of most of the SO_2 emissions is the fuel used to fire the kiln and dryers. The composition of the clays & plaster added to refractories and the amount of refractories produced affect the amount of SO_2 produced.

Pollutants emitted as a result of combustion in the calcining and kilning processes include sulfur dioxide (SO_2), nitrogen oxides (NOx), carbon monoxide (CO), carbon dioxide (CO_2), and volatile organic compounds (VOCs). The emission of SOx is also a function of the sulfur content of certain clays and the plaster added to refractory materials to induce brick setting. Fluoride emissions occur during the kilning process because of fluorides in the raw materials.

Volatile organic compounds emitted from tar and pitch operations generally are controlled by incineration, when inorganic particulates are not significant. Based on the expected destruction of organic aerosols, a control efficiency in excess of 95 percent can be achieved using incinerators.

Manufacture, Properties and Applications of Important Refractories

Some important refractories are described below :

FIRE CLAY REFRACTORIES

Manufacture :

Fire clays are hydrated aluminium silicates represented by Al_2O_3, SiO_2/H_2O. At higher temperature water is driven out. The common fire clays contain 25 to 45% Al_2O_3 and 50 to 80% SiO_2. Clays with higher Al_2O_3 contents are usually called higher alumina refractories. Fire clays are classified in several kinds of clays e.g. flint (hard clay), plastic clay, kaolin, etc. The term 'firebricks' refers to fireclay bricks. Flint clays are very hard and non plastic. Its plasticity, can be increased by addition of bond clays. Refractoriness of plastic clays can be increased by addition of more refractory clay. The desired properties of fireclay is achieved by mixing different clays in proper proportions. Fusion temperature of fireclay (varies from 1580 to 1750°C) is lowered by the presence of alkalies, sand, gravel, calcium silicate, iron oxide, magnesium silicate, iron silicate etc. The most harmful elements are the alkalies and basic elements which form double silicates of lower fusion point. Composition and fusion points of some refractory clays are tabulated in Table 2.1.

Table 2.1: Composition of Refractory Clay

Composition (%)	Clay Types		
	Plastic	Semi-plastic	Non-plastic (flint)
Al_2O_3	26-38	17-28	29-38
SiO_2	48-61	56-74	48-57
Fe_2O_3	0.6-2.5	0.6-2	0.6-1.6
TiO_2	0-1.8	0.8-1.5	0.5-2.0
CaO	0-1.3	0-2.75	0-0.06
Alkalies	0.15-0.90	—	< 0.5
Fusion point, °C	1600-1725	1600-1700	1650-1750

Weathered clay i.e. clay after mining is left in open thereby causing decay of organic matter present in it due to heat, cold and atmospheric moisture. This results in increase in plasticity of clay also due to production of some organic acids. It is crushed and ground in gyratory crushers and edge runners.

Specification/features of raw materials for fireclay refractories are tabulated in Table 2.2.

Table 2.2: Specifications of Raw Materials for Fireclay Refractories

Type of raw materials	Al_2O_3(raw) minimum %	Fe_2O_3(raw) maximum%	Alkalis %	Loss on ignition (LOI),%	P.C.E. Orton cone no, minimum	On calcination at 1450°C for 2 hours			Remarks
						Specific gravity	Porosity, %	Linear change, maximum %	
Plastic :									fine grain, soft, has long vitrification range, maximum grit is 2 and 5% for grade I and II respectively.
Grade I	30	1.5	2.0	14	30	2.70	4 6	12	
Grade II	33	2.0	1.5	14	32	2.75		15	
Semi-plastic	30	1.75	2.0	14	29	2.65	6	6	fine texture, has long vitrification range, soft
Non-plastic:									fine texture, hard, dense, non-flaky and less crumbly
Grade I	33	1.00	1.00	12	32½	2.75	4	2	
Grade II	32	1.20	1.25	12	31½	2.70	6	4	
Grade III	30	1.25	1.50	14	30½	2.10	10	5	

The fire clays can hold water mechanically in addition to the chemically held water. During firing, due to the expulsion of all this water and fluxing effect of the impurities, the product shrinks considerably and becomes out of shape & size. To overcome this difficulty the fire clay is mixed with materials like silica that expand and/or large amounts of materials which do not shrink. This latter material is known as grog. Grog is fire clay which has already been heated to 1300°C so that no further contraction in volume takes place. Broken and used fireclay bricks are also used as grog. A small amount of plastic clay is added to facilitate moulding. A mixture is made by grinding raw fire clay, grog and plastic clay with water. There are several ways of giving a final shape to the mixture. In colder and more humid climates they are dried in dryers.

Grog addition to fireclay is mainly as an anti-shrinkage element in the form of angular particles of various sizes to achieve better interlocking of grains. The main advantages of grog addition to clay are:

- reduced shrinkage on heating
- enhanced strength of fired refractories
- greater spalling resistance
- increased specific gravity and decreased apparent porosity
- increased production rate due to necessity of less addition of water to get a workable plasticity and lesser time required for drying the raw refractories.

Grog addition ranges from 20 to 90% in the clay mixture depending upon the properties of Raw clays and the desired properties of finished products. High percentage of grog necessitate the use of highly plastic clays for the development of proper bonds. Ordinary fireclay refractory contain 20-40% grog. Use of higher percentage of grog results in high heat duty and improved properties of the final product. High grog fireclay refractories may contain upto 90% grog.Size variation in the finished product are due to high shrinkages that take place on heating during drying and firing.

Quality of high grog products can further be improved by using :

- Fine dispersion and thorough tempering of highly plastic bond clay to ensure its proper distribution and effective binding by as thin layer of clay as possible.
- High grain density calcined material for improved strength and soundness of the product.
- Properly graded raw materials in the mixture to ensure highest packing density and reduction in the use of plastic clay for binding. Proper blending and uniform properties in the final product is ensured by a proper sequence of charging various raw materials in the mixture.

Finely crushed refractory clay is mixed with prepared grog with bond clay or calcined bauxite, if necessary either dry or wet. In wet mixing, the materials are mixed with water in a pug or pan mill. The materials mixture may be soaked in water for 30-36 hours or more before transferring to the pug mill so that the water is absorbed by the clay and distributed throughout the mass uniformly. The mixture after mixing may be allowed to mature for a few days in a cool celler before moulding to increase plasticity.

Bricks moulding is done either by hand moulding, pneumatic ramming or by mechanical pressing. Water content in hand moulding & machine moulding mixtures is 7-9% & 4-7% respectively. Higher moulding pressure is required for mixtures having lower moisture content whereas higher moisture containing mixes increase the slumping chances at the time of release from the mould. Hand moulded bricks are less dense compared to machine moulded and may not be of appropriate size. Machine moulding gives a higher production rate. Mechanical presses may be rotating table, friction screw, jolting, hydraulic and toggle type. Other moulding methods are extrusion, wire cutting and repressing. Special shapes like tyres and sleeves etc. are made by extrusion followed by repressing. Refractory mixture shaped by extrusion require 15-20% moisture in the mix and hence extruded shapes need special attention during drying to avoid crack formation & other defects. Pressure required for friction screw, toggle & hydraulic presses to produce complicated shapes (having uniform size, homogeneous structure and improved compressive strength) using a mixture containing 5-7% moisture is from 300 to 800 kg/cm^2.

Higher amount of fines and moisture reduces the effectiveness of deairation process. Under-pressing of bricks results in high porosity and low strength of the bricks whereas overpressing may cause crack formation in the bricks. Moulds and dies used for shaping fireclay bricks are slightly of oversize than the final size of the refractories as it shrinks on heating.

Moulded bricks are dried under shade or in well ventilated rooms. Steam heated floors are used for final drying of green hand moulded or extruded bricks after 1 to 4 days of natural drying. Drying is also done in continuous tunnel driers in which cars loaded with moulded bricks are pushed at a uniform rate and hot air at 150-250°C is introduced from the discharge end which leaves the drier at 60-80°C at charging end. Moisture lost by the bricks joins the outgoing air which increases humidity at the charging end of the drier and controls the drying rate in the initial stages. Humidity controlled driers heated by hot flue gases are used for bricks drying.

Depending upon the composition, moulding method, type and size of bricks, the total drying period ranges from 24 to 200 hours. Slow drying rate ensures uniform drying of both inside and outside of compact brick whereas faster drying rates make the bricks liable for crack formation due to uneven drying & development of strains in the refractory shapes.

The bricks are then fired in a kiln. At the lower temperature, water is expelled. At 980°C an exothermic reaction takes place due to the formation of mullite ($3Al_20_3.2SiO_2$) and the excess silica crystallizes as cristobalite. Usually the temperature is raised to 1400°C and maintained for a few hours to obtain uniform properties. For higher temperature use, the firing temperature may be 1500°C. The furnace is then closed and allowed to cool slowly. The total time taken is about a week.

Dried fireclay bricks are fired/burnt in batch or continuous type kilns which may be up-draught, down-draught or horizontal draught type. Depending upon the kiln/furnace temperature and changes that take place in the dried refractories, firing of fireclay bricks is divided into following five stages namely :

- steaming or smoking
- decomposition of clay
- oxidation or full fire stage
- incipient fusion
- annealing and cooling

In the first stage of firing called steaming, any moisture remaining in the dried brick is expelled upto 120°C which further continues as the kiln temperature increases. This stage of steaming and smoking lasts for about 20 hours.

During second stage, which also lasts for 20 hours (and begins at 350°C and extends upto 900°C), decomposition of clay occurs and combustibles, volatile matter & any molecular water present is driven out.

In full fire or oxidation stage (which lasts for 12 to 36 hours and extends upto 1350°C), various mineral transformations occur. At 950°C, exothermic mullite ($3Al_2O_3.2SiO_2$) formation takes place when alumina begins to combine with silica. Between 1350 to 1380°C, crystallisation of mullite is completed. The clay shrinks and hardens thereby no shrinking of refractories results on subsequent heating or in use.

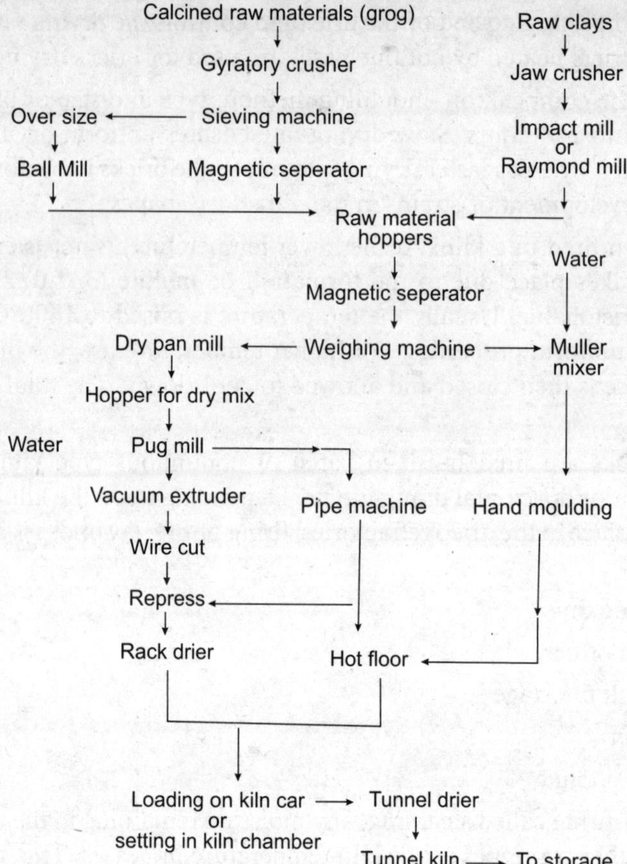

Fig.2.1: Flow sheet for wet plastic process of Fireclay Brick Manufacturing.

Fourth firing stage (which extends upto 1500°C) is accompanied by the incipient fusion of the impurities and filling up of the interstices between the solid particles by the fused mass. As the pores are filled in by vitrified mass, the refractories becomes less and less porous, bonding of the clay & grog particles take place resulting in the compact & hard product. Firing temperature is higher than that at which the bricks are supposed to be used. The bricks are soaked at highest temperature for some hours to achieve uniform properties.

Final stage of annealing and cooling involves very slow cooling of the fired bricks in the kilns itself. In batch type kilns, the fire is extinguished, the damper is closed and most of the heat is lost by radiation. This cooling takes the same time as the heating the goods to the maturing temperature.

Higher rate of firing are used for high grog refractories which have low shrinkage. High alumina refractories involve high firing temperature (1500-1680°C) depending upon the percentage of alumina. Firing temperature can be reduced by using some additives which on firing the refractories from liquid phases necessary to develop binding of refractory grains. However, such refractories have reduced refractoriness under load (RUL) values.

After burning, refractories are sorted and stored. Under fired bricks are fired again and overtired bricks are rejected.

Flowsheet for wet plastic process and dry process for manufacturing of fireclay bricks are given in Figs. 2.1 and 2.2 respectively.

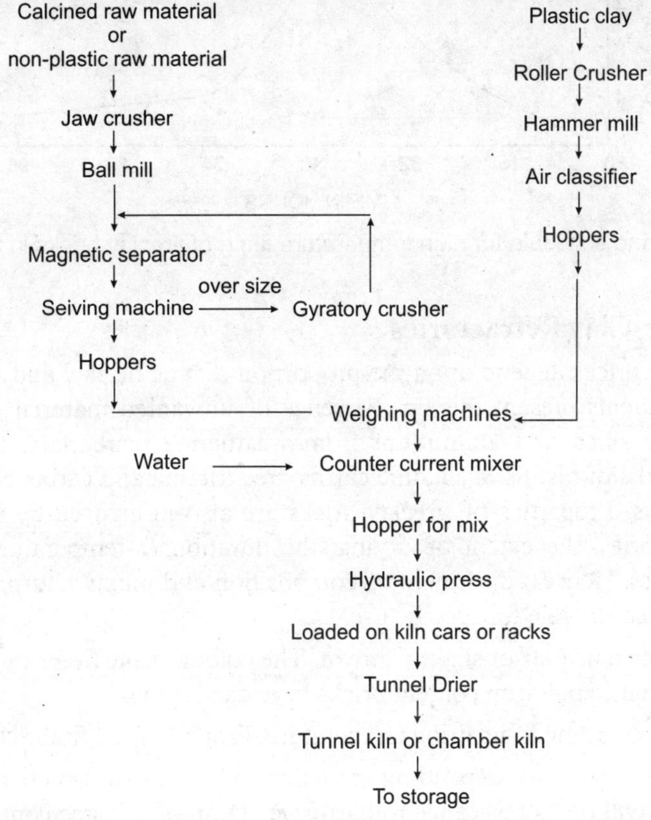

Fig.2.2: Flowsheet for dry process of fireclay brick manufacture.

In fact firing temperature of fireclay brick depends on its composition and it increases with the increase of alumina content, the maximum being 1600-1650°C for 80-95% alumina bricks. Two typical firing schedules for a low temperature and a high temperature tunnel kilns for firebricks are shown in Figs. 2.3 and 2.4 respectively.

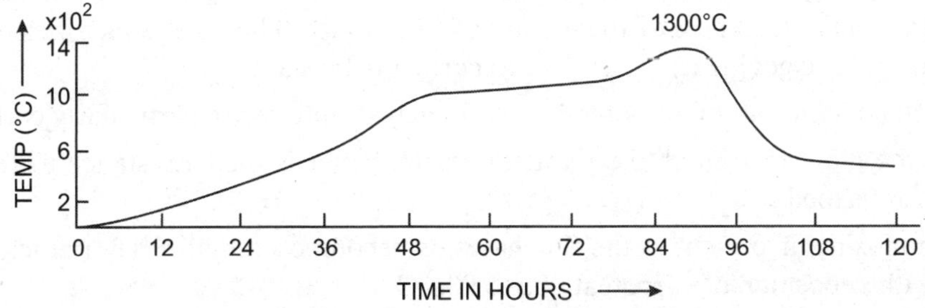

Fig.2.3: Firing schedule for low temperature firing of fireclay bricks in tunnel kiln

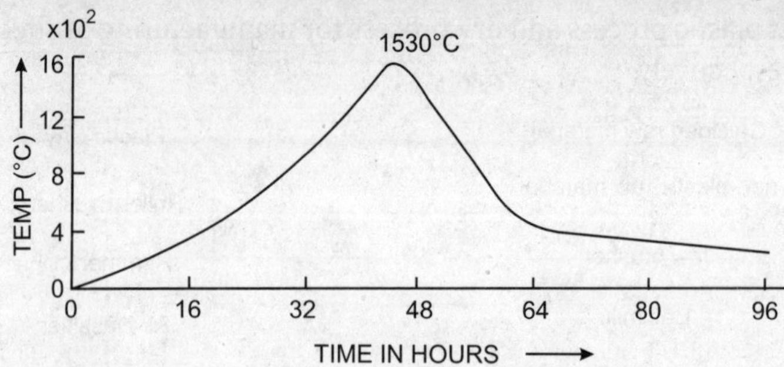

Fig.2.4: Firing schedule for high temperature firing of fireclay bricks in tunnel kiln

Properties of Fire-Clay Refractories

Properties of fireclay bricks depend upon the proportion & type of clay and amount of alumina, grog & other constituents present therein. Presence of unwanted material in the fireclay brick (upto 15%) like free silica and aluminium from weathering, carbonate, iron oxide, titanate, sulphide, sulphate, salts of alkalis & alkaline earths, free tritania and carbonaceous natter, reduce refractoriness of clays. Properties of fireclay bricks are also influenced by moisture content of the moulded refractories, the extent of drying, the duration & temperature of firing and the manufacturing process. Hence, adjustment of composition and manufacturing process is done to produce firebricks to be suitable for specific use.

Colour is generally pale buff or slightly brown. The colour of the firebrick depends on the iron content of raw materials. High iron content bricks have dark colour.

Hardness varies according to its firing temperature. Properly fired firebrick is as hard as steel.

Porosity varies from 8 to 24% depending upon the temperature to which it is fired. The greater the porosity, the less will be its resistance to corrosive action of furnace contents, fluxes etc. and also the less will be the effect of sudden changes of temperature. Amount and fineness of grog used also influences the porosity of these refractories.

Chemical Properties. At high temperature, fireclay refractories combine with alkalies such as soda, potash etc. Other bases such as lime, magnesia, iron oxide, ash etc. or salts such as chlorides, sulphates & carbonates from fusible aluminates, silicates and alumino silicates.

Resistance to slags. The resistance to slags and other corrosive materials is not due to its composition but due to closeness of the texture of the surface. Thus the refractories burnt at very high temperatures have got greater resistance to corrosion by slags.

Slags which contain sodium or potassium carbonate or nitrate are particularly corrosive.

The finer the grain texture of the refractories, the better is their resistance to the action of otherwise corrosive media.

Carbon monoxide, if present in the flue gases, is decomposed by fireclay refractories at high temperature with deposition of carbon at 400 to 900°C.

Strength or resistance to crushing. Crushing strength of bricks varies with temperature. An ordinary fireclay brick which has crushing strength as high as 950 kg/cm^2 when cold, has the following variations of crushing strength at different temperatures :

Temperature, °C	Crushing load, kg/cm^2
20	950
800	555
1000	475
1300	360
1500	65

There is a great reduction in crushing strength at about 800°C, which is very important in furnace technology.

Cold crushing strength of a well fired refractory is higher than that of a soft fired refractory. Normally, the cold crushing strength of fireclay bricks varies between 20 to 400 kg/ cm^2 depending upon their composition, texture and firing temperature.

Refractoriness. It is the resistance to fusion of the refractory under a steady rising temperature and in the absence of disturbing conditions. Fireclay refractories vary a lot in refractoriness depending upon the fireclay used for its manufacture. There are three grades of fireclay.

- 1st grade having softening temperature between 1770 and 1790°C.
- 2nd grade having softening temperature between 1670 and 1770°C.
- 3rd grade having softening temperature between 1580 and 1670°C.

So the refractoriness of a fireclay refractory corresponds to the softening point of the fireclay used.

Wide variation in refractories of the fireclay bricks varying from pyrometric cone equivalent (PCE) values of Orton 31 to 34 is dependent on the type of clay used, their alumina content & the type of grog used.

The maximum safe working temperature for fireclay bricks and other refractories is 1545°C.

Refractoriness under load (RUL). All fireclay refractories with a very few exceptions have softening point of 1350°C under a load of 3.5 kg/cm^2.

Fireclay bricks usually do not fail below 1350°C under a load of 2 kg/cm^2. RUL depends upon chemical composition and physical structure of the material. Impurities like iron, alkali etc. has a marked adverse effect on RUL.

Effect of heat. When fireclay refractory is heated, four distinct changes take place in it.

(a) ***Baking***. It is heating till the clay is decomposed and there is no more loss of water and it becomes porous. Baking takes place between 500 and 650°C.

(b) ***Vitrifying***. It is heating till some of the ingredients of fireclay are used and they fill the pores. Vitrification takes place between 1250 and 1400°C. In this stage kaolinite $Al_2O_3.2SiO.(2H_2O)$ is converted to mullite which is $3Al_2O_3.2SiO_2$.

(c) ***Squatting or Softening.*** It is heating till the ingredients are so much fused that the mass begins to loose its shape. It takes place between 1550 and 1770°C.

(d) ***Fusion.*** It is the stage when the material changes from solid to liquid state. The stage occurs above 1770°C. During fusion , mullite which is $3Al_2O_3.2SiO_2$ decomposes to form a mixture of corundum $(A1_2O_3)$ and mullite $(3Al_2O_3.2SiO_2)$.

The first two stages are generally complete when refractory (such as fire clay bricks or blocks) is fired after its manufacture. In order to complete these stages, the firing of the refractory in its manufacture should be as high as possible so that the volume changes, which are accompanied by these stages, are complete.

The 3rd and 4th stages of heating may take place when the refractory is actually used in the furnace.

Spalling or susceptibility to sudden changes in temperature. The disintegration of the refractory, when it is cooled too rapidly is called spalling. Many fireclay bricks etc. disintegrate if they are cooled very rapidly. Dense fireclay bricks or blocks disintegrate more rapidly on sudden cooling than the porous ones. So to increase the porosity generally grog, kieselguhr or saw dust etc. are added to fireclay; but with the increase of porosity, the resistance of the refractory to compression, abrasion and corrosion decreases. So the composition of the refractory is so adjusted that the porosity is not unduly increased.

The main causes of spalling in fireclay refractory are :

- low proportion of binding material;
- insufficient mixing;
- irregular or less pressure application during moulding and shaping ;
- irregular heating and cooling of the kiln when the refractory is fired.

Resistance to spalling can be increased by using a coarse grog, by increasing the porosity and decreasing the coefficient of expansion.

Fireclay bricks are the most commonly used refractories due to their cheapness. Its use in basic condition should be avoided as it is slightly acidic in nature. Due to its excellent resistance to thermal fatigue, fireclay bricks are used for regenerative furnaces and regenerators. These bricks are used for flues, chimney linings, steel castings, cement, pottery, petrochemical, steel and fertiliser industries. Other uses of firebricks are in iron blast furnaces, hot blast stoves, cupolas, recuperators and annealing, roasting, reheating & glass melting furnaces.

Specific Heat. Specific heat of fireclay bricks or block is as low as 0.25. It rises rapidly with rise in temperature and is 0.264 at 1300°C.

Thermal Conductivity. Thermal conductivity of fireclay varies from 0.8 to 0.95 (in C.G.S. units) from 300 to 1100°C. Fireclay refractory though a bad conductor of heat has a high power of radiation. So materials placed in fire chambers made of fireclay may be heated to a very high temperature, though the conductivity of fire clay is very low.

Electrical Resistance. Electrical resistance of fireclay bricks varies from 1.37×10^8 to 8.9×10^2 between 800 and 1500°C.

Durability of Fireclay Refractory. The most durable and best fireclay refractory should have high apparent density, high resistance to slags, high crushing strength and should have no volume changes when heated.

The properties of fireclay bricks are shown in Table 2.3.

Table 2.3: Properties of Typical Indian Fireclay Bricks

Type of fireclay bricks	Al_2O_3 %	Fe_2O_3 %	Refractoriness, °C	Apparent Porosity, %	Bulk density, gm/c.c.	Cold crushing strength, kg/cm^2	Thermal shock resistance	Permanent Linear Change		
								%	Tempe-rature, °C	Time, hrs
Medium heat duty	30-32	2-2.5	1655	22-25	1.95-2.05	200-250	Good	± 0.5	1350	2
High heat duty	38-40	2-2.5	1699-1717	23-26	2.0-2.10	170-230	Good	-0.5	1400	2
Super heat duty										
-I	42-44	1-1.5	1743-1763	15-16	2.15-2.25	400-600	Good	-0.5	1450	2
-II	42-45	1.5-2	1743-1763	18-20	2.10-2.20	250-300	Excellent	-0.5	1450	2

Table 2.3 (A): Composition and PCE of various heat duty typical fireclay bricks

S.N.	Fireclay brick types for various heat duty	% SiO_2	%Al_2O_3	% of other constituents	Py Pyrometric cone equivalent (PCE), °C
1	Super duty brick	49-53	40-44	5-7	1745-1760
2	High duty brick	50-80	35-40	5-9	1690-1745
3	Intermediate duty brick	60-70	26-36	5-9	1640-1680
4	High duty (siliceous) brick	65-80	18-30	3-8	1620-1680
5	Low duty brick	60-70	23-33	6-10	120-1595

Uses of Fire-clay Refractories. Fire-clay refractories are mainly used for the construction of furnaces, kilns, ovens, fire places, regenerator flues, boilers and many other like structures in which heat is generated or through which hot gases pass (e.g. blast furnace stoves). Because of its high co-efficient of thermal expansion, fireclay bricks are resistant to thermal fatigue/spalling. Coarse textured and high grog containing refractories have greater spalling resistance than those produced from fine textured & low grog mixtures.

Other uses of fireclay bricks include iron blast furnace lining, cupola, recuperators, annealing, roasting & reheating furnaces, beehive coke oven , glass melting furnaces, chimney and other furnaces which are allowed to cool frequently. They are also used in B.F. runner, steel casting and pig iron handling ladles. Besides, pottery, fertiliser, petrochemical and cement plants also use them extensively.

Salient features of Fireclay Bricks

Refractory fireclay consists essentially of hydrated aluminum silicates with minor proportions of other minerals. There are five standard classes of fireclay brick as per ASTM: superduty, high-duty, medium-duty, low-duty and semi-silica. These classes cover the range from 18 to 44% alumina, and from 50 to 80% silica.

A blend of clays is commonly used in the manufacture of high-duty and superduty fireclay brick. Flint clays and high-grade kaolin impart high refractoriness; calcined clays control the drying & firing shrinkages; plastic clays facilitate forming and impart bonding strength.

Superduty fireclay brick have good strength & volume stability at high temperatures and an alumina content of 40 to 44%. Some superduty brick have superior resistance to cracking or spalling when subjected to rapid changes of temperature. High firing temperature of fireclay bricks enhances the high temperature strength of the brick, stabilizes their volume & mineral composition, increases their resistance to fluxing, and renders them practically inert to disintegration by carbon deposition in atmospheres containing carbon monoxide gas.

Because of their greater resistance to thermal shock, high duty fireclay brick are used with better economy than medium duty brick for the linings of furnaces operated at moderate temperatures over long periods of time but subject to frequent shutdowns.

Medium-duty brick are used for conditions of moderate severity. Medium duty brick, within their serviceable temperature ranges, can withstand abrasion better than many brick of the high duty class.

Low-duty fireclay brick finds application as backing for brick with higher refractoriness, and for other service where relatively moderate temperatures prevail.

Most fireclay brick are made from blends of two or more clays. Some brick, of the low-duty class, are made of a single clay. The mixes for super duty and high duty brick commonly contain raw flint and bond clays, with or without calcined clay. In making brick of kaolin and various other clays, a large proportion of the mix is precalcined to control firing shrinkage and stabilize the volume & mineral composition of the product.

In making fireclay brick, the particles of ground clay must include a range of graded sizes, each in proper proportion. The moistened batch is then fed to a mechanically or hydraulically operated press in which the brick are formed under pressure.

In a modification of the power-press process, certain physical properties are enhanced by the application of a high vacuum during the forming of the brick. Brick made in this way have a more homogeneous texture and are harder, stronger, less porous & more dense than those made without vacuum. As a result, they are more resistant to impregnation & corrosion by slags and to penetration by gases.

The extrusion process is sometimes used for making special shapes. In making extruded brick, clays are ground in a dry pan, mixed wet or dry in a mixer, brought to the proper consistency in a pug mill, and extruded through the die of an auger machine in the form of a stiff column. The air is removed from the clay before extrusion by a deairing system within the auger machine chamber. The column is cut into brick by means of wires. The brick are then typically repressed to give them sharp corners & edges and smooth surfaces. Many intricate special shapes are formed in vertical piercing and forming presses, in which blanks from the extrusion machine are completely reshaped.

Brick formed by any of the processes described above are dried in tunnel or humidity driers. The temperature of firing depends upon the maturing temperatures of the clays, and upon the service for which the brick are intended. In firing the brick, several necessary ends are accomplished: free & combined water are driven off; iron & sulfur compounds and organic matter are oxidized,& the gases formed are eliminated; mineral transformations & changes in volume are affected; and the particles of clay are ceramically bonded together into mechanically strong brick.

Uses of Fire-clay Refractories. Fire-clay refractories are mainly used for the construction of furnaces, kilns, ovens, fire places, regenerator flues, boilers and many other like structures in which heat is generated or through which hot gases pass (e.g. blast furnace stoves). Because of its high co-efficient of thermal expansion, fireclay bricks are resistant to thermal fatigue/spalling. Coarse textured and high grog containing refractories have greater spalling resistance than those produced 20 Elements of Refractories from fine textured & low grog mixtures.

Other uses of fireclay bricks include iron blast furnace lining, cupola, recuperators, annealing, roasting & reheating furnaces, beehive coke oven , glass melting furnaces, chimney and other furnaces which are allowed to cool frequently. They are also used in B.F. runner, steel casting and pig iron handling ladles. Besides, pottery, fertiliser, petrochemical and cement plants also use them extensively.

SILICA REFRACTORY

The raw materials for silica refractories are the various forms of natural deposits of SiO_2 *e.g.* quartz, quartzite, ganister, sand, sandstone etc. The presence of an excess of basic oxides is undesirable as they lower the fusion temperature. One great difficulty with silica refractories is that silica occurs in several allotropic forms which are stable at different temperatures and have different specific gravities. Therefore, a raw material is chosen which can be easily converted into the more stable form. For this reason, common sands and pure quartzite are not very satisfactory. The raw material should have some building material preferably fine grained silica particles themselves.

In fact, quartz is the only stable form of silica stable at atmospheric temperature and quartzite is the most used naturally available form of quartz used for making silica bricks. Though, quartzite having 96-97% SiO_2 with < 0.3% alkalis are acceptable for silica bricks manufacture, however for super duty silica bricks , Al_2O_3 + TiO_2 + alkalis content should be < 0.5%, as higher percentages of impurities greatly reduces its refractoriness.

Bonding of silica grains is done by milk of lime which also serves as mineraliser and accelerates the formation of tridymite & cristobalite. Other mineralisers (like FeO, borax, Mg, Ba & carbonates, flourides & phosphates of lithium, sodium and potassium etc.) should be added to the raw materials for silica brick manufacture in minimum required quantity as they reduce the refractoriness of silica bricks. Lime clay mixture is also used for binding, however, higher percentage of lime reduces the abrasion resistance and strength of the silica bricks. To get a brick with hardened case and improved mechanical strength cellulose sulphite lye is used as a binder. Tar, gum, molasses etc. when used as binder increase the green strength of the brick. Resistance to slag attack of silica brick is increased by the addition of zircon, however compressive strength of zircon quartzite refractory is lower than quartzite or zircon material.

In order to produce silica bricks having high bulk density and improved properties, coarse (1-3 mm), medium (0.2-1 mm) and fine (-0.2 mm) grains of raw quartzite after its washing (to eliminate dust and impurities) and crushing/grinding (in jaw/gyratory crushers and ball mills) is mixed in proportions of 35-40%, 10-15% & 45-50% respectively.

Manufacture of silica bricks. For the manufacture of silica bricks, the graded material (e.g. quartzite) is mixed with required amount of binding material (i.e. lime) and water in a pan mill or revolving bottom counter flow mixer to produce a plastic paste. At higher temperatures, lime forms fusible calcium silicate which acts as a binding material. The mixture is then made into bricks by hand moulding or by machine pressing.

Moisture percentage in the semi dry mixture (called tempered mass) depends on the type of moulding and ranges from 4-5% for hydraulic pressing to 7-9% for hand moulding. While lower moisture content in the mix requires higher moulding pressure, the higher moisture content inceases the chances of slumping of the pressed bricks on release from the moulds. Because of non uniform ramming, hand moulding produces bricks of unequal hardness & density while machine moulding produces bricks of high hardness & uniform density. Higher moulding pressure produces bricks with lower thermal shock resistance because a highly compact mass on heating cracks due to insufficient gap between the expanding silica grains.

Toggle pressing or hydraulic pressing produces bricks having high bulk density, low porosity and high cold crushing strength while heavy sections are normally hand moulded. Friction presses and rotating table type presses are also used for mass production of standard shapes. Moulding rate depends on the shape & size of the bricks and may vary from 1 to 36 bricks/minute. Moulds/dies used for silica bricks is undersize to the extent of allowable expansion allowances as volume of silica increases on heating. The bricks are dried in the heated rooms.

Ordinary drying of silica bricks may take 48 to 96 hours while drying in hot room/floor requires less than 12 hours. Dried brick should contain < 0.2% moisture. Drying of bricks should be uniform throughout inside and outside of the bricks to avoid crack, formation. Tunnel driers using steam or hot flue gas from firing kiln are also used for uniform brick drying. When dried in hot rooms/floors, bricks need to be turned upside down very frequently to maintain uniform drying rate.

The bricks are dried in heated rooms. The dried bricks need to be handled with care, otherwise they crumble. They are then transferred to kilns which are heated by burning coal or gas. In the manufacture of refractory bricks on heating, mechanically and chemically held water is first removed. On heating further very often mineral transformations and volume changes take place. In the case of silica bricks another very important aim of heating is to convert quartz into cristobalite or tridymite which are the more stable allotropes of silica. If this change does not take place while firing, the volume will change during use and the structure will become weak. The temperature is slowly raised, in about 24 hours, to say 1500°C and maintained at this temperature for nearly 10-20 hours so as to allow quartzite to be converted into cristobalite which slowly changes to tridymite during cooling. Maintaining the temperature at 1500°C for a longer period is desirable but is not economical. Cooling has to be done still more carefully and takes 1 to 2 weeks. This results in the formation of a mixture of tridymite and cristobalite.

Heating and cooling rate of silica bricks during transformation should be very slow. Cooling of silica bricks below 600°C should be very slow to prevent its cracking and spalling. Microscopic examination of a well fired silica brick shows only tridymite and cristobalite.

Firing of silica brick is done usually in batch type down draught kiln. Sufficient gas is provided between the roof and the uppermost layer of silica bricks in the kiln as well as the close setting of the bricks in the kiln is avoided for accommodating its expansion on heating. Firing of silica bricks results in expulsion of water & organic matters present in it, vitrification of silicates formed by chemical action of lime/other bases with silica and the conversion of quartz to tridymite & cristobalite. Silica bricks firing can be done in tunnel kiln as well.

Firing schedule of silica bricks in batch type down draught kiln and in tunnel kiln have been shown in Fig. 2.5 and 2.6 respectively.

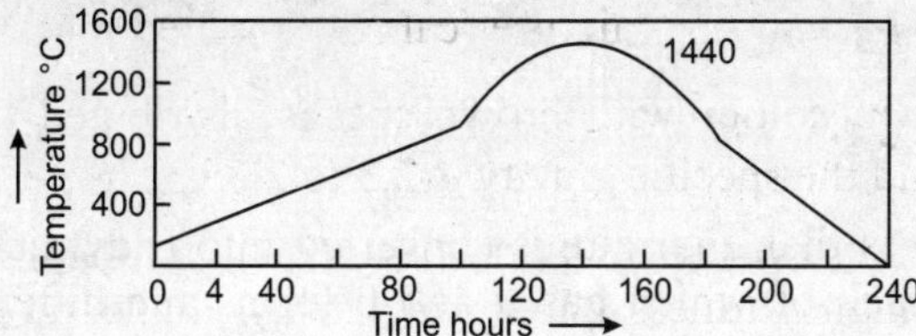

Fig.2.5: Firing schedule of silica bricks firing in.batch type down draught kiln.

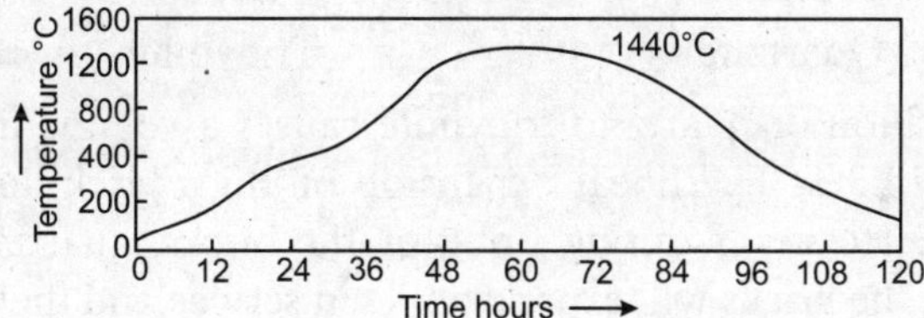

Fig.2.6: Firing schedule for silica brick firing in a tunnel kiln.

Various steps involved in the modern method of silica bricks manufacture is given in Fig. 2.7.

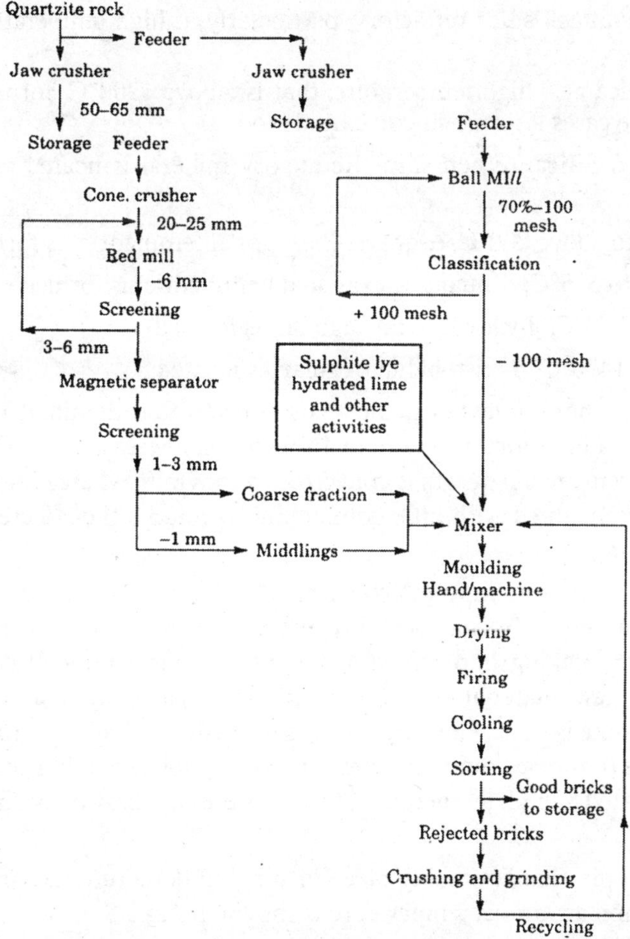

Fig.2.7: Various steps in the modern methods of Silica brick manufacture.

Properties of silica bricks

Silica bricks are yellowish in colour with brown specks distributed throughout the mass. The porosity is about 22-26% and the specific gravity is 2.3 to 2.35.

An important property is that quartzite is converted into tridymite at 870°C which changes to cristobalite at 1470°C, each of which has α and β forms and different specific gravities. The changes may be summarised as follows :

Sp.gravity	2.65	870°C	2.26		1470°C
Allotrope	Quartzite	⇔	Tridymite	⇔	cristobalite

Hence, the transformation of quartz to tridymite causes a volume increase of 17.2% and of quartz to cristobalite of 14.3%. The linear expansion of silica brick during firing (due to these transformations) is 3.5%. Hence, if during firing of the bricks, quartzite is not converted into tridymite and cristobalite, the bricks will expand while in service and the structure will fail.

Action of Chemical Agents. Silica is unaffected by any acid except hydrofluoric and phosphoric acid. Hydrofluoric acid attacks it in cold and phosphoric acid at 400°C. Bases and alkalis attack silica bricks at high temperature forming complex fusible silicates.

Alkalies and bases attack silica refractory, particularly at high temperature forming complex fusible silicates.

Carbon attacks silica at a high temperature, that is, above 1200°C, forming silicon & carbon monoxide and in some cases it forms silicon carbide too.

Action of Heat on Silica. When silica Refractory mineral is heated the following changes occur in it:

- On heating it to above 200°C, α-quartz (hexagonal tetrahedric) is formed.
- When heated to 575°C, α-quartz (hexagonal hemihedric) is formed.
- On heating to 870°C, β-tridymite (hexagonal) is formed.
- On heating to 1470°C, β-cristobalite (regular) is formed.

Cristobalite which is nore or less unstable tends to revert to tridymite if heating between 1470 and 1500°C is continued for a long time.About 16% volume increase (i.e., corresponding increase in linear expansion) occurs when quartz is converted to tridymite whereas conversion of tridymite into cristobalite is accompanied with little contraction in volume (i.e., decrease in thermal linear expansion correspondingly).

Natural silica (containing mainly quartz) which is unstable at high temperature undergoes transformation to its allotropic forms like tridymite and cristobalite involving very high volume changes. Hence, before making into refractory bricks, complete mineral conversion and hence stabilisation of silica raw material (quartz) by firing or calcination at high temperature for appreciable period of time is done. Constituents of the finished refractories (produced by reactions between solid particles of the raw materials) are rarely in complete equilibrium. Raw silica expands considerably when heated to high temperature, hence its pre stabilisation becomes necessary before making into bricks.

The various allotropic changes taking place in silica mineral (quartz) and the accompanying thermal linear expansion at various temperature is shown in Fig.2.8.

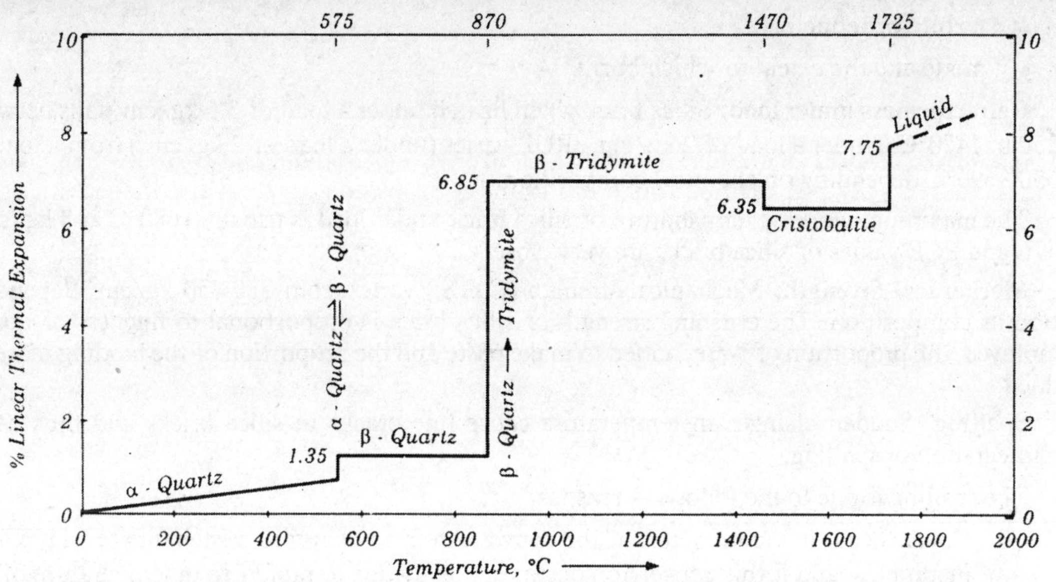

Fig.2.8: Allotropic changes in silica (quartz) mineral and assoicated thermal linear expansion.

On cooling all the above changes occur in reverse order upto 270°C and below 270°C the following changes take place:

(i) From 270°C to 180°C, β-cristobalite changes to α-cristobalite.

(ii) From 180°C to 150°C, β-tridymite changes to α-tridymite and then very slowly to quartz.

Cooling from 270°C to atmospheric temperature is accompanied by lot of volume changes and there is a danger of spalling, so this cooling is done very slowly.

Volume changes due to heating of silica refractory. Volume changes due to heating of silica are very important as the stability of silica refractory mineral depends upon this volume change when it is subjected to sudden change of temperature. These volume changes are due to the allotropic changes occurring in silica. These allotropic changes cause changes in specific gravity which in turn cause changes in volume.

The change from α-quartz to β-quartz from 200 to 575°C is accompanied by change in volume of 14 % (increase). At 870°C, when β -tridymite is formed, there is increase in volume of 14 to 16%. Conversion of tridymite into cristobalite is associated with little contraction in volume.

The conversion of silica into cristobalite, that is, heating of silica refractory upto 1470°C from room temperature is accompanied by an increase of 20% in volume and the reduction of specific gravity from 2.6 to 2.23.

The co-efficient of thermal expansion from 15°C to 940°C is about 0.0000051 inch per 1°C per inch while coefficient of expansion from 15°C to 1180°C averages about 0.0000064 inch per 1°C per inch

Refractoriness. The best silica refractory has softening temperature from 1715°C to 1760°C. The refractoriness depends upon the:

- proportion of lime soda, potash and alumina present in the refractory.

- texture of refractory.
- mode and the extent to which burnt.

Refractoriness under load. Silica brick when heated under a load of 3.5 kg/cm^2 fails between 1350 to 1470°C. Under a load of 2 kg/cm^2, RUL varies (under a load of 2 kg/cm^2) from a load of 1650-1690°C depending on chemical composition.

The maximum working temperature of silica brick under load is usually 1680°C at 3 kg/cm^2. RUL and PCE values of silica bricks are very close.

Mechanical Strength. Mechanical strength (C.C.S.) varies from 250-450 kg/cm^2 depending upon its composition. The crushing strength of silica brick is proportional to fineness of quartz employed, the proportion of water added to make paste and the proportion of the binding material added.

Spalling. Sudden changes in temperature cause fine cracks in silica bricks and thus cause disintegration or spalling.

The spalling is due to the following reasons:

(i) Cracks occur in the refractory if the conversion from quartz to cristobalite or tridymite is incomplete and if this conversion occurs suddenly due to rapid rise in temperature of the refractory. This type of spalling can be avoided by repeatedly raising the temperature of the refractory at a slow rate up to 800°C and then cooling it slowly.

(ii) Cracks occur due to the reversible expansion and contraction occurring too rapidly. This can also be avoided by heating the refractory slowly.

(iii) Spalling may occur due to the action of slags.

(iv) Stresses are created when silica brick is cooled below 250°C due to inversion of cristobalite.

In general, the spalling can be decreased in silica refractory by adopting the following means :

(i) By repeatedly heating the refractory to 800°C and then allowing it to cool slowly. By so doing, quartz is changed into tridymite rather than cristobalite.

(ii) By using a coarse grained refractory.

(iii) By increasing the porosity and decreasing the density.

Thermal Conductivity. It is higher than fireclay bricks. It varies from 0.003 to 0.004 cal/cm/sec at temperature of 700 to 1100°C.

Specific Heat. It is about 0.26.

Characteristics of a good Silica Brick/Refractory. A good quality of silica brick should have the following characteristics :

- Refractoriness should be as high as 1690°C.
- The percentage of silica should be as high as possible and the percentage of other oxides should be very low.
- It should have high crushing strength.
- There should not be any volume changes as far as possible when it is used.
- It should be burnt to such an extent that it should have high tensile strength when hot.
- It should have as little spalling as possible when subjected to sudden change of temperature.

Properties of some typical Indian silica refractory bricks have been tabulated in Table 2.4

Table 2.4: Typical Properties of Indian Silica Refractory Bricks

Type of Silica bricks used for	Minimum % of SiO_2	Maximum % of CaO	Maximum % of $Al_2O_3 + TiO_2$	Maximum % of Fe_2O_3	Refractoriness, °C	RUL under 2kg/cm²	Apparent porosity, %	Bulk density, gm/c.c.	Cold crushing strength, Kg/cm²	Permanent. linear change (PLC) at 1450CC for 2 hours, %
General Purpose	94	2.75	1.5	I	1680-1690	1670	25	1.8-1.9	250-300	+ 1
Coke oven	94	2.25	1.5	1.5	1680-1690	1650	17-24	1.7-1.9	250-300	+ 0.4
Super duty (as in glass tank furnace Crown)	97	2	0.5	0.5	1680-1690		24	1.75-1.90	300-400	+ 1
Furnace roof	94	2.5	1.3	1.0	1680-1690	1670	24	1.7-1.75	250-300	+ 1
Bessemer converter and steel transfer laddle	94	2.5	1.5	1.0	1680-1690	1650	24	1.7-1.75	250-300	+ 2

Uses of Silica Refractory. Silica bricks expand on heating. So silica refractory is used, for the following purposes:

- Where no shrinkage in refractory is required as in coke oven walls.
- Where high resistance to heat is required.

Silica refractories are used in the lining of furnaces for arches, crowns & higher parts of furnaces and kilns, that is, in portions which are the hottest in furnaces.

Due to their high heat conductivity, silica bricks are universally used in the construction of coke ovens viz. oven walls, top flue blocks, sole and roof of the ovens etc. In glass industry, silica bricks are used in roof and burner parts of the furnaces. Because of their excellent resistance to compressive stresses at high temperature, dome and upper checkerwork of hot blast stove (in blast furnace) are also lined with high density silica bricks having large specific surface area where stoves temperature remain high all through during their working and there is not much difference in dome temperature during on gas and on blast period in stoves. Due to their close fitting properties at high temperatures, high heat duty silica bricks are used in open hearth furnace roofs besides arches, crowns & other higher parts of different kilns and furnaces

Another application of silica bricks is in the construction of dome of blast furnace stove because of their excellent resistance to the compressive stresses at high temperature. High density silica bricks with large specific surface area is suitable for upper portion of hot blast stoves used to give higher blast temperature and where stove temperature remain high all through their operation.

Salient features of silica bricks.

Silica refractories are well adapted to high temperature service because of their high refractoriness, high mechanical strength & rigidity at temperatures almost up to their melting points, as well as their ability to resist the action of dusts, fumes & acid slags.

Among the important properties of silica brick are high melting temperatures, i.e. 1695 to 1710°C ; their ability to withstand pressure of 1.75 to 3.5 kg/cm^2 at temperatures within 28 to 56°C of their ultimate melting points , high resistance to acid slags; constancy of volume at temperatures above 650 °C and virtual freedom from thermal spalling above 650°C. At high temperatures, the thermal conductivity of most silica brick is somewhat higher than that of fireclay brick.

At temperatures below 650°C, silica brick have less resistance to thermal shock. They are readily attacked by basic slags and iron oxide at high temperatures in a reducing atmosphere.

Conventional quality silica brick have excellent service record due to its rigidity under load at high temperatures, high resistance to spalling above 650°C , high mechanical strength, resistance to abrasion, resistance to corrosion by acid slags and uniformity of size.

A light weight silica brick with a bulk density of 1,041 to 1,121 kg/m^3 is suitable for use up to 1650°C. At a mean temperature of 650°C , its insulating value is excellent.

Light weight silica brick are used largely for the insulation of silica brick constructions, especially the crowns of glass tanks. They are also ideal for the construction of tunnel kiln crowns, and their properties are conducive to arches having a wide span.

Certain superduty silica brick have been developed to meet the demand for a silica refractory that would permit higher furnace temperatures, give longer life and reduce maintenance costs. These brick contain no more than 35% alumina plus alkalies and titania.Superduty silica brick are used with excellent results in the superstructures of glass tank furnaces.

Semi Silica Bricks are made from siliceous materials having silica content of 75 to 92 %.These are made in a manner similar to silica bricks, are cheaper and find uses where service conditions are not so exacting that the more expansive silica bricks are needed.

Though having less refractoriness, semi-silica bricks have better spalling resistance than silica bricks. Semi-silica bricks are used as backing layers for silica bricks in coke ovens , kiln roofs etc. They are also used in the construction of reheating furnace roofs, regenerator chambers, furnace doors etc. where refractoriness coupled with resistance to slag action and spalling is desirable.

Siliceous refractory are based on pure dehydrated china clay, $A1_20_3 \cdot 2SiO_2$. The higher the proportion of $A1_20_3$ and the lower the proportion of fluxes, the higher the melting point will be. This is given by the formula:

Softening point, °C = $19 \times A1_2O_3 + 15 (SiO_2 + TiO_2) + 10 (CaO + MgO) + 6 (Fe_20_3 + Na_2O + K_2O)$ where, $A1_20_3$, SiO_2 are the percentages of oxides found by analysis (or in the refractory).

Semi-silica fireclay brick contain 18 to 25% alumina & 72 to 80% silica, with a low content of alkalies and other impurities. With notable resistance to shrinkage, they also have excellent load-bearing strength and volume stability at relatively high temperatures.

MAGNESITE REFRACTORIES

Magnesite refractories are the most important and widely used basic refractories. Magnesite is

$MgCO_3$ and occurs as a natural deposit but before it can be used as a refractory, it has to be heated strongly and the resulting product, MgO, is used as a refractory material.

In India, magnesite deposits are found in Tamilnadu, Karnataka and Uttar Pradesh. Dense and low porosity magnesite containing about 6% Fe_2O_3 (to achieve stability of magnesia at low temperature) is suitable for refractory making. Indian magnesites being purer (contains 1-2% Fe_2O_3), requires the addition of Fe_2O_3. Raw magnesite is dead burnt in shaft or rotary kiln at 1500-1700°C (to produce magnesia of crystalline form called periclase) and the lumps (in the size of 110-130 mm) are ground to < 5 mm size which is suitable for making good quality refractories. Ground magnesite is freed from dust, quartz, metallic iron (by magnetic separation) and other impurities.

An alternative for magnesite is dehydration of $Mg(OH)_2$ (produced from sea water by its treatment with slaked dolomite) to give magnesia [$Mg(OH)_2 = MgO + H_2O$]. 0.34 million litres of sea water requires 1.25 tonnes of dolomite (for precipitation) to produce one ton of magnesia.

Binding materials used for the manufacture of magnesite bricks include caustic soda, milk of lime, sodium silicate, molasses, tar, gum, portland cement, clay, feldspar, gelatinous silica, chrome iron ore, coal ash etc. depending upon the raw material and the properties required in end product. The best binding material is caustic-magnesia which produces less shrinkage in the brick on burning. Raw magnesite containing large amount of silica use milk of lime as the binder. A typical composition of dead burnt magnesite is as follows :

MgO = 87-90%, SiO_2 = 5-6%, $Fe_2O_3 + Al_2O_3$ = 1-2%
CaO = 1-2%, Loss on ignition = 1% maximum.

About 4-7% water is used for mixing with dead burnt magnesite to produce a moulding mixture having a typical seive analysis as -0.125 mm = 40%, 0.125-0.5 mm = 10% and 0.5-1.5 mm = 50%.

Drying of moulded bricks (usually moulded by hydraulic pressing using a pressure of 1000 kg/cm^2) is done at 20-30°C for about 3 weeks or more. Dried bricks should be carefully handled as they are friable and of low strength.

Magnesite ($MgCO_3$) is heated when most of CO_2 evolves at about 800°C giving caustic magnesia. Caustic magnesia is used as a cheaper refractory material. It is then heated to 1700°C till the decomposition is complete. Then no further shrinkage will take place in actual use. It does not absorb water or carbon dioxide from the atmosphere. Thus from magnesite, about 48% of dead burnt magnesite is obtained. Mineral or crystalline form of magnesium oxide (m.p. 2800°C) is calledpericlase. Periclase is crushed to the proper size & mixed with caustic magnesia or sulphide lye (obtained as a waste product from paper factories) or iron oxide as binding material and ground with water & made into bricks. The bricks are slowly heated to 1500°C, kept at that temperature for 8 hours and slowly cooled. Chemically bonded bricks which do not require firing are also made.

Properties of magnesite bricks. Magnesite brick can be used upto 2000°C without load and up to 1500°C under load(3.5 kg/cm^2). They are basic in nature but not as basic as dolomite and the reaction with other acidic bricks is not very rapid; therefore they can be used in conjunction with slightly acidic bricks.

Magnesite bricks cannot resist thermal shocks, loose strength at high temperature and are not resistant to abrasion. To overcome these defects, the bricks may be packed in metal boxes and used.

Magnesite bricks are fired in down draft chamber type kiln (similar to that used for firing silica bricks) or tunnel kiln (this gives a better control) at 1600-1800°C. Higher firing temperature reduces the after shrinkage of the bricks. Tunnel kilns require shorter firing cycle as compared to chamber-kilns which requires about 12 days firing cycle i.e. heating, soaking (for 8 hours) and cooling of the bricks. Cooling of'the bricks are done very slowly. The bricks are then sorted ; good ones are stored while rejected ones are recycled after crushing and grinding.

Composition of a typical Indian magnesite refractory brick is :

MgO = 86 % , Fe_2O_3 = 6 % , SiO_2 = 4.5 % , CaO = 2.5 % , Al_2O_3 = 1 %

As per Indian standards,magnesite refractory bricks should have :

MgO = >85 % , SiO_2 = < 5.5 % , CaO = < 2.5 %

Colour. Colour of magnesite bricks varies from grey to dark brown depending upon mainly its iron content. Bricks free from iron and other impurities are white.

Specific Gravity and Porosity. Their specific gravity is high (3.4 to 3.6) due to dense texture achieved by high moulding pressure. Presence of iron oxide increases the specific gravity while silica decreases the same. The extent of firing and presence of impurities also affect the specific gravity. The maximum apparent porosity of the bricks should be 24%.

Cold crushing strength (C.C.S.). C.C.S. of hydraulically pressed magnesite bricks is at least 350 kg/cm^2 , while the hand moulded bricks have a lower C.C.S. values.

Refractoriness and RUL. Refractoriness of a good quality magnesite brick is about 2000°C. Impurities like SiO_2, CaO, Al_2O_3, alkalies etc. reduce their refractoriness. RUL of magnesite bricks (under a load of 2 kg/cm^2) is about 200°C more than that of fireclay bricks and ranges from 125 to 1575°C. Magnesite bricks usually do not sag until very near the fusion point and remain rigid till they collapse just like silica bricks.

Thermal and Electrical conductivity. Since the heat conductivity of the magnesite bricks decreases with rise in temperature, they are useful for raising and maintaining high temperature. Electrical conductivity of these bricks is considerable at high temperature of about 1500°C.

Thermal spalling resistance. Magnesite bricks have low abrasion as well as thermal spalling resistance. Hence, these bricks are not able to resist the sudden temperature changes and disintegrate. The limitation of magnesite bricks can be minimised by :

- fine grinding of material before mixing and by using sufficiently high pressure during brick pressing.
- minimising iron oxide content in the brick as it provides better thermal shock resistance than the high iron magnesite.
- firing magnesite at high temperature for longer duration to ensure its complete conversion to periclase.

Resistance to slag attack. Being basic in nature, magnesite bricks resist the action of basic slags rich in lime and iron oxide. These bricks are not used in contact with acid slag. Bricks with high porosity and rough surfaces are liable to greater action of slags. Denser bricks resist slag attack more.

Uses of magnesite bricks. Magnesite refractories are used widely in the iron & steel industry in the form of bricks and as 'pea' magnesite (about 1 cm in diameter). Pea magnesite is used for making the bottoms of basic furnaces. The bricks are used in making basic furnaces, hot metal mixers, soaking pits, etc.

Magnesite bricks are used in those parts of a furnace which are required to resist the corrosive basic slag and which are not subjected to temperature fluctuation. These bricks are also used in furnaces used for raising and maintaining high temperature. However, the limiting factor that restricts the use of magnesite bricks is their low thermal spalling resistance, low RUL, high thermal conductivity and high cost.

Major consumer of magnesite bricks are L.D. converters, open hearth furnaces, hot metal (pig iron) mixers and electric arc & other steel melting furnaces. These bricks are also used in soaking pits and refining furnaces for gold, silver & platinum. Copper converters & fire refining furnaces; nickel, lead and antimony smelting & refining furnaces and cupro-nickel & other non-ferrous alloy melting furnaces also use magnesite bricks. Though costly, these bricks are preferred for lining hotter zones of lime and cement kilns for their increased resistance to the chemical action of hot clinker/calcine.

Induction furnace crucibles (used for melting ferrous and non-ferrous alloys) are constructed using clay bonded fused magnesia.

Metalcased magnesite refractories are used in open hearth and electric furnaces for steel melting. Steel case on heating soften and weld together resulting in monolithic lining with lowest permeability ensuring improved services. Tar or pitch impregnated/bonded magnesia is used in oxygen converters where a temperature of >1610°C is encountered or where continuous casting of steel is adopted. Fused periclase (MgO) is used as an electrical insulating material.

Magnesite paste is used for producing unfired pressed steel teeming nozzles (used for high speed top teeming of steel from large capacity ladles).

Plates/slabs for slide gates (used for steel pouring) are produced using bakelite impregnated periclase which has high heat & wear resistance and high strength. Fused periclase is also used as plate for slide gates.

Lining of copper converters & soaking pit bottoms are done using magnesite bricks while repair & construction of open hearth furnace bottom is done by pea size dead burnt magnesite.

Porous plugs and blocks used for bottom blowing of liquid steel with inert gases are made of corrundum periclase composite. Porous periclase tuyers are used for blowing low carbon steel with vacuum treatment for decarbonising. Magnesite tiles with periclase working layers and with addition of alumina is good material for stopperless casting in continuous casting plants.

Salient features of magnesite bricks.

Brick made with dead-burned magnesite are an important category of basic refractories. Magnesite brick are characterized by good resistance to basic slags as well as low vulnerability to attack by iron oxide and alkalies.

They are widely employed in applications such as glass tank checkers, sub hearth brick for electric arc furnaces, and sometimes as backup linings in basic oxygen furnaces. They are impregnated with pitch in the latter application. Magnesite compositions are also widely used to control the flow of liquid steel in continuous casting systems, either as the slide gate refractory or as a nozzle.

Various grades of dead-burned magnesite are available for the production of magnesite brick. They range from natural dead-burned materials, with MgO contents of 90% or less, to high purity synthetic magnesites containing 96% MgO or greater.

Since magnesia itself has an extremely high melting point of 2800°C, the ultimate refractoriness of a magnesite brick is often determined by the amount and type of impurity within the grain. In practice, the refractoriness of a dead burned magnesite is improved by lowering the amount of impurities, adjusting the chemistry of the impurities or both.

There are many types of magnesite refractories, both burned and chemically-bonded. They are divided into two categories on the basis of chemistry. The first category consists of brick made with low boron magnesites, generally less than 0.02% boron oxide, that have lime- to-silica molar ratios of 2:1 or greater. Often, the lime-to-silica ratio of these brick is intentionally adjusted to a molar ratio of 2:1 to create a dicalcium silicate bond that gives the brick high hot strength. Brick with lime-to-silica ratios greater than 2:1 are of higher purity than the dicalcium silicate-bonded brick. This greater chemical purity makes them more desirable for certain applications.

The second category of magnesite brick generally has lime-to-silica ratios between zero and one, on a molar basis.These brick may contain relatively high boron oxide contents (greater than 0.1% BO) in order to impart good hydration resistance. Sometimes, for economic reasons, these brick are made with lower purity natural dead burned magnesia with magnesia contents of 95% or less.At other times, the brick are made with very pure magnesites with MgO contents greater than 98% for better refractoriness.

CHROME-MAGNESITE BRICKS

Chromite brick's draw back is its lack of resistance to thermal shocks. Efforts to overcome this difficulty have led to the addition of 10 to 30% of dead burnt magnesite (MgO) to chromite. (In chrome magnesite the amount of chrome ore exceeds the amount of magnesite ; those in which magnesite predominates are termed 'magnesite chrome'.) The mixture is bonded by organic bonds (like molasses) and made into bricks & fired at 1350 & 1650°C. At a higher temperature the organic materials burn off and ceramic bond is formed.

To produce chrome magnesite basic bricks, ground chromite ore is mixed with powdered calcined magnesite (periclase) in the ratio of 70 : 30 in pug mills/edge runners with proper quantity of bond material. The material must contain > 18% Cr_2O_3 and > 30% MgO. Mechanical strength is imparted either by using chemical bonding agents or by developing bonding on firing. After moulding , the bricks (hydraulically or mechanically by applying a pressure of 850 kg/cm^2) into the desired shapes, they are dried under controlled condition at 60-80°C. After burning the dried refractories (in tunnel or batch type kilns) at 1500-1600°C, they are cooled slowly to avoid their disintegration.

Chrome magnesite bricks have the good qualities of chromite as well as of magnesite bricks. They can be used up to 1700°C and their refractoriness under load is 1600°C at 2 kg/cm^2. They are more resistant to thermal shocks than chromite bricks. Their chief defect is that they swell and burst in the presence of slags containing iron oxide.

The defects of chrome magnesite bricks have been further reduced by three improvements. The improved bricks are made by using coarse chromite mixed with fine calcined magnesite and using a chemical binding agent. The fine magnesite prevents slag penetration, and thus reduces the bursting of the bricks. The chemical binder eliminates and saves the cost of firing. The third improvement is the use of alternate layers of these bricks with layers of the bricks enclosed in boxes made of thin steel plates. In actual use, due to the high temperature, these boxes fuse together and provide a strong structure. As the bricks are enclosed in boxes these are known as Metalcase or metalclad or steelclad bricks.

Properties of chrome magnesite bricks

Colour. Depending upon the amount of impurities (mainly iron) present, the colour of chrome magnesite brick varies from grey to dark brown.

Specific Gravity. Its specific gravity is about 3.8 (as compared to 2.3 for silica bricks) which necessitates use of strong structure to withstand the total load of these heavy refractories.

Porosity. Apparent porosity of chrome magnesite bricks varies from 22 to 26%. Though denser brick can also be manufactured (by adjusting the firing temperature and time), but due to its higher specific gravity, chances of roof failure due to increased load of bricks increases. Apparent porosity of chrome magnesite bricks fired for 5 hours at 1350°C is about 24%.

Action of slag. Due to its basic nature, it resists the action of basic slags. However, absorption of iron oxide causes its disintegration due to formation of spinals (a needle like structure). Degree of disintegration of the chrome magnesite bricks depends on the extent of absorption of iron (as Fe_SO_4) which sets up strain due to growth of grains of chromite present in the brick.

Cold crushing strength. CCS of chrome magnesite bricks varies from 250 kg/cm^2 (for burnt quality) to 450 kg/cm^2 (for chemically bonded bricks used in roof lining).

Thermal spalling resistance. Though much lower than firebricks, its spalling resistance is higher than those of chromite or magnesite bricks.

Refractoriness Under Load (RUL). Its refractoriness is > 2000°. The value of RUL for chrome magnesite bricks under a load of 2 kg/cm^2 varies usually from 1530°C (for general purpose chemically bonded bricks) to 1600°C (for general purpose burnt refractories). RUL of these bricks suitable for roof linings in open hearth furnaces is about 1700°C , which is about 390°C higher than that of silica bricks under similar load conditions.

Uses of chrome magnesite bricks. Chrome magnesite bricks find extensive use in acid as well as basic open hearth furnaces in making the hearth, roofs, side walls and ports.

Chrome magnesite brick finds extensive use in the steel industry. It is used in the inner lining of basic oxygen steel making furnace (*e.g.* L.D. converter) and side walls of soaking pits (used for steel ingot heating). However, use of chrome containing refractories is avoided in those quality steel making furnaces where slightest chromium pick up by steel makes it out of specification.

The chrome magnesite bricks are also used in copper converters and copper smelting furnaces, lining of induction furnaces, side walls of electric arc furnaces, burner block of glass tanks & furnaces used for melting, remelting and refining of aluminium & aluminium alloys.

In the basic hearth, chrome magnesite bricks find their use in the construction of front & back walls from door sill level to roof, air & gas bridge walls under the ramp, bulk head walls,

inclined roof of air uptakes from bulk head roof level to the working platform level and then to slag pocket roof etc.

Burned chrome-magnesite brick may be of either the direct-bonded or silicate- bonded variety. The direct-bonded brands are used under more severe service conditions.

MAGNESITE CHROME (OR MAG-CHROME) REFRACTORIES

Manufacture of mag-chrome bricks. Magnesite chrome bricks are made from mixture of magnesite and chromite ores, the former predominantly present in the mixture. Clay, tar or organic binding materials (e.g. molasses) are used for bonding. The minimum percentage of MgO & Cr_2O_3 respectively in the mixture varies from 55% & 6% (in case of burnt and chemically bonded general purpose bricks) to 65% & 10% (in case of chemically bonded refractories for roof lining). Silica in magnesite chrome refractories (also called mag-chrome bricks) should be <5% as it reduces the refractoriness. In case of burnt mag-chrome bricks, bonding is developed on firing the shaped refractories while in chemically bonded bricks, the bonding agent itself binds the refractory grains and hardens due to chemical interaction between the refractory and bond material. Use of bonding agents (such as H_2SO_4, $BaSO_4$, $MgCl_2$, K_2CO_3, $MgSO_4$, Na_2CO_3 and Sodium silicates, chromates or orthosilicates) is restricted to minimum required to avoid their adverse affect on other properties of the brick.

A typical composition of magnesite chrome mass (which is used after aging to lime in open induction furnaces producing precision alloys) is as follow :

MgO = 65%	Cr_2O_3 = 14%	Fe_2O_3 = 6.5%	
Al_2O_3 = 4.5%	SiO_2 = 5%	FeO = 1%	
CaO = 3%	MnO = 0.5%	TiO_2 = 0.25%	K_2O = 0.25%

A size tolerance of ± 15% is allowed for mag-chrome bricks to cover the dimensional changes due to shrinkage and warpage during their manufacture.

Properties of mag-chrome bricks. Mag-chrome bricks have almost similar properties as chrome-magnesite bricks.

Porosity. The apparent porosity of mag-chrome bricks varies from 23% (for chemically bonded bricks for roof lining) to 24% (for general purpose burnt or chemically bonded bricks).

Refractoriness under Load (RUL). The minimum RUL at 2 kg/cm^2 for these bricks is about 1550°C.

Spalling Resistance. Mag-chrome bricks have better spalling resistance than chrome magnesite bricks and thereby these bricks have replaced long used silica bricks for roof lining of open hearth & other basic furnaces giving better life & higher working temperature.

These bricks withstand minimum 20 thermal cycles.

Cold Crushing Strength (C.C.S.). C.C.S. of mag-chrome bricks varies from 250 kg/cm^2 (for general purpose burnt bricks) to 450 kg/cm^2 (for chemically bonded bricks used in furnace roof lining).

Permanent Linear Change (PLC). PLC of these general purpose bricks at 1600°C for 2 hours varies maximum from 0.5% (for burnt bricks) to 1% (for chemically bonded bricks).

Resistance to slag action. Magnesite chrome bricks have got excellent resistance to the action of basic slag and iron oxide, however, presence of excessive FeO in slag causes their failure due to growth on account of expansion that occurs as FeO can be absorbed in the coarse grain spinals owing to capillary action before refractoriness is actually impaired. Refractories high in Fe_20_3 content may also involve growth due to carbon deposition and thereby disintegrate due to reduced spalling resistance which is further aggrevated by presence of higher amount of CaO. Hence magnesite chrome bricks should be low in CaO & Fe_2O_3 and their use should be limited to basic slags containing low percentages of FeO.

High specific gravity and high thermal co-efficient of expansion makes the arched furnace roof lined with magnesite chrome bricks succeptible to rise & fall with fluctuations in temperature resulting in crack formation thereby restricting their use under fluctuating temperature conditions. Such furnaces are therefore kept on gas heating even when fettling and other repairs are carried out between the tapping & recharging.

Uses of mag-chrome bricks. Magnesite chrome bricks are used for lining inclined roof of air uptakes, slag pocket roofs, port roofs, gas uptakes, back & front walls, burner zones etc. in basic open hearth furnaces.

Due to high resistance to the corrosive action of iron oxide, these bricks are used in soaking pits, reheating furnaces, electric steel making furnaces, cement kilns and roof lining of rever- beratory furnaces used for copper smelting.

Direct bonded cast mag-chrome bricks (containing chromite and periclase) are used for lining plants for vacuum degassing of steel.

Another use of the mag-chrome bricks is in metalcase form. These metalcased refractories, which have increased mechanical strength thereby having reduced handling problems & breakages are extensively used in the construction of suspended roof of open hearth furnaces and also in electric arc steel smelting furnaces. Metalcased bricks (i.e. burnt or chemically bonded bricks boxed in mild steel sheet cases) when used in furnace roof construction are set such that these are alternately placed with uncased refractory bricks. The steel cases on heating the furnaces get oxidised and form ferro-magnesium spinal (on reaction with chromite and magnesia) which fuses at 1600°C thereby resulting in the binding of the bricks and formation of an impermeable monolithic surface which is highly resistant to the corrosive action of molten slags & the flow of dust laden gases. Expansion on heating is taken care of by the slight curvature to the steel cases thereby improving the spalling resistance of these bricks.

Open hearth furnace lining using sintered magnesite & chromite guniting compounds containing silicate binders has helped in reducing its shutdown period for repairing and increased campaign time for its roof.

Salient features of mag-chrome bricks

Some magnesite-chrome brick are chemically bonded rather than burned. These chemically bonded brick do not have the high temperature strength, load resistance or slag resistance of burned compositions. They are widely used, usually as lower cost compositions to balance out wear profiles in various applications.

Chemically bonded magnesite- chrome brick are sometimes used with steel casing. In service, the steel oxidizes and forms a tight bond between the brick. The technique of steel casing has accounted for improved service life in many applications.

Fused magnesite-chrome grain brick have been developed that contain fused magnesite-chrome grain to offer improved slag resistance. Fused grain is made by melting dead burned magnesite and chrome ore in an electric arc furnace. The melted material is then poured from the furnace into ingots and allowed to cool. The resulting ingots are crushed and graded into grain for brick making. Brick made from this grain, are called "rebonded fused magnesite-chrome brick".

Fused magnesite-chrome grain has extremely low porosity and is chemically inert. In addition, brick made from this grain have a tendency to shrink on burning rather than expand, as is characteristic of many direct bonded magnesite-chrome brick. As a result of these features, the rebonded fused magnesite-chrome brick have lower porosity and superior slag resistance as compared to direct-bonded magnesite-chrome brick.

This type of brick is used in AOD's, degassers and sometimes in the more severe areas of non ferrous applications.

The fused grain brick used typically contain 60% magnesia. Some compositions contain a combination of fused & unfused materials for better spalling resistance, to lower cost, or to achieve a balance of properties that is appropriate to the particular application in which they will be used.

Some magnesite-chrome brick are made from coburned magnesite-chrome grain. Coburned grain is made by combining fine magnesia & chrome ore and dead-burning in a rotary cement kiln.The resulting grain is dense and exhibits a direct-bonded character.Like brick made with fused magnesite-chrome grain, brick made with coburned grain shrink in burning and thus can have lower porosity than certain classes of direct-bonded magnesite-chrome brick. Brick made with coburned grain find uses, such as in vacuum degassers in steel making and in certain nonferrous industries, such as primary copper & nickel production.

Magnesite-spinel brick have been more broadly used in recent years. The term "spinel"refers to the mineral MgO•AlO. Some magnesite-spinel brick are made by adding fine alumina to magnesia. On firing, the fine alumina reacts with fine magnesia in the matrix of the brick to form an in situ spinel bond.

One of the principal benefits of combining spinel and magnesia is that the resulting compositions have better spalling resistance than brick made solely with dead burned magnesite.

This feature results in the inhibition of peel spalling caused by temperature cycling & infiltration of constituents from the service environment.Spinel addition lowers the thermal expansion coefficients of magnesite compositions. This can reduce thermal stresses that contribute to cracking in certain environments.

Use of chrome free basic brick for environmental reasons has increased the importance of magnesite-spinel brick. Trivalent chromium (Cr^{+3}) present in magnesite-chrome brick can be converted to the hexavalent state (Cr^{+6}) by reaction with alkalies, alkaline earth constituents & other compounds that are present in some service environments. These factors have led to broad use of magnesite-spinel brick in rotary cement kilns. They have excellent spalling resistance, good thermal expansion characteristics & provide excellent service results in rotary kilns.

DOLOMITE REFRACTORIES

Dolomite represented by $CaCO_3$.$MgCO_3$, on calcination gives CaO.MgO. The difficulty in using it is that lime absorbs water & carbon dioxide at atmospheric temperature and thus causes the disintegration of the bricks. This can be prevented by either coating the bricks with tar so that during storage, contact with water and carbon dioxide is avoided, or stabilized by mixing serpentine (MgO.SiO_2) with dolomite & calcining, resulting in the formation of di or tri-calcium silicate which does not absorb water & carbon dioxide. After calcination, it is mixed with water, a silicate binder and made into bricks. The bricks are dried and fired at 1500°C for a day.

Major deposits of dolomite in India is in Orissa and Madhya Pradesh. They are also found in Bihar, Bengal, U.P., Rajasthan, Maharashtra, Gujarat and Karnataka.

Theoretically, dolomite contains about 30.5% CaO, 21.8% MgO and 47.7% $C0_2$, however, dolomite associated with calcite may have higher percentage of CaO. Ideally, dolomite used for refractory bricks manufacture should have about 45.6 mole% of $MgC0_3$ and 54.4 mole% of $CaCO_3$ and it should be hard, compact with uniform texture containing very little of iron, silica & alumina as they reduce the refractoriness of dolomite bricks.

Dolomite is dead burnt in rotary or shaft kilns at 1600°C in presence of some serpentine which stabilises the dolomite. MgO and CaO content in dead burnt dolomite is about 42% and 58 % respectively. Even dead burnt dolomite is hygroscopic in nature and hence to prevent its hydration, dolomite refractories are coated with wax or tar just after their firing & cooling. These bricks are generally stored in an air conditioned room maintained at low temperature and appropriate humidity. Furnaces like L.D. converter are then lined with these tar / wax coated dolomite bricks and once the furnace is heated, it is never cooled to atmospheric temperature, otherwise the furnace lining will disintegrate due to their hydration.

Prevention of hydration and stabilisation of dolomite. Dolomite is inactivated by inverting it into silicate or some other high fusion point compounds to prevent its hydration. Because of low fusion point (1540°C) of mono-calcium silicate, its conversion to dicalcium and tri-calcium silicate is preferred which have high fusion point of 2130 and 1940°C respectively. Out of a , b and y polymorphic forms of dicalcium silicate, a -form is stable at > 1410°C, b-form between 675 and 1410°C and y-form at < 675 °C. Formation of y-form is accompanied with 10% swelling resulting in explosion and thereby disintegration of the material. Addition of Fe_20_3 upto a maximum of 4%) to dplomite prevents the formation of y-dicalcium silicate. Addition of phosphates and boron also stabilises dolomite.

Calcination of dolomite. Dolomite and serpentine ($MgO.2SiO_2.2H_2O$) are mixed in a proper proportion and ground in a ball mill with a small quantity of water. The ground mixture is then calcined in a rotary kiln (rpm = 0.6–0.8) at 1600°C for a period sufficient for completion of conversion reaction. Total calcining and cooling time may be about 2.5 hours. Calcination of dolomite with serpentine leaves no free CaO but yield tricalcium silicate and periclase which form a very good refractory mixture. As the calcination temperature increases, first water& CO_2 are expelled and then formation of tricalcium silicate associated with some periclase and glass i.e. SiO_2 (depending upon amount of excess silica) occurs. Large size clinker formed on calcination is cooled, crushed and ground. The desired percentage of coarse (1-3 mm.) medium (0.2-1 mm) & fine grains (-0.2 mm) in the ground dead burnt dolomite mass are usually 55-57%, 5-15% & 20-40% respectively.

Manufacture of dolomite refractory bricks. Properly size graded dead burnt dolomite is mixed with bonding material and 4% of water to a semi dry mass. Binders for dolomite refractories include milk of lime, tar, magnesium silicate, lime, plastics, basic slags, magnesium chloride etc. Since presence/addition of clay results in the formation of ortho-silicate accompanied by 10% volume increase, their presence in the mixture is always avoided. The mixture is then moulded to the desired shape by a hydraulic press employing a pressure of about 700 kg/cm^2 to give strong bricks. Bricks are then dried in a stream of dry clean air at < 60°C to minimise their cracking. Burning of bricks is then carried out in down draught kilns at > 1450°C for about 24 hours. Lower firing finishing temperature and shorter soaking periods result in breaking of tricalcium silicate to dicalcium silicate and lime which undergoes hydration when exposed to the atmosphere and thereby causing the disintegration of refractory bricks/shapes.

Raw dolomite is also used for making dolomite bricks. Raw dolomite after crushing & grinding is mixed with 20% tar and bonding is developed by addition of <5% iron oxide. Silica, alumina, chromite etc. are then added as mineraliser. Mixture is then hand or power press moulded into the desired shapes and stored in sheds where they get hardened. The bricks are then fired in a down draught kiln by first slowly heating to attain 350-400°C in 20-35 hours and finishing the firing at 1200°C. Because of comparatively lower finishing temperature of 1200°C (as compared to 1450°C earlier), the bricks produced are of inferior quality. Hydration and deterioration of these bricks during storage is controlled by giving them a covering of wax or tar thereby avoiding direct contact with air.

Mineral bonded dolomite refractory bricks are superior compared to tar bonded because of higher firing temperature. Tar bonded dolomite bricks are extensively used in L.D. converter, however, pitch bonded dolomite shows a better resistance to hydration.

Properties of dolomite bricks. Dolomite refractories are more basic and less refractory than magnesite. Furnaces using unstabilized bricks should not be cooled during use because lime would slake and the bricks disintegrate. Dolomite bricks can withstand a load of 3.5 kg/cm^2 at 1650°C. They are not very resistant to thermal shocks.

Colour, Specific gravity and Porosity. Depending upon the percentage of iron and other impurities in dolomite bricks, their colour varies from white to buff. The specific gravity and apparent porosity of dolomite refractories are 3.2 and 23% respectively.

Cold Crushing Strength (C.C.S.). C.C.S. of dolomite refractories containing tricalcium silicate is above 400 kg/cm^2.

Refractoriness and RUL. Depending upon the structure and composition, the refractoriness of dolomite refractories varies from 1540 to 2130°C. Dolomite bricks having monocalcium, dicalcium & tricalcium silicates in their structure fuse at 1540,2130 & 1900°C respectively. RUL of dolomite bricks (containing periclase and tricalcium silicate) under a load of 2 kg/cm^2 is 1650 °C.

Thermal spalling resistance. Dolomite bricks containing tricalcium silicate in their structure has a lower thermal spalling resistance compared to magnesite bricks. However, dolomite bricks containing dicalcium silicate in their structure has a better thermal spalling resistance as dicalcium silicate is less succeptible to temperature variations than magnesium silicate. Dicalcium silicate changes to its y-form when cooled to <675°C accompanied by 10% swelling and hence results in small explosions & disintegration of the material. Formation of y-form of dicalcium silicate in dolomite bricks can be prevented by the addition of Fe_2O_3 upto 4% which does not significantly reduce the refractoriness of the dolomite bricks.

Resistance to slag attack. Though having better resistance to molten steel, dolomite bricks, however, have lower resistance to iron oxide and lime slag attacks. Dolomite bricks having dicalcium silicate in their structure have greater resistance to slag attack than magnesite bricks. Lime present in non-stabilised dolomite brick is very reactive hence, the use of non-stabilised dolomite bricks or dolomite bricks containing free lime must be avoided otherwise when the furnace is cooled for repair, the free lime present will absorb moisture (as it is hygroscopic in nature) resulting in disintegration of dolomite brick lining. Stabilisation of tar dolomite brick is carried out by its tempering in tempering kiln.

Uses of dolomite bricks. Stabilized dolomite bricks are replacing magnesite bricks because they are much cheaper. Dolomite refractory is widely used in basic furnaces like L.D. converter.

Outer most wear out lining of L.D. converter is of tar dolomite bricks which give the converter a life of 150-180 heats under Indian conditions. Ramming mixture for electric furnaces and basic open hearth furnaces are prepared by mixing dead burnt dolomite with tar. Dolomite refractories (1-10 mesh size) are used as fettling material for patch repairs of these furnaces after metal tapping and also while they are still hot. Dead burnt dolomite (alone or mixed with coke breeze) is used for closing the tap hole of basic open hearth furnace. Hot metal desul- phurisation (external to blast furnace by lime addition) ladles are lined with dolomite bricks because they have better resistance to soda slag than fireclay bricks. Because of their better resistance to thermal shock and slag attack than fireclay sleeves, unfired sleeves of stabilised dolomite are used for making upper parts of stopper rods particularly when holding times are long. '

Dolomite brick are available in burned and carbon-bonded compositions. The carbon-bonded varieties include both pitch and resin-bonded versions. Some of the carbon-bonded products contain flake graphite and are somewhat analogous to magnesite-carbon brick. Dolomite brick are widely applied in applications as diverse as argon-oxygen decarburization vessels (AOD's), rotary cement kilns and steel ladles.

HIGH ALUMINA REFRACTORIES

Refractories of this group are not used so extensively because they are costly. As a general rule the refractoriness increases with increase in the Al_2O_3 content of alumino silicate refractories.

Compared to fireclay refractories, high alumina refractories give better service under severe condition due to their high abrasion, erosion, slag & spalling resistance; good volume stability, high refractoriness and higher load bearing capacity. Alumina content in high alumina refractories vary from 45 to 95%. Composition of some high alumina materials/clays used for the manufacture of such refractories are tabulated in Table 2.5.

Table 2.5: Composition of Some High Alumina Materials

Composition, percentage	Type of high alumina materials			
	Kyanite	Sillimanite	Bauxite	Diaspore
Al_2O_3	58-62	61	58-62	64
SiO_2	34-37	36.5	1.2-2	25
Fe_2O_3	0.3-0.8	1.1	1.4	0.75
TiO_2	0.88-1	0.75	3.4	1.2
CaO	0	0.6	0	0
Loss on ignition	2.3-3.3	—	25-30	9

Fusion point of all the above high alumina materials are >1800°C and the alkali content except CaO) is almost nil.

Raw materials for high alumina refractory bricks which are chemically or ceramic bonded is mostly precalcined for giving lower firing shrinkage. Dry pressing is normally used for moulding the mixture of high alumina material & granular grog, though hand moulding and pneumatic ramming is also done to produce some special shapes. High alumina refractory bricks may be bauxite, sillimanite, mullite or corundum base. Bauxite is the common raw material for the manufacture of high alumina refractories which gives higher alumina content and thus improves their refractoriness as the alumina/silica ratio increases. A 72% alumina in the mixture produces refractories whose structure completely shows mullite. Firing a mixture of bauxite or any other form of alumina with required quantity of quartz sand/kaolin (silica bearing material) produces synthetic mullite. Mullite and α-alumina are the main phases present in mullite refractories which imparts it excellent thermal spalling resistance along with high corrosion and deformation resistance at high temperature.

Kyanite and sillimanite are the most important refractories of this group both represented by the formula Al_2O_3, SiO_2. Both change on heating to mullite ($3Al_2O_3,2SiO_2$) and silica. In the case of kyanite this change takes place at 1310°C and is accompanied by a great increase in volume whereas in sillimanite the change to mullite takes place at about 1600°C without any appreciable increase in volume. Therefore, it is necessary to strongly heat kyanite and convert it to mullite and then make bricks and finally fire in the usual manner. Sillimanite can be made directly into bricks and fired as usual. In both cases, plastic clay is used as binding material.

Kyanite deposits are found in the Chotanagpur division (Bihar) and sillimanite in the Khasi Hills Assam), Rewa and Bhandara (M.P.) and as beach sand in Kerala.

Bauxite suitable for refractory manufacture is found in Kutch, Bhavnagar and Saurashtra. Diaspore found in Jammu is suitable for refractory manufacture.

Due to the formation of mullite, both kyanite and sillimanite are almost ideal refractories. Mullite contributes to high refractoriness (can be used upto 1800°C), high refractoriness under load, high chemical resistance and low co-efficient of expansion, therefore they have high degree of resistance to thermal shocks. They are more or less neutral refractories and can be used in most environments. Both the materials and sillimanite in particular, have high mechanical strength.

They are costly and hence are not used in very large quantities.

Properties of high alumina refractories. With increasing alumina content, the fusion point of high alumina refractories increases. As for example, the refractoriness (PCE) of refractories containing 50% alumina is 1763°C (Orton 34) whereas that of > 95% alumina containing refractories is 2000°C (Orton 42). Properties of some typical Indian high alumina refractories are tabulated in Table 2.6.

Table 2.6: Properties of Typical High Alumina Refractories

Type of high alumina refractories	Al_2O_3, %	Fe_2O_3, %	Refrac- toriness, °C	Apparent porosity, %	Bulk density, gm/c. c.	Cold crushing strength (C.C.S.), kg/ cm^2	Permanent linear change (PLC)		
							%	Temp., °C	Time, hours
Kyanite base —									
Grade I	58-60	0.5-1	> 1804	22-25	2:2-2.4	300-400	- 0.5	1500	2
II	61-63	0.5-1	> 1804	22-25	2.3-2.9	250-300	- 0.4	1500	2
III	61-63	0.5-1	> 1804	15-16	2.45-2.55	500-700	+ 0.5	1600	2
Mullite and Corundum base —									
Grade I	70-74	0.4-0.6	> 1850	16-20	2.45-2.55	600-800	- 1.0	1700	5
Grade II	84-86	0.5-1	> 1850	18-20	2.70-2.80	500-800	- 0.5	1600	5
Grade III	94-96	0.3-0.5	> 1850	20-22	3.5-3.65	600-800	±0.3	1600	2
Bauxite base —									
Grade I	68-71	2-2.5	> 1804	20-23	2.65-2.75	300-400	+ 2.5	1600	2
Grade II	84-86	3-1	> 1850	18-20	2.70-2.80	500-800	+ 1.5	1600	1

Thermal shock resistance of all the above mentioned high alumina bricks are excellent.

Uses of high alumina refractories. Kyanite and sillimanite refractories are widely used in glass making furnaces.

High alumina refractory bricks are also used in blast furnace stoves (in dome portion)& other regenerative furnaces, cement & lime rotary kilns, electric arc furnace roof, ladles and glass melting furnaces. Due to their non-wetting characteristics, the chemically bonded high alumina refractory bricks are specially used in alumina melting and holding furnaces. Porous high alumina lances for blowing steel with inert gases in ladles and to use chamotte ground dense refractories for lining blast furnace shaft are the other uses of high alumina refractories under development.

Blast furnace hot blast stove wall and air ducts are lined with mullite-corundum type high alumina refractories (containing > 78% alumina) made by mixing kaolin with high grade chamotte (containing alumina powder of 10-60 micron size), briquetting and firing at 1750°C for 80 hours. Natural sillimanite blocks (a high alumina refractory) is used in glass industry.

Salient features of high alumina refractories

High alumina brick refers to refractory brick having an alumina (Al_2O_3) content of 47.5 % or higher and distinguishes them from brick made predominantly of clay or other alumino silicates which have an alumina content below 47.5%.

According to the ASTM convention , high-alumina brick are classified by their alumina content as: 50%, 60%, 70% and 80% alumina classes.They contain their respective alumina contents with an allowable range of plus or minus 2.5% from the respective nominal values. The 85% and 90% alumina classes differ in that their allowable range is plus or minus 2% from nominal. The final class, 99% alumina, has a minimum alumina content rather than a range, and this value is 97%.

There are several other special classes of high alumina products such as :

- **Mullite brick** — predominantly contains the mullite ($3Al_2O_3 \cdot 2SiO_2$) which, on a weight basis, is 71.8% Al_2O_3 and 28.2% SiO_2.
- **Chemically bonded brick** — usually phosphate-bonded brick in the 75 to 85% Al_2O_3 range. An aluminum orthophosphate ($AlPO_4$) bond can be formed at relatively low temperatures.
- **Alumina-chrome brick** — formed from very high purity, high-alumina materials and chromic oxide (Cr_2O_3). At high temperatures, alumina and chromia form a solid solution which is highly refractory.
- **Alumina-carbon brick** — high-alumina brick (usually bonded by a resin) containing a carbonaceous ingredient such as graphite.

A brick classified as a 50% alumina product has an alumina content of 47.5 to 2.5%. Chemically, such brick are similar to superduty fireclay brick which can contain up to 44% alumina. Brick within the 50% alumina class are often upgraded versions of fireclay brick with the addition of a high-alumina aggregate. Compositions of this class are designed primarily for ladles. These 50% alumina class brick have low porosity and expand upon reheating to 1600°C- desirable features for ladle applications since they minimize joints between brick, giving a near monolithic lining at service temperature. These bricks have low thermal expansion and good resistance to spalling. Many high-temperature industries use them as backup brick.

50% alumina products based on high purity bauxitic kaolin, and other ingredients in the matrix, provide exceptional load-bearing ability, alkali resistance and low porosity. These qualities make such brick an excellent choice for preheater towers and calcining zones of rotary kilns.

The 60% alumina class is a large, popular class of products. These brick are used in the steel industry, as well as rotary kilns. Brick in this class are made from a variety of raw materials. Some are produced from calcined bauxitic kaolin and high purity clay to provide low levels of impurities. As a result of firing to high temperature, these brick have low porosity, excellent hot strength & creep resistance, and good volume stability at high temperatures.

70% Alumina Class brick is the most frequently used class because of its excellent and cost-effective performance in multiple environments. Applications include the steel-industry and cement & lime rotary kilns. Most brick in this class are based on calcined bauxite and fireclay. Brick are usually fired to fairly low temperatures to prevent excessive expansion in burning which causes problems in final bauxite to form mullite. The brick typically undergo large amounts of secondary expansion when heated. This is advantageous in reducing the size of joints between brick and providing a tight vessel structure, e.g., a rotary kiln.

A higher cost and higher quality alternative to producing a 70% alumina brick is represented by brands based on high purity calcined bauxitic kaolin. These brick have superior high temperature strength & refractoriness and significantly lower porosity than typical products based on calcined bauxite. Due to their more homogeneous structure, they show somewhat less expansion on reheating than bauxite based products.

80% alumina bricks are based primarily on calcined bauxite with additions of various amounts of other fine aluminas and clay materials. They are usually fired at relatively low temperatures to maintain consistent brick sizing. Most brick in this class have about 20% porosity, good strength and thermal shock resistance. Because they are relatively inexpensive, perform well and are resistant to most slag conditions present in steel ladles, they are used extensively in steel ladle applications.

90 and 99% alumina bricks contain tabular alumina as the base grain and may include various fine materials such as calcined alumina, clay and fine silica. As these brick generally have low impurity levels, alumina and silica typically make up 99% of the chemical composition.Usually, the only mineral phases present are corundum and mullite. Properties such as high hot strength, creep and slag resistance benefit from this purity level.

ZIRCONIA REFRACTORIES

Zirconia refractories are usually classed as neutral refractories but they are not quite resistant to acid slags and hence are on the border line between neutral & basic refractories. Baddeleyite (which is the chief source of zirconia i.e. ZrO_2) has a monoclinic structure upto 1000°C , a tetragonal structure between 1000 and 1900°C and a trigonal structure above 1900°C. Presence of lime or magnesia transforms it into stable cubic structure but it reduces the refractoriness of the bricks produced from it. Stable cubic zirconia (prepared by mixing zirconia with MgO or 5% CaO & heating to 1700°C) after crushing, grinding & sizing is mixed with bond materials (like H_3P0_4, CuO, zirconates of magnesium dioxide) in edge runners. The mixture is then shaped by hand moulding, dry or stiff pressing under a pressure of 560 kg/cm^2 or by slip casting. Firing temperature of the bricks varies from 1550 to 1700°C depending upon the properties required in the product.

The raw material is ZrO_2, which occurs as zirconite, zirconia and zircon mineral. The mineral undergoes volume changes on heating & cooling and is stabilized by adding MgO/CaO and firing at 1700°C. The binding material is colloidal zirconia (prepared by drying zirconium titrate solution and taking the residue into solution & again drying). Alumina can also be used as a cheaper binding material. The mixture is made into bricks, dried slowly and fired at 1750°C or at an even higher temperature for still better bricks.

Properties of zirconia bricks. Zirconia bricks can be used upto 2000°C and specially prepared bricks can be worked upto 2600°C. Refractoriness under load is 1600°C at 3.5 kg/cm^2. They resist the basic sags but are not satisfactory for use with acidic slags. They are resistant to thermal shocks.

Pure zirconia (fusion point: 2677°C) reacts with H_2 & N_2 above 2200°C but below this temperature is stable in both oxidising & reducing atmosphere. Because of its low thermal conductivity even at high temperature, it is very useful as an insulator. Due to its high electrical conductivity at high temperature, it is used as inductors in high frequency induction furnaces. It has a low spalling resistance due to succeptibility of zirconia to structural changes on heating thereby obviating its use in those places where fluctuating temperature conditions exist. Ceramics made from zirconia-hafnia (60 mole% $Zr0_2$ + 40 mole% HfO_2 typically) solid solutions have low vapor pressure and excellent strength & thermal spalling resistance.

Uses of zirconia bricks. Zirconia bricks are very expensive and are used only when the temperature is very high as in high frequency electric furnaces.

Zirconia does not react with any metal (except titanium) and hence is used in receptacles used for refractory metals & alloys at high temperatures. Small articles like burner tips, sheath for thermocouples, crucibles, ignition boat etc. are made by slip casting of zirconia. Nozzles used in continuous casting steel plants are made of zirconia. Zirconia is also used for lining high temperature ceramic kilns, thermal barrier coatings for jet engines and crucibles & furnaces used for melting precious metals, enamels, glazes, special glasses etc. Zirconia-hafnia ceramics is used as tiles for supporting high melting alloys heated to 2000-2300°C.

SALIENT FEATURES OF BASIC REFRACTORIES

Basic refractories are so named because they exhibit resistance to corrosive reactions with chemically basic slags, dusts & fumes at elevated temperatures.

Basic refractories generally fall into one of the following five compositional areas:

- Products based on dead burned magnesite or magnesia.
- Products based on dead burned magnesite or magnesia in combination with chrome-containing materials such as chrome ore.
- Dead burned magnesite or magnesia in combination with spinel.
- Dead burned magnesite or magnesia in combination with carbon.
- Dolomitic products.

One of the more important types of magnesite brick are those that have low boron oxide contents and dicalcium silicate bonds. These chemical features give the brick excellent refractoriness, hot strength and resistance to load at elevated temperatures. Another category of magnesite brick contains a higher boron oxide content to improve hydration resistance.

Basic refractories containing chrome are an important group of materials due to their excellent slag resistance, superior spalling resistance, good hot strengths and other features. Historically, silicates in the groundmass or matrix formed the bond between the chrome ore and periclase in the brick. However, the advent of high purity raw materials in combination with high firing temperatures made it possible to produce "direct bonded" brick, where a ceramic bond between the chrome ore and periclase particles exists. These direct bonded brick exhibit superior slag resistance and strengths at elevated temperatures.

Magnesite-spinel brick have become important due to a desire to replace chrome-containing refractories because of environmental concerns. Brick made with spinel & magnesite have better spalling resistance and lower coefficients of thermal expansion than brick made solely with dead burned magnesite. These features minimize the chance of the brick cracking during service.

Basic brick containing carbon include pitch impregnated burned magnesite brick with carbon contents up to 2.5%, pitch bonded magnesite brick containing about 5% carbon and magnesite-carbon brick which contain up to 30% carbon. Development of the more corrosion resistant magnesite-carbon brick has resulted in decreased consumption of pitch impregnated & pitch bonded magnesite brick. In addition, magnesite-carbon brick have replaced magnesite-chrome brick in applications such as electric arc furnaces.

Dolomitic products are an important class of refractories that offer a good balance between low cost and good refractoriness for certain uses.

The principal raw materials used in the production of basic refractories are dead-burned & fused magnesites, dead-burned dolomite, chrome ore, spinel and carbon.

One result of this effort has been the development of technology to address specific wear mechanisms by employing special additives in the refractory composition. These additives generally constitute less than 6% of the total mix. Examples of these special additives include zirconia, which is sometimes used to improve the spalling resistance of burned basic refractories. As carbon has become an important constituent in the formulation of composite magnesite-carbon refractories, metallic additives, such as powdered aluminum, magnesium or silicon are used to improve hot strength and oxidation resistance. Small boron carbide additions also can improve the oxidation resistance of certain magnesite-carbon compositions. These compositions are used in special applications such as bottom blowing elements of basic oxygen furnace.

ZIRCON REFRACTORIES

Zirconium silicate (a refractory compound of zirconium) exists in nature as 'Zircon' mineral. It is also produced by heating tetragonal zirconia with silica at 1460°C. Zircon is stable upto 1750°C but at higher temperature, it decomposes yielding zirconia and silica glass or cristobalite (which is a stable form of silica above 1440°C). Zircon and zircon flour are mixed with the binder (such as sodium silicate, phosphoric acid, ground zircite etc.) to a stiff paste which is press moulded to the required shape. Slip casting is also used for moulding zircon refractories which are dried and fired at 1600°C.

Refractoriness of zircon refractories is 2000°C and **RUL** is 1600°C under a load of 2 kg/cm^2. It has a high spalling resistance and low co-efficient of expansion. Zircon refractories are used in remelting furnaces for aluminium because they are not wetted by molten aluminium or alumina. They are also used in boilers as they offer excellent resistance to the action of coal and oil ash. Zircon nozzles are used in continuous casting of steel. Zircon-clay masses give excellent rate of sintering and a high compressive strength. Use of zircon refractories in contact with some phosphates, iron oxide and fluorspar should be avoided as they react with these substances.

BAUXITE BRICKS

Bauxite is the ore from which aluminium is extracted. Pure Al_2O_3 is a highly refractory material (m.p. 2010°C) but ordinary bauxite contains iron and silica as impurities and gives bricks fusing at about 1000°C but properly prepared bricks can be used upto 1750°C.

Besides containing iron oxide and silica, which lower melting point, bauxite contains water of crystallization. It is heated to 600°C to remove water and can then be used directly for making inferior grade bricks. Usually the calcination temperature is 1400 to 1500°C which prevents further shrinkage in the bricks. The calcined material is crushed & mixed with plastic refractory clay and made into bricks & fired at 1500°C.

In fact, bauxite (which is hydrated alumina - $Al_2O_3.H_2O$ or $Al_2O_3.3H_2O$) which contains silica & iron as impurities loses its water of crystallization when heated between 900 to 1100°C and yields y-alumina (sp. gravity -3.65). On further heating, it changes to corundum (sp. gr.- 3.99) which is more stable a -form of alumina. Bauxite used for making refractories bricks should contain >58% alumina, < 5% SiO_2, <3% Fe_2O_3 and < 3% TiO_2.

Properties of Bauxite Refractories

Shrinkage. Bauxite refractories show a great shrinkage even at high temperature. This is a great drawback. This shrinkage defect can be eliminated by making the refractory from fused bauxite.

Chemical Reactions. Bauxite being basic in character is not attacked by basic slags. Bauxite refractories are attacked by lime. This property of bauxite of not being attacked by basic slags makes it superior to fireclay refractories which are subject to such attacks.

Refractoriness. Bauxite refractories are best resistant from 1790 to 1880°C.

Refractoriness under Load. Bauxite refractory when heated under a pressure of 3.5 kg/cm^2 softens at 1500°C.

Maximum working temperature. Bauxite refractory can be heated up to 1700-1750°C.

Spalling and cracking is quite common in bauxite refractories.

Mechanical Strength. Bauxite refractories have got high mechanical strength. A bauxite refractory having up to 25% of fireclay does not soften below 1500°C under pressure of 3.5 kg/cm^2.

Resistance to Abrasion. If bauxite is sufficiently heated beforc making it into a refractory, it has got very great resistance to abrasion.

Porosity. Bauxite refractory is highly porous and this decreases its durability when used in furnaces.

Uses of bauxite refractories. They are used for making the bottoms of basic Bessemer and open hearth furnaces and as lining for cement & other kilns and furnaces working at high temperatures.

CHROMITE BRICKS

Chromite refractories are important neutral refractories. The raw material is chromite represented by ($Cr_2O_3.FeO$). It contains $MgO.Al_2O_3$ which acts as binding agent for the bricks. Cr_2O_3 content of the ore varies from 30 to 60%. Some of the Indian ores contain as high as 2% Cr_2O_3. The presence of large amounts of silica and iron oxide reduces the fusion temperature, hence these are undesirable.

Chromite ore (which is found in India in Bihar, Maharashtra, Tamilnadu, Andhra Pradesh and Orissa) suitable for chromite refractory manufacture should contain at least 40% Cr_2O_3; and must be hard, lumpy & have fine crystalline structure. A typical composition of chromiteore is as follows :

Cr_2O_3 = 45%, Iron oxide = < 18%, Silicates of magnesium = 10-15%
CaO = < 1%, SiO_2 = 3-5%, Al_2O_3 = 7-10%
TiO_2 and alkalies = < 1%

After crushing and grinding the chromite ore, it is mixed with water and bond material (usually heated tar though organic substances are also sometimes used) to prepare a plastic paste. Dry mixing is also done sometimes. Hand or machine moulded bricks are slowly dried to avoid uneven drying and crack formation. Bricks are then fired in batch or continuous kilns and the firing is finished at 1400°C. Heating rate is kept low to control swelling which may result, in the disintegration of bricks in the kiln if they are closely packed.

Properties of chromite bricks. Chromite bricks can be used up to 1800°C and their refractoriness under load is 1430°C at 3.5 kg/cm^2. At the atmospheric temperature, the crushing strength is 350 to 550 kg/cm^2. Chromite bricks are resistant to basic and moderately acid slags, but are affected by the slags in a reducing atmosphere. They are not resistant to thermal shocks. They have moderate thermal conductivity.

Chromite refractories burst in presence of iron oxide. Presence of silicates of magnesium is undesirable as it lowers the refractoriness. Properties of chromite bricks vary depending on the chrome ore composition and type of bond material used. Refractoriness of chromite bricks with very low silica content can be as high as 1900°C. However, in practice they are rarely used above 1500°C. Lime bonded chromite refractory bricks are more refractory than clay bonded ones. Chromite bricks having excess silica have lower refractoriness under load (RUL). They usually fail at 1400-1450°C under a load of 2 kg/cm^2. It has a low thermal conductivity.

Depending upon the composition, the specific gravity of chromite bricks varies from 3.8 to 4 and the porosity from 24 to 27%. Cold crushing strength (C.C.S.) varies from 350 to 550 kgf/cm^2.

Chromite bricks containing free silica often crack at 200 to 300°C. However, thermal spalling resistance of chromite bricks free from silica is better than magnesite and silica refractories because its sensitivity to thermal shock increases with increase of free silica.

Uses of chromite bricks. Chromite bricks find extensive use in steel making and copper refining furnaces. They are particularly useful in separating acid & basic refractories and are good for any use so long as not subjected to thermal shocks, but they are expensive.

Chromite bricks have wide applications in steel and non ferrous industries. It is used in the bottom of soaking pits in slabbing mill of steel plant. However, its high cost prohibits its liberal use. This brick is also used as a separating layer between acid and basic linings in basic open hearth furnaces to prevent their chemical reaction with each other.

CARBON/GRAPHITE REFRACTORIES

Graphite is a crystalline form of carbon and its natural deposits are mixed with calcarious and siliceous materials. The natural deposit is known as plumbago. Coke has been used as a cheap substitute for graphite but is not as satisfactory as the latter.

Carbon occurs in nature in both amorphous or crystalline form. In graphite, carbon atoms are arranged in a hexagonal framework. Graphite is less prone to oxidation compared to other forms of amorphous carbon due to its lower porosity (<5%). Ash content in graphite ranges from 7.5 to 12.5%. Composition of graphite is given below which is found in Karnataka, Bihar and Orissa.

SiO_2 = 42-54%	Al_2O_3 = 15-20%	Fe_2O_3 = 13-28%
MgO = 3.5-8.5%	CaO = 1-4.5%	K_2O = 2.2—2.6%
Na_2O = 0.1-2.5%	TiO_2 = 0.9-1.4%	

Bonding materials for carbon refractories (which are made from coke or graphite) are usually tar or pitch (free from ammonia & light oils), however, milk of lime, fire clay, sodium silicates etc. are also used. To produce carbon refractory bricks, coke (containing < 8% ash) is crushed/ground to 1-2 mm size and stored in dry place. About 10 kg of coke powder is spread on a clean floor and hot tar or pitch is poured on it. It is then mixed with shovels quickly to get a mixture of uniform texture avoiding lump formations. The mixture is then moulded in wooden or iron moulds which are oiled before use. After removing the excess material (by a hot rod of iron or stamps), it is allowed to cool and then mould is pulled up leaving the brick on the baseboard. Bricks are cooled in ventilated rooms. They should not be exposed to sunlight as it softens the tar and prevents hardening/setting of moulded shapes.

For obtaining superior products, graphite after grinding is treated with hydrochloric acid to remove the mineral impurities. It is mixed with a binding material like plastic clay, sodium silicate or tar and made into crucibles. Carbon bricks, which are also extensively used, are made from coke of very low ash content and employing tar as the binding material. During firing, contact of the bricks with air is minimized by filling the space in between the bricks with a mixture of sand and powdered coke.

Firing of carbon refractories is done out of contact with air in a completely air tight down draught kiln. The bricks to be fired are packed with a mixture of sand and coke in cupboards of the kiln to prevent their distortion during initial heating. In the kiln, heating is done indirectly & slowly to 250°C at which tar is converted into coke and the refractories harden. Temperature is finally increased to 1300°C and maintained for 50-70 hours at which burning of the brick is completed. The kiln is then cooled down very slowly to room temperature and the kiln is then opened & unloaded. Cooling time almost equals heating time. The total cycle time from initial firing to the opening of kiln is about 200—250 hours. Defective refractories are crushed and recycled.

Fireclay refractories coated with tar or carbon deposited on them, under controlled conditions, are also used. The carbon makes them more resistant and increases the life.

Refractory shapes from graphite fireclay mixture is made in the same way as from coke, however, graphite is purified (to remove silicious & calcareous raw materials) before use by grinding and washing or by froth floatation. Purified graphite is then mixed with fireclay to the extent of 15 to 25%. Graphitised fireclay bricks are produced by soaking fireclay refractories in hot tar. Molten tar fills the pore of these refractories. Soaked refractories on slow heating to 250°C in reducing atmosphere get graphitised by carbon deposition on the pore surfaces as far as the tar penetration inside the brick has taken place.

Properties of carbon/graphite refractories. Graphite is infusible and sublimes at 3600°C. It is not attacked by any slag or chemical but in the presence of oxygen, at a higher temperature, it oxidizes. It is resistant to thermal shocks and has high thermal & electrical conductivity. It is light grey in color.

Porosity of carbon refractories varies from 20 to 25% and the specific gravity from 1.5 to 1.9. They are light grey in colour and give a ringing sound when struck. Refractoriness of these bricks are very high and they do not show any deformation under a load of 2 kg/cm^2 at 1750°C. Cold crushing strength (CCS) of carbon refractories is about 350 kg/cm^2. It has high spalling resistance.

Since, carbon blocks are generally not wetted by melts, its use in the blast furnace hearth prevents skull formation, if the blast furnace hearth runs cold and hence helps in regulating its working. Carbon refractories, however, oxidise and burn when exposed to air on heating.

Uses of carbon/graphite refractories. Carbon bricks find the most extensive use in lining the hearth and the bosh of the blast furnace. They are also used in other furnaces and equipment where the temperature is high and the atmosphere is non oxidizing.

Carbon refractory blocks are used for lining blast furnace hearth (due to their capacity to withstand heat, load & chemical action during iron making) and bosh walls as also cupola hearths. They may be used in the lower stack wall of the blast furnace also. Due to presence of reducing atmosphere in blast furnace, life of carbon refractories is prolonged because the metal oxidation is minimum in this furnace. High thermal conductivity of carbon lining helps in easy temperature control in the lower parts of the blast furnace. Iron and slag runners are also lined with carbon refractories which are made to accurate sizes as they undergo very low shrinkage. Corrugated shapes are used to develop strong non permeable joints which prevents metal/slag penetration, loosening of joints and floating of carbon blocks.

Graphitised fireclay (which is produced by soaking fireclay refractories in hot tar at 150°C for 24 hours followed by slow heating in reducing atmosphere upto 250°C resulting in its graphitization) is useful for stoppers, nozzles and laddle bricks used in positions which are more prone to slag attack (either due to high casting temperature or because of the chemical composition of metal/slag).

Carbon is used as a ramming mixture with tar for the construction of electrolytic cells e.g. aluminium reduction cell.

Tar/Pitch impregnated carbon refractory bricks have better resistance to oxidation & slag attack, higher compressive strength, reduced porosity & gas permeability (due to closing of pores) as compared to non impregnated ones.

Use of rammed carbon packing linings ensures better heat transfer of water cooling shells in blast furnaces resulting in reduced temperature of refractory hot face. Coke with pitch binder is used as filler material for packing blast furnace between carbon lining & cooling plates and also between the fireclay lining & the furnace shell.

Graphite crucibles of various sizes (having an apparent porosity of 30 to 40% and bulk density of 1.4 to 1.5 gm/c.c. depending upon the size) are used for melting cast iron, aluminium, bronzes, copper, brasses etc.

CERAMIC-GRAPHITE REFRACTORIES

A new class of carbon refractories called ceramic graphite refractories such as magnesia graphite and alumina graphite are finding their use in steel making processes due to their high resistance to spalling & action of basic slag at high temperature coupled with reduced specific refractory consumption. Alumina graphite bricks are used in slide gate teeming & continuous casting systems in the form of pouring tubes, sliding plates and nozzles etc. Magnesia graphite refractories (have

<10% porosity) are now replacing alumino silicate refractories for lining steel ladles used as secondary treatment vessels for production of high quality steels. The composition of magnesia graphite bricks is, C = 5-20%, CaO = 2.2-2.4%, SiO_2 = 0.7-1.5%, Fe_2O_3 & Al_2O_3 = 0.3-0.4% each and balance is MgO. Its C.C.S. after tempering varies from 323 to 633 kg/cm^2. Its bulk densities after coking & tempering are 2.76-2.99 & 2.80-3.04 gm/c.c. respectively. Magnesia graphite bricks are extensively used in high powered electric arc furnaces and basic oxygen furnaces due to their high fusion point & resistance to basic slags.

OTHER HIGH REFRACTORY MATERIALS (SPECIAL REFRACTORIES)

There are many specialised service conditions for which, though the quantity of the material needed is small, resistance to very high temperature is needed. Materials for this come from pure metallic oxides, carbides, borides, nitrides silicides and sulphides. Since, impurities lower the melting points, great care should be taken to use pure materials and not to use any low melting binding agent. The shape is to be given by special techniques, like the combined effect of heat and pressure.

OXIDES REFRACTORIES (ThO_2, BeO, Al_2O_3, SiO_2, MgO etc.)

The highest melting oxide is thoria (m.p. 3070°C) but its use is very limited due to its cost and difficulty in fabricating it. Widely used oxides are magnesia (2800°C), zirconia (2680°C), beryllia (2570°C), alumina (2040°C) and silica. The pure oxide is finely ground with water and a temporary organic binder such as starch, wax, high boiling alcohol, etc. On firing the final shape, the temporary binder is destroyed and the flue particles sinter together. Solid surface reactions take place and gradually new tiny crystals are produced which hold the shape.

Very fine grained pure oxide of thoria ($Th0_2$) is fired at 2000°C to make it very dense. Very expensive and basic in nature, thoria has a high resistance to basic slag. Being radioactive in nature, it requires careful handling. It has a poor thermal spalling resistance. Its main use is in making crucibles used for melting some high purity metals like gold.

Though magnesia ware is difficult to fabricate, it is resistant to the oxidising atmosphere and basic slags upto 2500°C and is very strong/rigid. Crucibles made of magnesia is used for melting pure metals like nickel, platinum etc.

Beryllia resists the action of alkalis but is attacked by acids and oxides like magnesia & alumina forming fusible slàgs. Inhalation of beryllia dust may prove fatal. Beryllia has a high thermal conductivity, good thermal spalling resistance, inert to the action of carbon, stable under nuclear irradiation condition and a low electrical conductivity. Its usage include radiation shields in carbon electrical resistance furnace, as a moderator in nuclear reactors, making crucibles for melting beryllium & zirconium etc.

High purity alumina obtained by dehydration of bauxite is very hard & strong has good mechanical properties and a good wear resistance. Alumina crystals (obtained by treatment of alumina with sodium carbonate followed by crystallisation of (3-Al_2O_3 from the melt) are fused & cast as corundum and fired at 1700°C after fine grinding & moulding. This corundum is used in the laboratory as alumina tubes & melting crucibles where high temperature (upto 1900°C) and solid-gas reaction is involved. But the corundum is attacked by basic slag, iron oxide & flourine gas and has a poor thermal spalling resistance ; hence it is used only where alternative cheaper material is not available.

Alumina is used for making pyrometer & electrical tubes, spark plug, nose pieces for missiles, rock cutting drills, abrasives and roller bearings used at higher temperature.

Pure/fused form of silica has a high thermal spalling resistance and a very low co-efficient of thermal expansion continuous use of silica ware at > 1350°C temperature ensures it a long life ; however it is attacked by alkaline compounds. It is used for making tubes, dishes, thermocouple sheaths, crucibles etc.

Fusion points and specific gravity of some pure refractory oxides in increasing order of fusion temperature is tabulated in table 2.7.

Table 2.7: Fusion Point and Specific Gravity of Some Pure Refractory Oxides

Refractory oxides	Fusion point, °C	Specific gravity
SiO_2	1715	2.32
MnO	1780	5.40
TiO_2	1850	4.24
BaO	1917	5.72
NiO	1950	6.80
V_2O_3	1977	4.87
Al_2O_3	2050	3.97
Cr_2O_3	2075	—
BeO	2550	3.03
CaO	2570	3.23
ZrO_2	2677	5.56
MgO	2800	3.58
HfO_2	2810	9.68
UO_2	2875	10.96
ThO_2	3070	9.69

Each of the oxides has some special property. For example, alumina is the hardest of all and is useful where wear resistance is required. Beryllia has high resistance to thermal shocks and is not reduced by carbon or carbon monoxide upto 2000°C. In spite of the high melting point of magnesia it cannot be used above 1700°C in a reducing atmosphere and above 2200°C in an oxidising atmosphere. As zirconia undergoes polymorphic changes , its resistance to thermal shock is low above 1250°C. However, it can be stabilized by adding calcium or magnesium oxide.

SUPER REFRACTORIES

Pure oxides are used for making so called 'super refractories' which are suitable to be used in contact with molten metals in oxidising conditions because of their high melting point. To produce these oxide super refractories, chemically purified material is sintered/fused in electric furnaces followed by its grinding to fine powder. Plasticity to the mixture is imparted by using some organic plasticiser. Shaping of the mixture is done by slip casting extrusion, hot pressing or hydraulic pressing. On firing these oxide refractories (at high temperature and controlled atmospheric &

temperature conditions), the organic material present therein burns off and recrystallisation on heating imparts high pack density to the bricks.

These refractory oxides shapes/bricks are very dense, highly resistant to the chemical action if slags, have high cold crushing strength, have stability at high temperature and are comparatively inert to the H_2, N_2, cracked NH_3 & carbonaceous or sulphurous atmospheres. At high temperature, they show high electrical resistance and also resistance to abrasion, erosion & corrosion. However, their use is restricted to special purposes due to their non availability, high production cost and special fabrication methods to shape them. Zirconia, thoria, beryllia and oxides of Mg, Al, Ce etc. are important refractory oxides.

CARBIDE REFRACTORIES

Generally, silicon carbide or carborundum is used for manufacturing carbide refractories.

Carbides of various elements such as tantalum carbide, zirconium carbide, titanium carbide , boron carbide etc. are useful as refractories for special purposes. They are very costly, so their use is restricted only to special purposes.

SILICON CARBIDE (SiC)

Silicon carbide or carborundum is one of the very important refractories. It is produced in electric furnaces by heating together up to 1300-2200°C, the following mixture:

Sand (SiO_2)	2–54%
Coke (C)	35%
Saw dust	7–11%
Salt	1.5–4%

The following reaction takes place :

$$SiO_2 + 3C \rightarrow SiC + 2CO$$

Composition of carborundum produced is generally the following :

Silicon	65%
Carbon	30%
Impurities	5%

Raw Materials for Silicon Carbide Refractories. Raw materials used are:

Carborundum (SiC). It is crushed and ground to the required size; suitable proportion of different sizes from 16 mesh to 100 mesh are mixed together depending upon the kind of refractory to be produced.

Binding materials such as lime, plaster of Paris, aluminium sulphate, dolomite, kaolin or felspar,glue, mineral oil, tar, molasses, resin or glycerin.

Manufacture of silicon carbide refractories. Ground carborundum and binding material are mixed with a little water in a pan mill and the mixture obtained is moulded into bricks by hand or by presses.

The moulded bricks are burnt in electric furnaces up to 1400°C to 2000°C depending upon the property desired.

Silicon carbide refractories are known as first quality refractories and contain at least 85% SiC. Grains of SiC have a thin silica film on their surfaces. Proper bonding is achieved by a reaction of the bond material and this thin film forming on adherent interface. Depending upon the type of bond in them silicon carbon refractories are classified into following five major groups :

- semi silicon carbide
- oxide of clay bonded
- nitride bonded
- oxy nitride bonded
- recrystallised

Semi-silicon carbide refractories are essentially fireclay refractories to which lot of SiC grains have been added. Clay/oxide bonded refractories are the largest carbide refractories used now a days in which SiC grains are >85% (except where alumina or zircon is added). The bond is developed due to silicon & silicates formed by added clays and oxides (e.g. those of Ca, Fe and Ba).

Silicon carbide refractories bonded with silicon nitride (Si_3N_4) are more oxidation resistant but are more prone to failure during thermal cycling. Oxy-nitride bonded silicon carbide refractories (having a silicon oxy-nitride bond) has more improved properties than those of oxide bonded compositions and is also highly resistant to oxidation growth. It is strong, highly oxidation resistant, stable in many chemical environments and bonds strongly to the SiC grains. It also offers a higher degree of protection to the SiC grains than that achieved by any other bonds.

Properties of silicon carbide bricks

- It is from dark grey to blue black in colour.
- Hardness and roughness is very high.
- True specific gravity varies from 3.17 to 3.21. It is light in weight.
- Melting point is about 2500°C and thus has high refractoriness but starts decomposing at 2200°C. It is chemically inert to acid slags.
- It has high thermal conductivity and RUL.
- Very low coefficient of expansion and high thermal shock resistance.

Uses of silicon carbide bricks. Usefulness of carborundum refractories lies in its hardness, low coefficient of expansion, great strength and toughness. Generally it is used for making saggars and for furnaces where it is not subjected to influence of reducing gases.

Silicon carbide refractories are used for metal melting furnaces like cupolas, aluminium remelting furnaces, crucible furnace linings, crucibles for melting non-ferrous metals; for radiant tubes, muffles etc. in heat treatment plants; for some smelting & refining furnaces including aluminium reduction, zinc smelting and also for muffles, combustion chambers& ceramic recuperators. Because of its high hardness, skids used in reheating furnaces are made out of it. These are also used for making nozzles for rocket motors, nose pieces for rockets, electrical resistor bars/spiral (used in the construction of high temperature furnaces) etc. Rotary furnaces producing ultramarine employ silicon carbide refractories. Use of SiC in presence of oxygen or iron oxide particularly at high temperatures is avoided as they attack it.

SIALONS

They are the refractory materials of oxide/monoxide compounds having high density and high strength. A solid solution of alumina in silicon nitride is a specific example of this group of ceramic materials from which its name Si-Al-O-N (Sialon) has been derived. Sialons have poor sintering characteristics and hence to get low porosity bricks hot pressing at a high pressure of 180—300 kgf/cm^2 and at 1700—1750°C in graphite moulds is required. Sialon is also produced by hot pressing a mixture of silica and aluminium at 1400°C in nitrogen atmosphere. Sialons have good oxidation resistance and resistance to the corrosive action of high temperature molten metals like iron, copper, zinc, aluminium, steel etc. They are also not eroded/corroded by sulphuric acid, borax, alkalis, hydrochloric acid etc. Hence, Sialons are suitable for storage and transportation of the above materials.

ELECTROCAST REFRACTORIES

Refractories produced from raw materials like alumina, chrome-alumina, alumina-zirconia, alumina-silica etc. by fusion in an electric arc furnace and then casting it in the desired shape/size followed by its annealing.

Electrocast mullite refractory is superior in various physical & chemical properties compared to alumino-silicate refractory produced by firing at high temperature as complete mul- litisation does not occur during firing without actual fusion of the whole mass of silica and alumina which is possible in electric fusion only. Electrocast mullite used upto 1800°C is neutral and withstand the attack of slag, molten metal & thermal shock. As a result they are suitable for use in lining molten steel handling/transfer ladles, open hearth slag pocket, reheating furnace hearth, upper portion of regenerator checkers and those parts of heat treatment, reheating & annealing furnaces which are subjected to thermal shocks. Ground electrocast mullite is rebonded to make burner blocks, tiles, saggers, grooved heating element carrier shapes used for special purposes.

Electrocast high purity alumina fused at about 2000°C has high abrasion resistance and hot strength. They have very high refractoriness, chemical stability and are inert to oxidising & reducing conditions in the furnaces. As a result they are used for :

(i) lining combustion chambers in acetylene production

(ii) petroleum oil gasification reactors

(iii) catalytic reforming and cracking units in oil refineries

(iv) lining carbon black reactors

(v) lining domes of chemical reactors in fertiliser plants

(vi) blast furnace stack

(vii) L.D. converters

(viii) hearth construction of slab/billet reheating furnace

(ix) lining rotary kilns

(x) vacuum degassing chamber of special alloy steels etc.

Electrocast corundum and mullite-corrundum masses have high strength, heat resistance & wear resistance properties. They are used in slide gate plates for molten steel pouring. It is also used for lining the equipment for refining and vacuum treatment of steel. Electrocast alumino-silicate castables are used for lining door and stand (ascension) pipes of coke ovens.

Electrocast fused products of MgO, $A1_2O_3$, Cr_2O_3 (with little Fe_2O_3) are used in lining such parts which are liable to wear in absence of thermal shock viz. slag belt and hearth/bottom of continuous melting furnaces used for sulphide materials. It has a very low open porosity (0.7 to 1.8%), high compressive strength (1050 kgf/cm^2) and a high deformation temperature of 1750°C under a load of 2 kgf/cm^2.

Crucibles used in vacuum electric induction furnaces for melting precision alloys are made of electrocast magnesite, corundum and zirconia. Electrocast zirconia is the best out of the three for above purpose as it undergoes less reduction with the metal or less contamination by oxygen comparatively. Electrocast periclase is used in slide gates for steel teeming.

The main consumers of electrocast refractories of various compositions are modern glass tank furnaces, cyclones & ducts carrying dust laden particles at high velocity and skid rails m steel slab reheating furnaces.

CERMETS

A combination of ceramics and metal/metal alloy is called 'cermet'. Cermet is a special type of refractory which combines the good property of both the ceramics and metals or their alloys. It is produced by pressing & sintering of powdered ceramic and metallic materials to the compact shapes by powder metallurgy techniques. Ceramic materials are brittle, have low thermal spalling resistance & have high strength at high temperature while metals & their alloys are ductile but loose strength on heating. Cermets have good properties of both. Ceramic materials normally used in the production of cermets are mainly high purity oxides (e.g. alumina, silica, chromite, magnesia, zirconia, beryllia etc); high purity carbides (of titanium, silicon or tungston) and borides (of aluminium, titanium or zirconium). Metals used in cermet production are normally iron, cobalt, stainless steel, nickel, manganese, chromium, aluminium, beryllium etc. In cermets, the metal phase acts as a bonding material for ceramic phase grains. Cermets being very hard & strong are suitable for fabrication of wear resistant tools (e.g. cutting off wheels, rock-boring drill bits) and those parts of space crafts & missiles which require high strength at high temperatures and under high wear conditions (e.g. TiC-Ni and Al_2O_3-Cr composition cermets).

Oxide cermets (e.g., $A1_2O_3$,-Cr) has high resistance to thermal fatigue and is used in missiles & space crafts. Carbide cermets are suitable for cutting and boring tools.

Uranium dioxide (UO_2) dispersed in stainless steel matrix is a cermet which is used as a nuclear fuel. It has good mechanical strength at high temperature, good corrosion resistance and good stability under nuclear irradiation conditions.

SILLIMANITE AND ALLIED MINERALS

The following are the natural occurring minerals under this heading. They are all used as refractories:

(i) Sillimanite, $Al_2O_3SiO_2$. *(ii)* Mullite, $3Al_2O_3.2SiO_2$. *(iii)* Kyanite, $Al_2O_3.SiO_2$.

SILLIMANITE

It occurs naturally in Assam and Rewa in India. It has got the composition Al_2O_3.SiO (fibrolite). It occurs as long needle shaped crystals. Crystals are orthorhombic.

Manufacture of Refractory materials from naturally occurring Sillimanite.

- Sillimanite occurs in nature in slabs and blocks. These blocks & slabs are cut into smaller blocks and bricks.
- When it is in small pieces, it is ground & mixed with fireclay and then molded into refractories.

Properties of sillimanite. It has a high softening point, that is above 1800°C.

- It is perfectly stable upto the softening point.
- It is stable in oxidising as well as reducing atmosphere.
- Coefficient of expansion is as low as 0.45 x 10^{-5} and this enables it to withstand sudden changes in temperature.
- It has no volume changes when heated.
- It has got neutral reaction.
- It is highly resistant to corrosive action of many slags and is resistant to abrasion.
- It has got great strength even upto its softening temperature.
- It has very low electrical conductivity.
- It has high thermal conductivity, that is 10'B.T.U./hr./sq.ft/°F/inch.
- Its hardness is from 6 to 7 Mhos.
- Its specific gravity is 3.23.
- Pure sillimanite is colourless , but generally it is brown coloured due to the presence of iron in it.

Uses of sillimanite. It is used for making blocks, bricks, crucibles, saggars, refractory fittings for electrical goods and tubes for surface combustion.

KYANITE

Its composition is $Al_2O_3.SiO_2$ (Disthene). It occurs in India in Rewa and Assam. Its crystals are triclinic.

It has all the properties and advantages of sillimanite but it is slightly softer than sillimanite as its hardness is 5 to 7 Mohs while sillimanite has hardness from 6 to 7 Mhos..

Decomposition temperature of kyanite is 1350—1380°C.

The other defect in kyanite is that it has got more volume expansion per cent than sillimanite. The volume expansion is 16% in case of kyanite and 6% in cases of sillimanite.

MULLITE

Its composition is $3Al_2O_3.2SiO_2$. It occurs rarely in nature. It is formed when kyanite, an- dalusite or sillimanite is heated. It is formed in burnt fireclay also.

Mullite is very hard. Its specific gravity is 3.16. Its decomposition temperature is 1810°C. It is highly resistant to alkalies but is slowly attacked by molten glass.

So all these properties make it one of the very good refractories. In order to make a good refractory of fireclay, its composition is generally so balanced by adding a little bauxite (Al_2O_3), some accelerators such as alkali, silicates, magnesia etc., that it conforms to the chemical com position of mullite. Then it is fired at 1350—1380°C so that mullite, $3Al_2O_3.2SiO_2$ is formed.

FAILURE OF REFRACTORIES

The most common cause of failure of a refractory is chemical reaction with the environment in which it is worked. It should certainly not be in a chemically dissimilar environment e.g. an acid refractory should not be used in a furnace using basic fluxes & slags and vice versa. In spite of using proper fluxes there may be impurities which may react with the lining. In the chemical reaction, the porosity of the refractory plays an important part. The more porous it is, the greater will be the depth to which the slag will penetrate and destroy the refractory. Porosity also affects the reactions with the gases. The deposition of carbon from carbon monoxide in fireclay refractories in a blast furnace is an important cause of failure of the refractory. Deposition of carbon sets up high internal stresses resulting in the disruption of the bricks. The higher the temperature , the faster the chemical reaction takes place. Sometimes rise in temperature beyond the safe limit quickly brings about the destruction of the refractory. Lastly, the agitation or movement of the materials is important. If there is agitation on the surface, new surface is quickly exposed to chemical attack and destroyed.

Attack on refractories by slag or metal, is unavoidable. It can be minimised by deoxidation with aluminium metal. The quantity, basicity and fluidity of the slag, are its important factors influencing the reaction with refractories. The reaction products are low melting calcium- alumino-silicates and ferrites. Pure mullite is more resistant to slag attack than mullite with vitreous silica as intrusion in the steel castings, if present in a larger amount. Thus, the Mn/Si ratio in steel is important as far as the corrosion of refractories is concerned. Erosion of refractories, generating scorified refractory particle, is more undesirable than corrosion by reaction yielding glassy intrusions. Low carbon steel produces scorified refractory particles as instrusions, and thus reduces the life of the ladle lining, while the life is fairly high with medium carbon steel. Again, steel with 1% C, 0.3% Mn, 0.3% Si and 1% Cr, does not erode the refractory lining but gives a smooth glassy furnace. Refractory life also depends upon the dwelling time and operating temperature.

The behavior of some oxides and elements towards quartizitic and high alumina refractories, have been tabulated in Table 2.8.

Some more reactions in actual performance of refractories are described below :

- Bursting expansion is caused by a solid solution of ferrous chromite and magnesite in chrome of the composite basic brick. Magnesioferrite is free from bursting expansion.
- Dusting due to transformation of C_2S can be prevented by addition of P_2O_5, B_2O_3 or Cr_2O_3.
- Iron oxide in Foresterite or Periclase bricks, should be low in order to avoid formation of eutectics in the checker bricks of glass melting tank furnaces.
- Corrosion of refractories by glass is due to the diffusion of alkali ions into the solid phase.
- A 50°C rise in the operating temperature of glass melting, will reduce the life of refractories in general by 50%. A higher temperature is essential for faster working and improved quality of glass.
- Free silica and free iron should be avoided in a reducing atmosphere.

- H_2S releases active H_2 at 1050-1100°C in a near stoichiometric atmosphere.
- V_2O_5 in presence of 2% Na_2O, lowers the refractoriness of both silica & firebrick by 20°C but in the presence of 5% Na_2O by 85°C & 40°C for silica & firebrick respectively.
- The chemical analysis of refractories is not of much importance, unless there is a chance of direct reaction between the refractories and the products of the process, provided the physical properties are satisfactory.
- 4-5%; Fe_2O_3 lowers the refractoriness of silica brick by about 30°C but do not have any appreciable effect on firebrick; CaO, however, has a detrimental effect on firebrick.
- Silica refractories form a low fusible manganese silicate due to the slagging reaction of cupronickel and nickelsilver.

The probable reactions, described in Table 2.8 will help in selecting a suitable refractory for specific applications.

The second important cause is "spalling". The cause of spalling may be thermal, mechanical or structural. Thermal spalling may be due to unequal expansion or contraction caused by the difference in temperature at different parts or by rapid changes in temperature. Mechanical spalling is mostly due to carelessness in loading the furnace or in the removal of materials from the furnace thereby damaging the refractory. Structural spalling takes place because of the change in composition of the refractory due to reactions with slags, etc. with the result that its coefficient of expansion changes. Thus different parts expand and contract to a different extent. Spalling of a refractory can be reduced by proper design, construction and operation of the furnace.

Abrasion may be listed as the third factor. Abrasion means mechanical rubbing away of the material. In the furnace, there is a movement of materials and often hard & solid material slides along the refractory. There are heavy molten slags and metals flowing. Dust bearing gases often move at high speeds. All these rub off the refractories and cause its failure.

Table 2.8: Reactions with Refractories

Effect of following oxides and elements	Quartzitic refractories	High-alumina refractories
SiO_2	Without effect	Forms mullite which, however, has a very high melting point.
B_2O_3	Forms an eutectic	Forms $AlBO_3$
P_2O_5	Forms an eutectic	Forms $AlPO_4$
SO_3	No effect	No effect
F	Together with other components forms an eutectic	Forms low melting compounds
Cr_2Os	Practically no reaction (CrO would react strongly at 1500°C with SiO_2 to form Cr_2SiO_4; Cr_2O_3 reacts with iron silicate (forming $FeCr_2O_4$, spinel) which melts easily, is decreased, and as a result, the slag may become more viscous again	Practically no reaction
TiO_2	Forms a high-melting eutectic (in the presence of CaO, forms titanate)	Forms high-melting Al_2TiO_5

Al_2O_3	Forms high melting mullite	No effect
Fe_2O_3	It itself has no effect, in a reducing atmosphere a severe effect due to the formation of Fe_2SiO_4, which produces low-melting eutectics	Little effect in itself, in a reducing atmosphere formation of hercynite $FeO.Al_2O_3$ takes place
C	Can reduce SiO_2 to volatile state or cause silicizing of steel	Probably without any effect
CaO	Reacts to form calcium silicates, which in themselves have high melting points, but together with other components, particularly Al_2O_3 and Fe_2O_3, forms FeO eutectics	Reacts to form calcium aluminates, $3CaO.Al_2O_3$; $CaO.6Al_2O_3$, etc. which in themselves have high melting points, but together with other components, e.g. especially Fe_2O_3, form eutectics, $4CaO.Al_2O_3.Fe_2O_3$ (brown-millerite)
MgO	Reacts to relatively high-melting compounds, as long as there is no CaO or FeO present	Reacts to form spinel phases
FeO	Forms low-melting mixtures	Reacts to form spinel phase (hercynite)
MnO	Forms Mn_2SiO_4 at 1251°C and eutectics at 1345°C	Forms spinel, which melts at ca. 1850°C. The eutectic between MnO and Al_2O_3 lies at 120°C (galaxite)
ZnO	Zn_2SiO_4 (willemite); in combination with other components, e.g. FeO, low-melting compounds	Reacts to form spinel phases, gahnite
PbO	Forms low-melting compounds and eutectics	—
CuO	Forms low-melting compounds and eutectics	—
Na_2O and K_2O	Form very low-melting compounds and eutectics (K_2SiO_3 or Na_2SiO_3)	Form very low-melting compounds ($NaAl_2O_3$, KAl_2O_3) and eutectics, leading to a scaffold formation in blast furnace stack; $2K_2O.Al_2O_3.2SiO_2$ or $2Na_2O.Al_2O_3.2SiO_2$ may also form
H_2 and steam	Vaporization of SiO_2	Vaporization of Al_2O_3
CO and CH_4	C deposition on alymino-silicate refractories containing iron compounds; Zn supports the reaction but S compounds and NH_3 retard it; this is the cause for carbon monoxide disintegration in the blast furnace stack ($2CO = C + CO_2$)	

CHOICE OF REFRACTORIES

The choice of a suitable refractory depends on a large number of factors classified into the following groups :

Factors related to refractories. The properties of the refractories available need to be studied.

Factors related to the furnace design. The refractory is to be used in a certain furnace. The choice of refractory will depend upon the type of the furnace, conditions of heating (e.g. a coke oven is kept continuously at a high temperature for years but a cupola is daily cooled) and loading, degree of insulation, provision for thermal expansion made in the furnace, whether there are moving parts or not (as are in rotary kilns and in tunnel kiln cars) etc.

Factors related to Operation. The most important factor in this group is the chemical nature of the materials (like the ores, fluxes, fuels etc.) in contact with the refractory, temperature in different parts of the furnace & temperature fluctuations, fluidity of the metals & slags, velocity of gases in contact with the furnace lining and abrasion due to moving charge, gases & the molten metals.

Cost factors. While considering the initial cost of different refractories their life during use should not be overlooked.

It may be seen that selection of a suitable refractory is quite complicated. Much data is available from the working of the refractories in similar furnaces elsewhere. These will help in actual selection. Out of the large number of factors only a few are so essential that by comparison other factors are of minor importance. For example, in some cases exposure to high temperature is the chief problem. In other cases high load has to supported, in only a few cases the refractory has to withstand both high temperature and high load. In some cases high insulating property is needed, in other, high thermal conductivity. Greatest weight, in the selection should be given to the more essential properties for the particular use.

The cost of handling the refractory, transportation, brick laying cost, are the same whether superior or inferior bricks are used. If an inferior refractory is used and it has a shorter life, it will have to be replaced more often. This will involve additional handling costs, in addition to loss in production. Therefore, the true cost of refractories should be considered on the basis of its effect upon the overall cost of production per unit output. Table 2.9 gives the summary of properties of refractory bricks.

Table 2.9: Summary of Characteristic Properties of Refractory Bricks

<table>
<tr><th colspan="4">Type of bricks</th><th>Characteristics</th></tr>
<tr><td colspan="4">High-Alumina</td><td rowspan="3">High refractoriness, which increases with increasing alumina content. High mechanical strength at high temperatures. Excellent to fair resistance to spalling. Greater resistance (than fireclay brick) to corrosion by most basic slags and fluxes.</td></tr>
<tr><td>Al_2O_3%</td><td>50</td><td>60</td><td>70</td></tr>
<tr><td>PCE</td><td>>34</td><td>>35</td><td>>36</td></tr>
<tr><td colspan="4">Fireclay</td><td rowspan="6">The chemical and physical properties of fireclays vary between wide limit; hence, fireclay bricks with widely varying combination of properties are available. This fact accounts for their suitability for service under widely different operating conditions. Most fireclay bricks have relatively good spalling resistance and thermal insulation value. Fair resistance to acid slags and fluxes, lower resistance to basic slags and fluxes. »</td></tr>
<tr><td colspan="2">Type</td><td colspan="2">PCE</td></tr>
<tr><td colspan="2">LHD</td><td colspan="2">15</td></tr>
<tr><td colspan="2">MHD</td><td colspan="2">29</td></tr>
<tr><td colspan="2">HHD</td><td colspan="2">31.5</td></tr>
<tr><td colspan="2">SD</td><td colspan="2">33</td></tr>
<tr><td colspan="4">Silica</td><td>High refractoriness and resistance to abrasion. High mechanical strength at high temperatures. Thermal conductivity at high temperatures is about 25% greater than that of fireclay brick. High resistance to corrosion by acid slags. Fair resistance to attack by lime, magnesia and iron oxide. Readily attacked by basic slags and fluorine. Not subject to thermal spalling at temperature above 650°C; poor resistance to thermal spalling at low temperature.</td></tr>
<tr><td colspan="4">Magnesite</td><td>Extremely high refractoriness and high thermal conductivity. Great resistance to corrosion by basic slags; poor resistance to slags containing high percentages of silica. The chemically bonded and metal encased bricks have marked resistance to spalling.</td></tr>
<tr><td colspan="4">Chrome</td><td>High resistance to corrosion by basic slags and moderately acid slags & fluxes. In general, basic slags do not adhere to chrome bricks. Under certain unusual conditions, iron oxide is absorbed and causes a damaging expansion. Thermal conductivity is lower than that of magnesite brick but higher than that of fireclay brick.</td></tr>
</table>

Magnesite Chrome	Mechanical strength and stability of volume at high temperatures. Excellent resistance to spalling. High resistance to corrosion by basic slags.
Chrome Magnesite	Mechanical strength and stability of volume at high temperatures. Excellent to good resistance to spalling. High resistance to corrosion by basic slags.
Foresterite, 2MgO. SiO_2	High refractoriness. Excellent strength at high temperatures. Marked resistance to corrosion by alkali compounds. Fair resistance to most basic slags; attacked by acid slags.
Carbon brick	Extremely inert under reducing atmosphere in acidic and basic slag, very good volume stability, high mechanical strength & thermal conductivity.

Basic refractories in general are less resistant to thermal shock than fireclay and high alumina refractories. However, resistance to spalling of some basic bricks of special composition is equal to that of super duty fireclay bricks.

Basic bricks of any class may be metal encased and internally plated. Such bricks have much better resistance to spalling than bricks without the casings. Most metal encased bricks are magnesite chrome compositions.

Indian resources of refractory minerals has been given in table 2.10.

Table 2.10: Indian Resources of Refractory Minerals

Refractory Minerals	Places of Occurrence	Remarks
Fireclay	Kumardhubi (West Bengal), Jharia and Hazaribagh (Bihar), Sambalpur (Orissa), Shimoga (Karnataka), Jabalpur (Madhya Pradesh), Rajkot (Gujrat)	—
Silica	Hazaribagh, Dhanbad and Gaya (Bihar); Hyderabad (Andhra Pradesh); Ratnagiri (Maharashtra)	—
Bauxite	Palamu and Ranchi (Bihar), Bilaspur, Mandla, Jabalpur and Balaghat (Madhya Pradesh), Poonch and Raisi (Kashmir), Kutch and Saurashtra (Gujrat), Sambalpur (Orissa), Belgaum and Bababudan hills (Karnataka), Salem and Nilgiris (Tamilnadu), Kolaba and Kolhapur (Maharashtra)	Refractory bricks grade bauxite should be low in Fe_2O_3 and TiO_2
Magnesite	Singhbum (Bihar), Almora (U.P.), Salem (Tamilnadu), Kadakola (Karnataka)	—
Chromite	Jaipur and Keonjhar (Orissa), Ratnagiri (Maharashtra), Salem (Tamilnadu), Kittoburu and Jojohatu (Bihar), Krishna (Andhra Pradesh), Chickmaglur and Chitaldurg (Karnataka)	Refractory bricks grade chromite should be of the following composition (percentage): Cr_2O_3 = 30-40, $Cr_2O_3 + A1_2O_3 = 60$, SiO_2 = 10 (maximum)
Dolomite	Sambalpur and Birmitrapur (Orissa), Nagpur (Maharashtra), Kurnool (Andhra Pradesh), Shimoga (Karnataka)	Refractory grade dolomite should contain at least 35% $MgC0_3$, maximum 1% silica and maximum 1.5% ($FeO + A1_2O_3$)
Graphite	Dhenkanal and Bolangir (Orissa), Trivendrum (Kerala), East Godavari (Andhra)	—
Kyanite	Lapsa-Buru, Ranchi and Singhbhum (Bihar), Jaipur (Rajasthan), Hasan (Karnataka)	Lapsa-Buru deposit is the largest in the world

Refractories used in Various Furnaces/ Industries

Requisites of refractory for various furnaces/processes are discussed below.

IRON AND STEEL INDUSTRY

Iron and Steel industry is the major consumer of refractories with about 70% of refractories being consumed by them.

The different areas of the Iron & steel manufacturing processes are exposed to different temperatures, slag and sulphurous gases. The diversity in the operating conditions of different equipment's demands different grades for different areas of applications .

Steel industry uses refractory for diverse applications in:

- Blast Furnace
- Coke Ovens
- Torpedo Ladles
- Basic Oxygen Furnace
- Soaking pits
- Electric Arc Furnace
- Reheating furnace
- Twin Hearth Furnace
- Annealing furnace
- Secondary Refining Ladles
- Stainless Steel Vessels
- Slide Gate Refractories
- Purging Refractories
- Tundish Refractories
- Induction Furnace

Blast furnace is the focus of any integrated steel plant. Blast furnace is used to reduce the iron ore to iron. The charge consisting of the sinter, lumps of iron ore, coke and limestone *etc* is fed from the top in different ratios. Air heated in the blast furnace stoves, is applied from the bottom of the furnace. The hot blast comes in contact with the descending charge in furnace and the iron ore gets reduced to iron due to reducing conditions on account of CO_2 and CO in the furnace. CO provides further heat and a very high temperature is developed because of which the iron gets melted which, along with the slag is collected in the hearth from where these are tapped separately from different tap holes.

The working hearth is required to withstand high temperature, temperature cycling, and impact from the charge. The sub hearth of the furnace has an extended lifetime in comparison with other components of the lining, and is only replaced after severe metal penetration through the working hearth or after major water leakage.

Hearth linings are normally constructed with high quality 42-62% Al_2O_3, Mullite, conventional carbon block. Current furnace designs favor Carbon/Graphite block with super micro-pores. Monolithic linings based upon high MgO powders are rammed and contoured into the sidewalls.

Blast Furnace (B.F.) Proper. Service conditions are different in different sections of B.F. Temperature is highest at the hearth level, lower in bosh section and corrosion is caused by the slag. Hence hearth and bosh refractory should:

- be highly resistant to corrosion by slag.
- have stability of volume at high temperature to resist penetration of metals into the joints. Hence super duty fireclay bricks are used for lining.

In the lower in wall section—high refractoriness and resistance to abrasion is required.

In upper in wall and top section, resistance to abrasion by descending stock is required, so the refractory used should be strong and tough to resist wear.

Free carbon deposited within the temperature range of 370-540°C in the presence of catalytic Fe_2O_3 or alkalies deposited on the lining or condensed in the pores cause destruction in wall and top section. Brick should not contain particles of ferric oxide.

Carbon blocks are used in bottom hearth walls.

Carbon refractories are used in the blast furnace hearth mainly due to their high thermal conductivity, non wetting characteristic to molten metal, volume stability, low thermal expansion, resistance to alkali & silica, increased strength with increasing temperature and high thermal shock resistance. The performance of carbon refractory block in the blast furnace hearth depends upon:

- carbon quality
- furnace design
- efficiency of cooling
- environment inside the furnace

Carbon refractories are also used in the hearth wall, bosh wall and even in the lower stack wall apart from the hearth bottom.

In some blast furnaces, fused cast refractories have been used in the bosh and stack areas. Use of castable refractories in the blast furnace bosh, lower stack, upper stack and in hot blast system have exhibited superior performance.

B.F. Stoves. Bricks for lining of combustion chamber and dome should have high:

(i) refractoriness & resistance to spalling.

(ii) Strength, density and resistance to alkaline fumes & dust.

Super duty fire clay bricks for upper courses and high duty bricks in lower courses are used.

With increase in the capacity of the blast furnaces (upto 10000 tons/day) and the hot blast temperature upto (1300°C), radical changes in design and use of refractories have been introduced. Use of high alumina refractories in blast furnaces and hot blast stoves is of recent advent. The introduction of carbon refractories in blast furnace has a similar trend. Fused cast refractories have been used in the bosh and stack areas of blast furnace in many countries. Use of castable refractories in the blast furnace bosh, lower stack, upper stack and in hot blast system have shown superior performance.

In a blast furnace, normally high heat duty (HHD) or super duty (SD) firebricks are used in the stack where the temperature varies from 200°C at the top to 1500°C at the lower portion. The neck of the stack may be made with castable covered with wear resisting plate. The hearth and hearth bottom may be made of super duty firebricks, high alumina and/or carbon blocks/ bricks.

In the blast furnace, purer refractory material is required for superior performance due to slag and alkali attack as the low porosity or fusion cast refractories is not the solution. Fusion cast bricks with smaller & uniform crystals of corrundum is better in bosh & stack for giving scaffold free surface and minimum rate of wearing. The throat of the blast furnace is made of dense abrasion resistant firebrick or castable refractories. The carbon bricks used in the blast furnace hearth bottom are positioned not preferably by keying but are anchored by positioning a portion of the huge block under the side wall brick work so that they can not float. Life of the carbon blocks depends not only on its quality but also on construction, design, cooling system, environment inside the furnace and burden preparation. However, the required properties for these bricks/blocks are maximum thermal conductivity, good abrasion resistance, almost zero permeability, very low apparent porosity (< 5%) and maximum oxidation resistance.

Iron and steel industry is the largest consumer of fireclay, silica and basic refractories.

Hot Metal Mixer. Mixer temperature is < 1350°C, hence hard burnt high duty or super duty fireclay is used for lower part of lining. At higher operating temperature, a belt of the brick of mullite type at the metal line resists the washing action of metal and corrosion by layer of slag.

Hot metal mixers are used for storing as well as homogenizing (in respect of silicon content of iron) the pig iron before charging it into steel making converters/furnaces (e.g. L.D. converter, open hearth furnaces *etc.*). Despite low operating temperature (< 1350°C) of the mixer, the refractories in it pose problems. The common problems are excessive joint erosion resulting from washing by molten iron and corrosion by slag. Also, the spout needs to have a streamlined construction. Brick lining used in the hot metal mixer should be denser, more volume stable, hard fired super duty bricks which has a prolonged life by minimising joint erosion. Critical areas around spouts should be lined with high alumina bricks. High duty blast furnace fire brick can be used for lining the other areas of the mixer.

Acid Bessemer Converter. Outer lining is of fireclay bricks next to the steel shell and inner lining is of siliceous material. The bottom or tuyer section is built with special fireclay brick shapes. Space between individual tuyer is rammed with a mixture of plastic fire clay and crushed fire brick.

Basic open hearth furnace (BOF)

Roof - Silica brick is used in the roof due to following properties:

- high refractoriness
- its ability to resist the action of dust and fume.
- high mechanical strength and rigidity upto the melting point.
- constancy of volume at red hot conditions.
- freedom from spalling.
- lightness in weight.

Table 3.1: Refractories for All-basic Open Hearth Furnace

Furnace area	Refractory type	Operating temperature, °C maximum
Hearth bottom	Magnesite brick, magnesite and dolomite ramming masses	1650
Banks	As above, but mainly dolomite	1650
Frontwall	Chrome-magnesite, metal encased chrome- magnesite brick, magnesite up to sill level	1650
Backwall	As above with slope covered with dolomite, magnesite, chrome refractories	1650
Main roof Port roof	Magnesite-chrome brick/chemically bonded metal encased	1680-1650
Gas port	Steel clad chrome-magnesite brick	1680
Uptakes	chrome-magnesite brick	1595
Slag pocket	Silica brick, fireclay brick, chrome magnesite bricks	1540
Upper checker	HHD/SD fireclay, silica, forsterite or magnesite bricks	1425 (gas) 1150 (air)
Lower checker	HHD fireclay brick	1000 and below
Checker chamber	wall-mag-chrome	1400-1500
Flues to stack	MD fireclay brick	750
Stack	MD fireclay bricks	650

The open hearth furnace which is used for steel making is an all basic furnace. Its various parts are shown in Fig. 3.1 and the operating temperature in the various zones and the refractory requirements are tabulated in Table 3.1.

Refractories for L.D. Converter. The L.D. converter vessel may be lined with tar dolomite, tarmagdol or tar magnesite bricks. Such bricks may be tempered. A typical refractory lining pattern of L.D. converter (100-130 tons capacity) used in Indian steel plant is described below:

The refractory lining of L.D. converter consists of two parts viz. Converter Body, and Bottom.

Converter Body Refractory Lining. Converter body is given a permanent lining and above it a wear lining.

(a) **Permanent Lining.** The full height of the converter body is given a permanent lining of chrome-magnesite brick of thickness 115 mm.

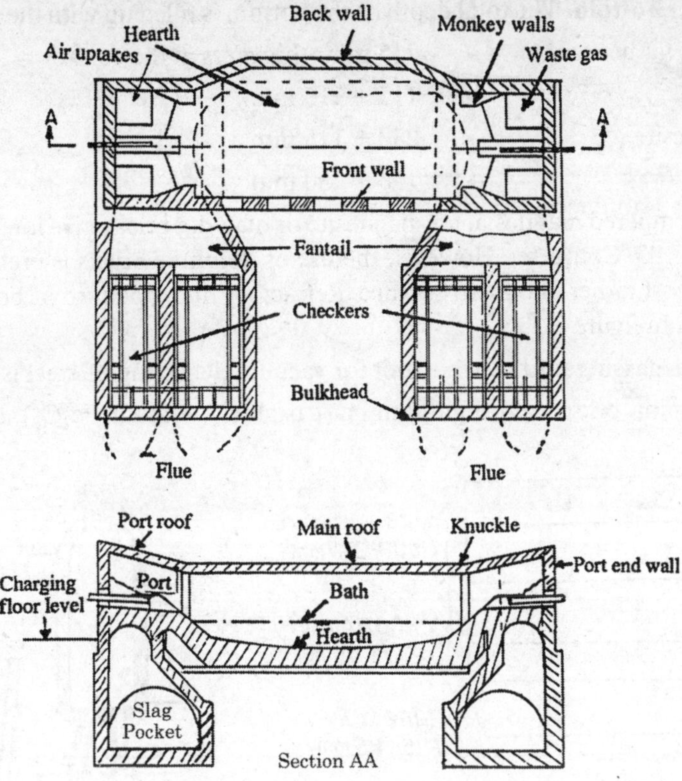

Fig.3.1: Open hearth furnace.

(b) **Wear out Lining.** Varying thickness of wear lining of tar dolomite brick or periclase- spinnel is given just after the permanent lining depending upon the position of the converter body. The converter body is divided in three parts. Top conical portion, barrel and bottom conical portion.

(i) **Top and bottom conical portion** — Here the thickness of lining is 125 + 550 mm.

The total length of the top conical portion is covered by 13 rings out of which top most 2 rings are of periclase spinnel brick and the rest 11 rings are of tar dolomite brick.

Similarly, bottom conical portion is covered by 9 rings out of which bottom most 2 rings are of periclase-spinnel, and the rest 7 rings are of tar dolomite brick.

The height of each ring is 125 mm.

(ii) **Barrel** — Here, the lining consists of a ramming mass of tar dolomite of thickness 50 mm just after the permanent lining.

Then a lining of tar dolomite brick of thickness 125 + 125 + 550 mm at the two trunion sides, and of 125 + 550 mm thickness at charging and tapping sides are given. The total height of the barrel is covered by 32 rings of 125 mm height.

(c) **Tap hole lining** — The tap hole consists of a block of cross section 690 x 690 mm with a circular cross section of 125 mm diameter at the centre. The intermediate portion is filled up with mag chrome brick.

L.D. Converter Bottom. The total depth of the bottom is filled up with the following materials:

Mica crumbles (at bottom) → 115 mm (thickness at the centre)

Fireclay → 115 + 115 mm

Chrome-magnesite → 230 + 115 mm.

Mag-Chrome → 150 + 230 mm

Pitch bonded tempered magnesia, tar magnesite or mag-dot bricks give longer life than sintered dolomite bricks in L.D. Converter. However, the use of dolomite bricks is preferred because of its appreciably lower cost rather than performance. Refractory lining pattern of bottom lining of L.D. Converter is shown in figure 3.2.

Vacuum steel degassing unit. Such a unit for vacuum degassing of steel is shown in Fig.3.3.

Only high alumina bricks and high alumina castables (without Fe_2O_3) are used in vacuum degassing units.

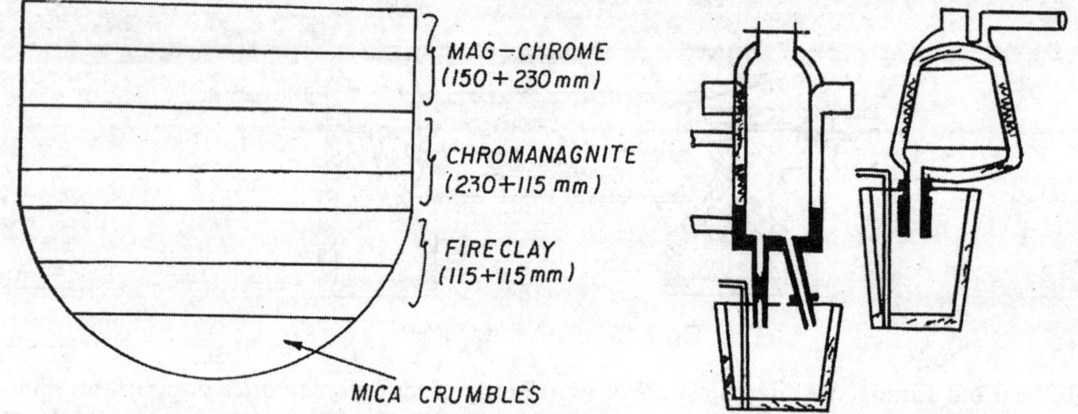

Fig.3.2: Refractory lining of bottom of L.D. Converter.

Fig.3.3: Vacuum steel degassing unit

Continuous Steel casting units. Steel is cast directly in the form of slabs by continuous casting unit. Zircon nozzles are used in continuous casting of steel.

In conventional method of steel ingot casting in moulds from the laddle, two practices are in vogue. In one, stopper rod is provided to block the hole in the laddle bottom while not pouring the steel while in other method slide gate system is provided to pour the steel in the moulds. In steel teeming laddle using stopper rod, fireclay refractories are used for sleeves, stoppers & nozzles while in slide gate system high alumina or magnesite slide gate plates & nozzles are used. The slide gate system is shown in Fig. 3.4.

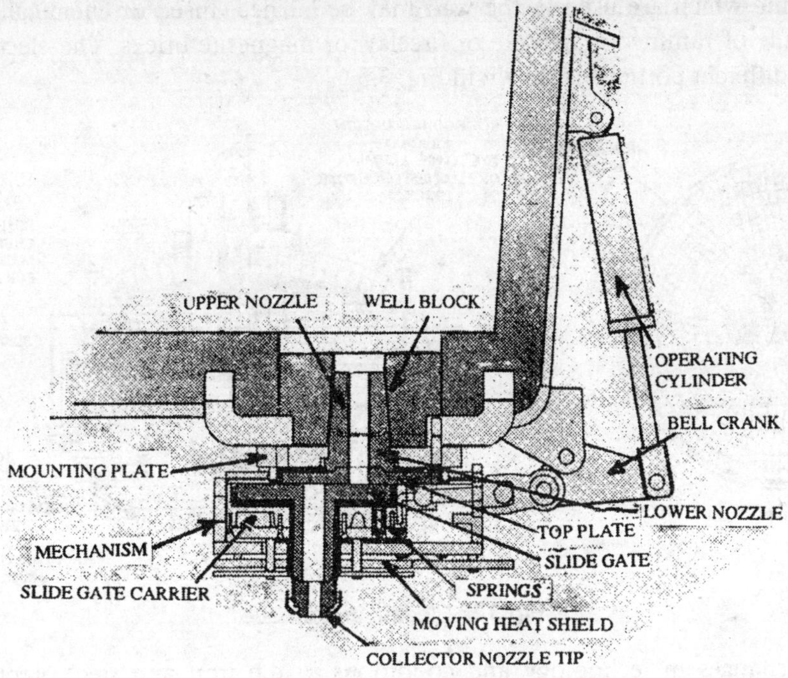

Fig.3.4: Slide gate system

Steel Teeming Ladle. Steel ladle is used to receive molten steel from the L.D. converter where after addition of deoxidiser or alloying elements, it is teemed into ingot moulds through a nozzle and stopper or through slide gate system in the bottom of the ladle.

The refractories used for lining the steel ladle should have:

- abrasion resistance to molten steel .
- resistance to the slag layer on the top of molten steel .
- low thermal conductivity to prevent the formation of skulls .
- tight joints to prevent adhesion of the skull to the bottom and sides of the ladle .
- good resistance to thermal spalling due to sudden heating when the molten steel is poured into the ladle .
- freedom from particles of the refractory mixing in the molten steel and forming inclusion.

Usually high heat duty fireclay bricks are used for lining of steel teeming ladles. However, the current development trend is to use aluminium silicate monolithics for the purpose.

Refractories for by-product coke ovens. Quality requirements of coke oven refractory bricks have become more stringent as the modern coke ovens are taller (7 metres tall), operate with higher coking temperature (1150 to 1300°C) and high coke chamber volume (47 m^3). Most of the construction of coke ovens is of silica bricks. Regenerators below the coke deck level is of fireclay refractories.

Electric arc furnace for steel making. In the electric arc furnace, the roof may be silica, 70-90% alumina or mag-chrome with delta portion rammed with high alumina monolithic.

Mag chrome which are used in the wall may be burned, fused or chemically bonded. The bottom is made of rammed magnesite or fireclay, or magnesite bricks. The electric arc furnace with lining in different portions is shown in Fig. 3.5.

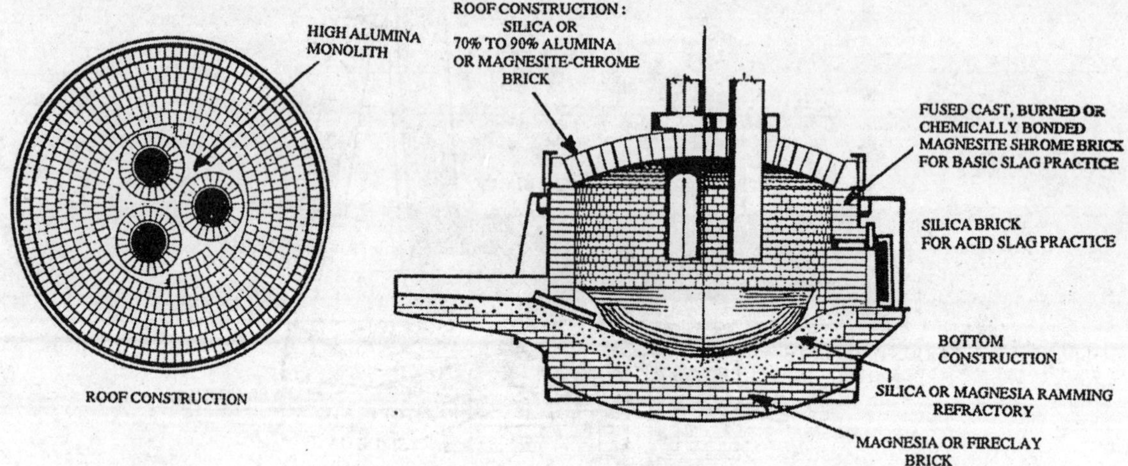

Fig.3.5: Electric arc furnace

With the changes in technology and conditions in the iron and steel plant furnaces, the refractory utilisation trend is also changing. The development trend in the use of refractories in iron and steel plant is tabulated in table 3.2.

Table 3.2: Development Trend in use of Refractory in Iron and Steel Industry

Areas of application	Future development trends
Blast furnace	Corundum, silicon carbide and magnesia bricks
Hot blast stoves	Mag-chrome and magnesia checkers
Torpedo ladle for hot metal	Andalusite, tar bonded alumina-silicate bricks
Basic oxygen converter	Preventive gunning, magnesia-graphite bricks with internal reinforcements
Electric arc furnace	Magnesia-graphite bricks
Steel ladle	Basic lining, aluminium-silicate monolithics
Continuous casting plants	Oxides ceramic, tundish with basic lining

The general practice of refractories used in different furnaces of iron and steel industry are given in Table 3.3.

Acid open hearth furnace. For complete super structure, (except the hearth) super duty silica brick is used. Working hearth is made by sintering layers of ground gannister or silica sand into place upon a sub-bottom made of silica brick. Small proportion of fireclay or slag is added to promote sintering.

Soaking Pits. Soaking nits are used for heating the steel ingots (varying in weight typically from 16 tons to 32 tons each and charging upto 8 ingots in a soaking pit). Soaking pits brick lining must have:

- good load bearing capacity
- high abrasion resistance
- resistance to iron oxide attack (particularly at the slag line).

Table 3.3: Refractory used in Iron and Steel Industry

Furnaces	Maximum operating temperature °C	Main refractories used
Blast furnace	1500-1600	Fireclay and high-alumina refractories and carbon blocks
Coke ovens	1100-1500	Fireclay and silica bricks
Open hearth furnace:		
• Basic roof	1650-1700	Magnesite, mag-chrome and chrome-mag bricks
• Silica roof	1600-1650	Silica bricks
Bessemer converter	1500-1650	Fireclay and silica bricks
LD converter	1650-1700	Dolomite and magnesite bricks
Electric arc furnace	1550-1700	Magnesite, mag-chrome, ganister or corundum bricks
Reheating furnace	1000 - 1500	Fireclay and high-alumina refractories
Soaking pit	1200 - 1400	
Ladle	1300-1500	Fireclay refractories (including bloating type).
Pit-house	1300-1400	Fireclay refractories (including bloating type) as also some special types like, graphite and magnesite stopper and nozzle.

The top of the soaking pit walls and the soaking pit cover lining must be highly resistant to spalling. Pit walls normally employ more siliceous bricks as these bricks meet the above requirement better and require less maintenance than do fireclay bricks. Slag line maintenance is a problem but 70% alumina bricks give good results. Super duty bricks, semi silica bricks, castables and ceramic fibre lining are particularly suitable for cover lining. The reasons for use of wide variety of refractories in soaking pit is because of wide variation in operating conditions.

Annealing furnace. Combustion chamber, walls and roof are made of high duty fireclay bricks.

Heating and Forging Furnaces

(*a*) **Hearth**. Plastic chrome ore ramming mixture and hard burnt chrome magnesite bricks which provide excellent resistance to attack by scale at high temperature is used in hearth.

(*b*) **Side walls.** High duty and super duty fireclay brick, and semi silica bricks are used.

(*c*) **Roof**. High duty fireclay brick is used.

Refractories used in the boiler. Temperature in the boilers for steam generation (for any purpose whatsoever) may be upto 1400°C but it can be as high as 2000°C in its combustion chamber. Firebricks are used upto a temperature of 1200°C but for higher temperature kaolin bricks, mullite or even chromite and silicon carbide are used. Refractory failure in boilers are due to rapid temperature fluctuation, vibration and fuel ash attack. In case of marine boilers, mortar joints should be of air setting type as it plays an important role.

Refractories used in glass industry

The glass industry furnaces are categorized by:

- High operating temperature
- High Insulation of crown
- High pull rates and low melting Rates
- Corrosive vapors and Batch dust

The bonded refractories are utilized in melter, refiner, regenerator, dog-house and ports of furnace.

Melter/Refiner: Refractories are needed for superstructure, crown and bottom. The important characteristics of the refractories in these areas are:

- High Alkali Vapor Resistance
- High hot strength
- High Corrosion Resistance

Crown: The furnace crowns are heavily insulated to reduce the energy loss. The important characteristics of refractory to meet this requirement are –

- Very Low flux factor <0.5; high hot strength
- Very low residual quartz <0.5%; good volume stability at service temperature
- High alkali vapor resistance
- Close dimensional tolerances required for laying with zero joint concept.

Regenerators: Regenerator in single or multi pass design is a common energy saving measure adopted. The basic requirement of the glassmaker from a regenerator is to maximize the heat recovery from outgoing gases apart from a long trouble free campaign life. This coupled with the increase in melting rates and use of alternative fuels has increased the severity of the working conditions in regenerator, subjecting the refractories to adverse chemical attack and high temperatures.

To achieve optimum energy saving and long campaign life, the refractories for such areas should be selected for:

- High thermal shock resistance
- High hot strength
- High creep resistance at service temperature
- High heat transfer

Checker Work: Regenerator checker system is a most vital component of a glass tank furnace from the point of energy saving. To increase the efficiency of heat transfer and to withstand the onslaught of the batch dust, high temperature and thermal cycling, the refractory products are characterized by:

- High thermal shock resistance
- High resistance to the chemical attack of batch dust
- High heat transfer

The needs of the different regions of the regenerator are different with condensation zone being the most critical zone.

Zirconia is useful refractory material for glass furnaces primarily since it is not easily wetted by molten glass and because of its low reaction with them.

Glass melting tank furnace is used to produce high temperature to fuse an appropriate mixture of silica, soda and lime into a molten mass *i.e.* glass. It is a continuous regenerative furnace fired usually by producer gas or furnace oil. The longitudinal and transverse sections of a glass melting tank furnace are shown in figures 3.6 and 3.7 respectively.

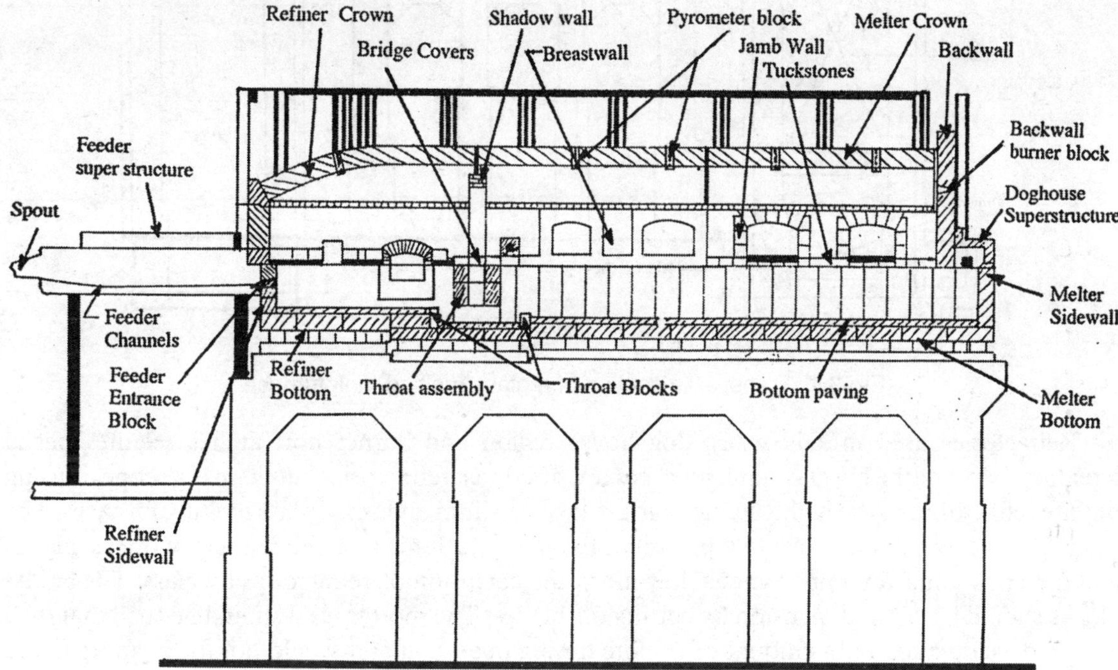

Fig.3.6: Longitudinal section of a glass melting tank furnace

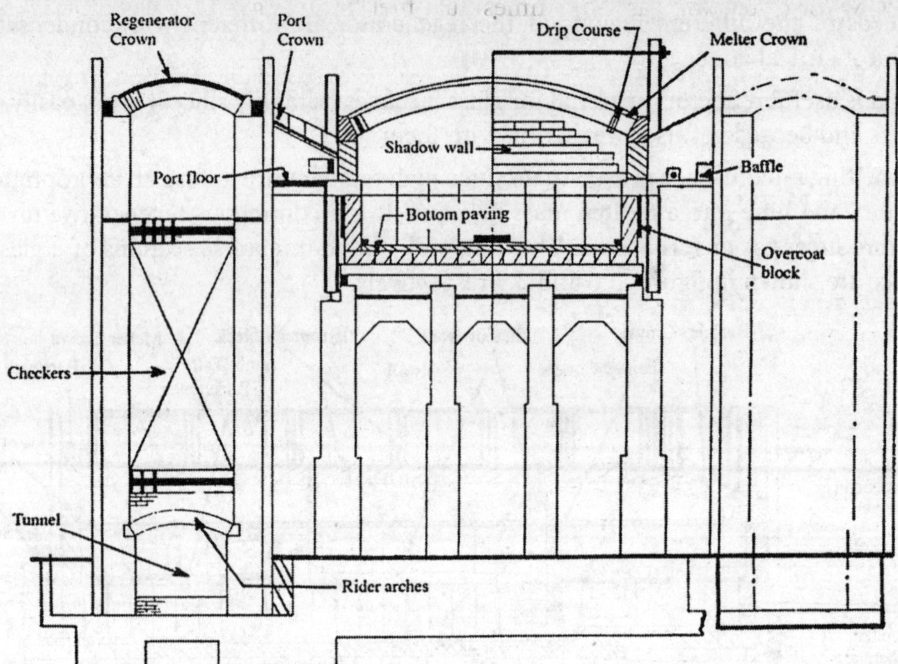

Fig.3.7: Transverse section of a glass melting tank furnace

Refractories used in side walls, dog house region and burner port arches require special attention. Most of the big glass tank furnaces use fused cast refractories but there is a scope of using mullite refractories in the highly vulnerable areas of medium and small glass melting furnaces. The desired refractory should have low porosity (about 15%). Surface of the refractory bricks/blocks and the brick joints are important as they affect the corrosion of refractories by glass. The brick/ blocks should be cut to dimension by cutting off blades. The mortar used should be such that they should develop maximum amount of mullite during operation and would not form cracks in the joints.

Corrosion of the refractories by glass is due to the diffusion of alkali ions into the solid phase present in the refractories. In case the fuel used in the glass melting furnace happens to be furnace oil then there is a possibility of presence of V_2O_5 along with alkali vapor in the flue gas. V_2O_5 in the presence of Na_2O lowers the refractoriness of both silica and alumino-silicate refractories by upto 85°C depending on their amount. A 50°C rise in the melting temperature of glass reduces the refractory life by 50%. To minimise the heat loss and to utilise the maximum amount of heat through the regenerator of the glass melting furnace, high temperature resistant insulating materials with very low thermal conductivity are used. Refractories used in the construction of glass melting tank furnace have improved significantly during last many years.

Refractories used in rotary kilns for calcination of limestone and dolomite

In the rotary kilns used for calcination of limestone and dolomite, the raw material is fed into the kiln at its elevated end and the burner pipe is at the other end so that the material moves in the opposite direction to that of flue gases. The fuel used may be tar, furnace oil, LSHS, pitch creosote mixture (PCM) or coke oven gas. Sometimes dual fuel firing (e.g. coke oven gas and tar) is also done. The calcined material is discharged to a rotary cooler underneath (which is water jacketted) which cools the lime and dolo produced by blowing air. Dimension of a typical rotary kiln is as follows:

Diameter = 3.6 m, Length = 75 m, Speed of the kiln = 1 rpm Inclination of the kiln to the horizontal axis = 3.5°

Typical dimension of rotary cooler are as follows:

Diameter = 2.5 m, Length = 38 m, Speed = 3 rpm Inclination of the horizontal axis = 3.5°

A proper thermal regime is maintained in the kiln to ensure proper calcination of limestone or dolomite. Calcination of dolomite require higher temperature than that for limestone. According to the requirement of chemical processes, different zones having different tempera-tures are developed in the kiln. The temperature in different zones are different and may vary from 600 to 1150°C for limestone kilns and from 800 to 1700°C in case of dolomite calcination rotary kilns. The important properties required in the refractories used in rotary kilns are resistance to abrasion, spalling and chemical action. However the degrees of these resistances required vary from zone to zone. The temperature is lowest in the material charging end and highest at the burner end.

For calcination of dolomite following zones are maintained as shown in table 3.4.

Table 3.4: Temperature Zones in Dolomite Calcination Kilns

Zones	Length of Kiln, metres	Temperature, °C	Process involved
Pre-heating	10-15	800-900	Moisture removal from the material.
Decomposition	30-35	900-1200	Liberation of CO_2.
Exothermic burning	8-10	1200-1300	Silicate and Calcium alumino ferrite formation.
Calcining	8-10	1300-1700	Partial melting of aluminates and ferrites and material
Cooling	4-6	1300-800	Cooling of calcined dolomite

For calcination of limestone following zones are maintained as shown in table 3.5.

Table 3.5. Temperature Zones in Limestone Calcination Kilns

Zones	Length of Kiln, metres	Temperature, °C	Process involved
Pre-heating	35-40	600-700	Moisture removal
Calcining	30-35	700-1150	Liberation of CO_2
Cooling	4-6	1150-600	Cooling of lime

Refractory lining of kiln and cooler

The full length of the kiln is lined with refractory bricks. Burning zones of kiln are lined with chrome magnesite bricks, while cooling and pre-heating zones are lined with fireclay bricks. In cooler, the shell is lined with fire clay bricks, 8M length of shell from its discharge end is not lined.

Refractory lining pattern in Dolomite Kiln: (from burner end)

(a) 0 - 2.3 M	–	Fire clay
(b) 2.3 - 55 M	–	Chrome magnesite
(c) 55 - 75 M	–	Fire clay

Lining pattern in Lime Kiln : (from burner end)

(a) 0 - 2.3 M	–	Fire clay
(b) 2.3 - 32.5 M	–	Chrome magnesite
(c) 32.5 - 75 M	–	Fire clay

Refractories used in Cement Industry

In cement industry, the kiln, the preheating system, and the cooler operate in a very high temperature range and therefore lined with refractories to protect steel structure against heat, abrasion and chemical attack.

Cement Rotary Kiln is the heart of any cement plant directly influencing the plant productivity. Nature of stresses is different in different areas, necessitating different quality of refractories for different zones.

Common refractory materials used in cement industries are:

- Dolomite
- Magnesia-Chrome Spinel
- Magnesia-Alumina Spinel
- Fireclay and
- High Alumina

The cement kiln clinker area is usually provided with Dolomite refractories. Advantages of Dolomite refractories for cement are well established . It is the most compatible material with the cement kiln clinker as it provides excellent coating stability, has very good resistance to thermal shock and alkali attack in varying kiln operating conditions. The refractories with Zirconia enrichment are used for crack arresting .

The Transition zone/Preheat zone/Preheater Precalinator areas are usually provided with Magnesia-Alumina spinel variety with high thermo-elastic & thermo mechanical property and various grades of high alumina bricks (45-80% Al_2O_3) with high strength.

Cement clinkers are produced in rotary kilns during cement manufacture either by dry or wet process. Normally, high alumina or basic refractories are used for lining the rotary kilns used in cement plants. The main zones in the rotary kiln are: drying, preheating, transition, burning and cooling. There are further sub-divisions in these zones. The temperature and operating conditions in these zones are quite different. Sometimes chains are fitted in the drying zone to break and remove the dried up clots of slurry fed into the kilns. The temperature in various zones varies from 180 to 1600°C or more.

The schematic diagrams of rotary kilns for dry and wet process of cement manufacture indicating the various zones with operating conditions are shown in Figs. 3.8 and 3.9 respectively.

The quality of refractories used in cement rotary kilns depends upon the size & capacity of the rotary kilns and also upon the process followed. Largest Indian kiln is of 4.5 m dia (capacity = 1200 tons/day) whereas those used abroad are upto 8 m dia (capacity = 5000 tons/day).

Smaller cement rotary kilns were earlier using refractory having 55% alumina content but with increase in size, presently 70% alumina or sometimes even upto 80% alumina refractory bricks are used. These high alumina refractories are mainly used in the burning/clinkering zone. In other zones, dense firebricks having 30-45% alumina are used which are able to resist abrasion. Mullite phase present in the high alumina bricks resists the chemical reaction. Hence higher is the mullite content, higher is the resistance to the chemical reactions occurring in the rotary kiln. Phase diagram of $CaO-Al_2O_3-SiO_2$ reveals the area of composition of alumino-silicate refractories resistant to the action of calcium silicates. Refractories used in cement kilns should have the capacity of forming a coating over them which protects the bricks from further corrosion and abrasion. It should be noted that refractories with 85% or more alumina do not form this coating.

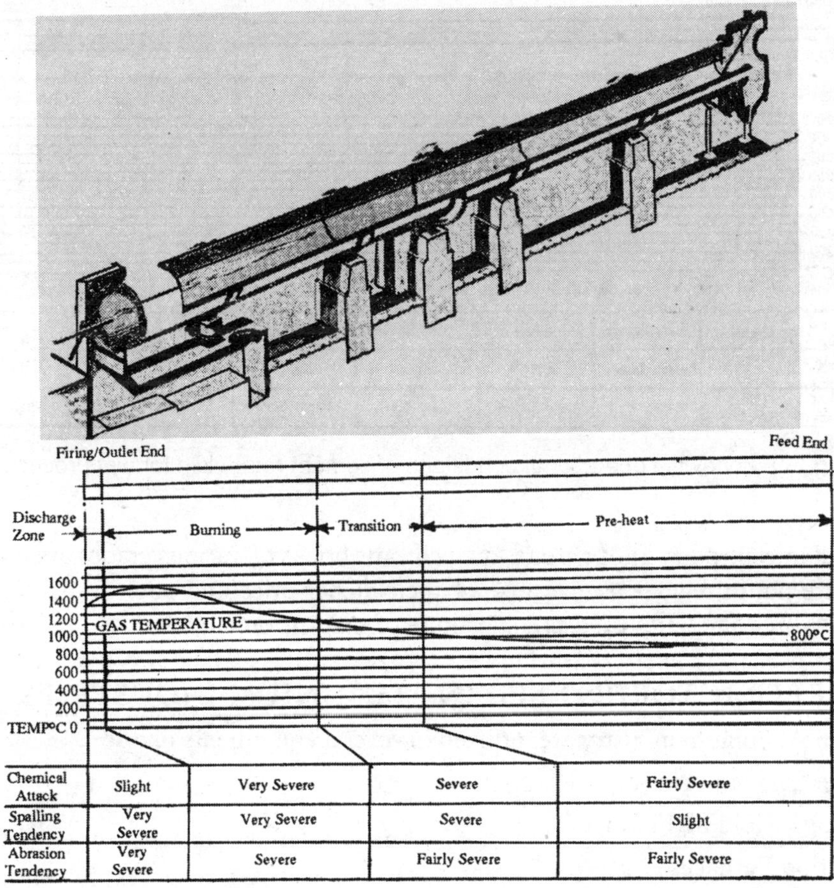

Fig.3.8: Zonewise operational conditions of cement rotary kiln for dry process.

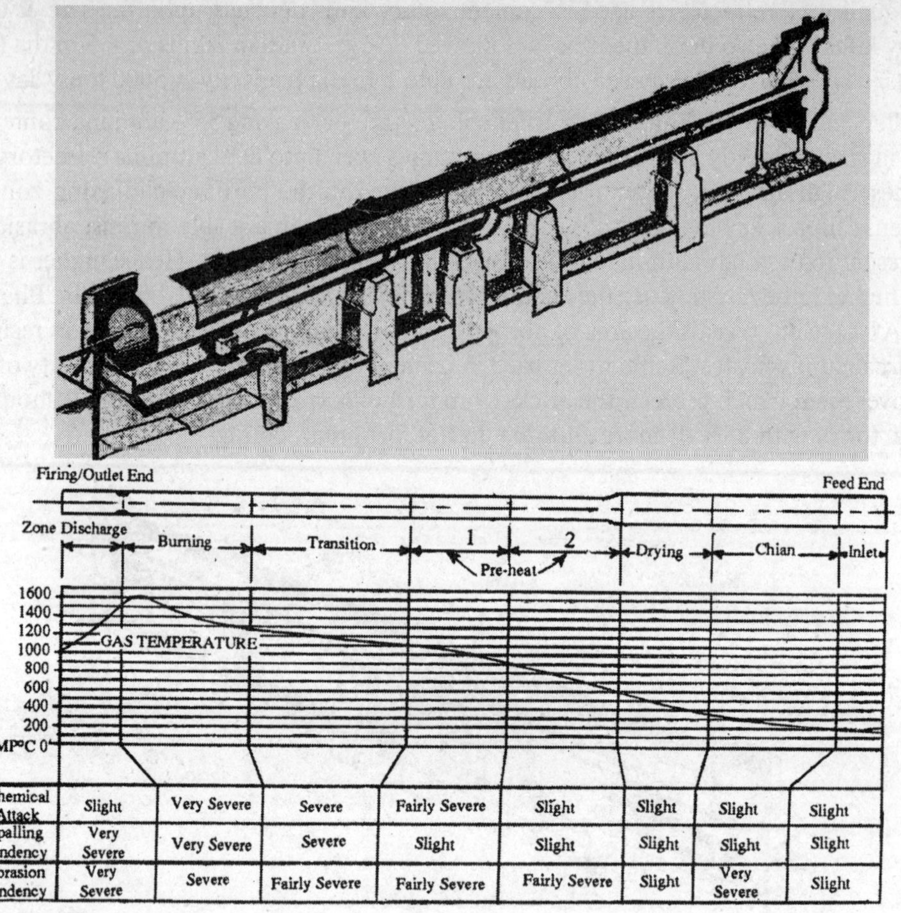

Chemical Attack	Slight	Very Severe	Severe	Fairly Severe	Slight	Slight	Slight	Slight
Spalling Tendency	Very Severe	Very Severe	Severe	Slight	Slight	Slight	Slight	Slight
Abrasion Tendency	Very Severe	Severe	Fairly Severe	Fairly Severe	Fairly Severe	Slight	Very Severe	Slight

Fig.3.9: Zonewise operational conditions of cement rotary kiln for wet process.

In larger diameter rotary kilns abroad, mag chrome bricks of various grades are used. In India, basic bricks are not in regular use because of their higher price and poor availability. Stabilised dolomite bricks are sometimes used as a substitute of mag chrome bricks.

REFRACTORIES USED IN ALUMINIUM INDUSTRIES

Refractories in Aluminum industry are utilized in several equipments like :

- Anode Baking Furnaces
- Melting / Holding Furnaces
- Induction Furnaces
- Reduction Pots
- Ladles and Launders
- Most common refractories utilized in Aluminum Industry include:

- Super duty Fireclay & High Alumina Bricks
- Phosphate Bonded High Alumina Bricks – These are ideally suited to molten aluminum ladle/furnace applications due to their unique chemistry, including special aggregates and additives.

The special requirements for refractories in Aluminum industry are:

- Resistance to Carbon/Aluminum penetration
- High Strength
- High Corrosion resistance

Range of products for Aluminum contact applications includes 'Insulating, dense, Low Cement and Ultra Low Cement Castables'. These products are not only reaction resistant to Aluminum contact but also totally non-wetting.

Pot shell is essentially a refractory-lined vessel composed of an outside steel shell and layers of refractory linings. The interior layer of the refractory lining is exposed to the process (molten metal, slag, corrosive gases). The linings between the shell and working lining are often referred to as the "safety or insulating linings." Insulating linings are used to limit heat loss and to maintain the vessel shell temperatures at reasonable levels.

Typically, the pot shell lining consists of several layers refractories blocks or bricks joined together using some monolithic material of varying qualities.

Aluminium industries have facilities for extraction of aluminium metal, production of alumina (by calcining aluminium hydrate made from bauxite) and carbon electrode block manufacturing. Maximum use is of dense superduty fireclay bricks (having 42-45% alumina). However, bauxite calcining rotary kilns and aluminium melting/holding furnaces require 85% alumina bricks. The refractories used in electrolytic reduction cell (for reduction of alumina to aluminium) should have resistance to abrasion, thermal spalling, dusting, attack by molten aluminium and mechanical shock. Refractories used in contact with molten aluminium should not have free ferric oxide or silica for eliminating the following thermit reactions.

$$4Al + 3SiO_2 = Al_2O_3 + 3Si, \qquad Fe_2O_3 + 2Al = A1_2O_3 + 2Fe$$

In the melting and holding furnaces in the aluminium industry, both the phosphate bonded bricks and monolithic refractories (containing 85% alumina) are mostly used. Even mortar used in those furnace are phosphate bonded 85% alumina. In the bauxite calcining and alumina calcining rotary kilns, mullite bricks or phosphate bonded 85% alumina bricks are used in the firing zone while in the preheating & cooling zones, dense superduty fireclay refractories and abrasion resistant castable refractories are in use. Monolithic refractories of varying alumina contents are in use in all types of melting furnaces of aluminium.

Dense, low porosity, spall resistant, superduty fireclay bricks (with 42% alumina) is used in carbon anode baking furnaces and petroleum coke calcining rotary kilns. These bricks must be resistant to slagging and abrasion.

Refractories used in various furnaces of aluminium industries are summarised and tabulated in table 3.6.

Table 3.6: Refractories used in Aluminium Industry

Furnace	Operating temperature, °C	Refractories in use
Alumina calcining	1000 – 1400 1150	Superduty fireclay dense fireclay, mullite or abroasion resistant castables
Bauxite calcining	1600-1700 1300-1400 550-600	72% or 85% alumina bricks dense fireclay chrome-mag
Petroleum coke calcining	1350	Abrasion resistant castable or superduty fireclay
Carbon anode baking	1500	Wall - superduty fireclay bricks, Flue - 60% alumina end – dense fireclay
Melting and holding (reverberatory)	800-850	Metal contact -85% alumina bricks, roof – plastic or castable, super structure – Superduty fireclay bricks or 75 – 85% alumina bricks
Rotary	—	40% alumina dense bricks
Induction	—	Graded quartzite or rammed siliceous clay
Electrolytic	900	Firebricks, carbon/silicon carbide

REFRACTORIES USED IN SOME OTHER NON-FERROUS INDUSTRIES

Refractory used in Copper Industry

In copper industry, "Direct Bonded Magnesia Chrome Brick" is the most popular choice . The bulk of copper is produced using flash smelters followed by reverberatory furnaces. Flash furnaces have major benefits over reverberatory furnaces, particularly in fuel efficiency, and increased yield.

The other type of copper smelting include; Isa Furnace, Pierce Smith Converters, Anode refining Furnace, Rotary Holding Furnace, Slag Cleaning Electric Arc Furnace, *etc.*

Primary smelting vessels, whether they are flash furnaces, reverberatory furnaces or electric arc furnaces; all subject the refractory lining to attack by slag and sulphurous gases. The slag composition is normally high in Fe_2O_3 & SiO_2, both of which attack magnesia in the refractory to form magnesio-ferrite, and Forsterite, whilst the sulphur present in the vapor phase also attacks magnesia to form magnesium sulphate. All reactions in the refractory system are accompanied by volume expansion, leading to crack formation behind the working face, which are then lost during temperature cycling.

Direct bonded magnesite -chrome refractories are the preferred materials in contact with copper slags, owing to their greater tolerance to acidic slags.

The copper converter carries similar slags to those found in the smelting process, together with sulphurous gases, and consequently similar modes of refractory attack take place. Preferential wear frequently takes place at the tuyere line of the converter owing to the combination of slag attack and thermal shock. High hot strength chrome - magnesite with good thermal shock resistance is a prerequisite.

There is little slag present in the Anode furnace, but the refractory lining is in contact with highly penetrative molten copper and cuprous oxide. Penetration of the refractory material with copper leads to disruptive failure owing to crack formation at the limit of penetration. Reaction between copper oxide and silica in the refractory leads to the formation of low melting point compound which exacerbate the formation of cracks & loss of the working face. The working environment of the Anode furnace demands chrome - magnesite material with high hot strength to minimize copper penetration and disruptive loss of the working face.

The slag cleaning furnace has special requirements depending upon the slag practice. It is necessary to maintain a frozen slag layer on the refractory lining, therefore, a good slag and penetration resistance is required. Direct bonded Magnesite - chrome materials are preferred.

In copper industry, magnesite bricks are also used for the metal contact areas in smaller furnace, converter anode furnace and remelting furnace. The main factors affecting refractory lining in copper converter (operating at 1200-1500°C) are reaction with faylite, FeS & ZnS and resistance to slag, spalling & abrasion. Due to formation of $MgSO_4$ ($MgO + SO_2 + ½O_2 = MgSO_4$) in magnesite bricks, there is a chance of crack development. It may as well spall due to migration of SiO_2, Fe_2O_3 and Cu_2O while mag chrome or chrome mag brick shows better performance due to formation of Cu_2O. Cr_2O_3 spinel.

In lead and zinc extraction industries, the common requirement is of high heat duty and superduty fireclay bricks. However, limited quantities of silica, magnesite and high alumina refractories are also required.

The refractories used in exothermic or rotary furnace for desulphurisation process of Zns (operating under reducing atmosphere at 1100-1300°C) should be resistant to the corrosive action of zinc vapor, high in mechanical strength and low in gas permeability. The rotary furnace (used in Waelz process) is lined with basic bricks; periclase spinel being preferred over chrome mag bricks. Such furnaces are also used for making lead, tin, mercury and antimony. SiC is used for retorts and trays in the production and refining of zinc. Due to severe slagging action in rotary kiln for smelting zinc ore, dense, low porosity, firebricks with good thermal spalling resistance are superior to 70% alumina bricks. Blast furnaces or reverberatory furnaces used for smelting galena (lead ore) at 1200°C requires highly abrasion resistant refractories. The lining should not react with PbO which may be aggrevated due to presence of antimony.

In nickel industry, fusion cast mag chrome bricks are used in the heavy wearing region of flash smelter. In the blast furnace (used for smelting nickel), firebricks are used in the shaft and magnesite in the hearth bottom. In reverberatory furnace used for nickel smelting, magnesite ramming mass is used over firebrick base in the hearth. Sometimes, electric arc furnaces are used for nickel smelting which has a silica roof while other areas are built of carbon.

Due to their non wetting characteristics to molten silver, copper & aluminium, the aluminium nitride refractories are used in their smelting and vaporising furnaces.

Tin is produced from stannous silicate or calcium stannate in a reducing atmosphere at 1200°C. Here, the reaction of molten metal and slag is resisted by either carbon or basic bricks.

The roaster used in the production of magnesium (which operates at only 500-550°C) is lined with firebricks while for thermic reduction in the electrolyzer, bricks used are magnesite, carbon or high alumina refractories.

Carbon industries mostly use dense superduty fireclay bricks. **Petroleum industries** require refractories with higher fusion point for resisting the fluxing effect of vanadium, sodium, sulphur *etc.* present in petroleum products. Refractories required in **steel heat treatment furnaces** (operating at < 1450°C) should be resistant to thermal spalling, abrasion and slag/scale attack. Refractories with high Al_2O_3 and low Fe_2O_3 content are used in **naphtha reforming (by steam) reformers** in fertiliser plants.

Ceramic fibres now a days are used in almost all the furnaces where there is no molten metal or molten glass. Even in furnaces treating molten metals, they are used as backing materials while in other furnaces they work as working hot face as well. Ceramic fibres are used on the car deck of tunnel kilns, firing pottery/refractory items, heat treatment furnaces *etc.*

Refractories used in some non ferrous metals industries is tabulated in Table 3.7.

Table 3.7: Refractories used in some Non-ferrous Metals Industries

Non-ferrous Metal Industry	Furnace	Operating temperature, °C	Refractory in use
Copper	Multiple hearth or fluidized bed roaster	800-900	MD fireclay
	Reverberatory (smelting)	1250-1500	Magnesite, Silica, SD fireclay
	Converter	1100-1250-1500	Magnesite, Chrome-mag
	Holding/anode		Magnesite, Silica,SD fireclay
	Remelting		Magnesite
	Wirebar (reverberatory)		Silica, SD fireclay
Brass (copper alloy)	Induction (loop)	800-1000	Quartzite with clay bond or Calcined clay with clay bond
	Reverberatory/Rotary	—	Silica, magnesite, SD fireclay
Cupronickel	Induction (loop)	1400	Fused alumina or Magnesite with clay bond
Zinc	Roaster		Dense fireclay
	Sintering		Dense fireclay
	Reduction		—
	Retort	1000	Fireclay with SiC
	Rotary (Waelz process) (recovery plant)		Firebrick, Chrome-mag or other Basic spinel
	Blast furnace:		
	Shaft	1300	Firebrick
	Hearth	960-1100-1200	Chrome-mag
	Pure zinc-smelter	1400	No refractory/Silica/SiC; Cast iron roof
	Reverberatory		Firebrick
	Low frequency induction		Aluminious brick

Lead	Sintering		SD/HD fireclay
	Blast furnace	200 at bell,1200 at tuyere	hearth-HD Fireclay, Magnesite, shaft-Firebrick
	Lead kettle	450-200	HD/MD fireclay, Graphite retort for zinc and noble metal recovery
	Softening furnace (reverberatory) [Dross (reverberatory)]	750-	HD fireclay, Magnesite/ Chromite-mag, High alumina
	Cuppellation cupola	Very high temperature	Mullite roof, Magnesite and Firebrick
	Lead oxide Recovery	—	HD fireclay, Magnesite HD/SD fireclay

In all the applications of the refractories, the first and foremost of their properties is their refractoriness *i.e.* the stability to withstand the operating temperature. The next important property is the chemical reaction with the process materials particularly liquid and gas.

Table 3.8 gives the list of typical bricks used in various furnaces in metallurgical industries.

Table 3.8: Typical Refractories used in Various Furnaces

Furnace	Refractories used
By product coke ovens:	
Oven bottom	Chrome magnesite
Oven walls	Silica bricks
Regenerators	Fireclay bricks
Oven top	Fireclay + insulating bricks.
By product plant saturator	Acid resistant bricks.
Hot metal (pig iron) mixer:	
roof	fire bricks
bottom	magnesite
end walls	magnesite.
L.D. Converter (BOF):	
Converter body	Permanent lining of chrome magnesite bricks followed by wear out lining of tar bonded dolomite bricks
Bottom of L.D. converter	Mica crumble followed by fireclay brick, chrome magnesite bricks and finally outermost layer is of magchrome bricks.
Hot metal ladle	fire bricks
Beehive coke ovens	fire bricks
Rotary kiln for limestone/dolomite calcination:	
Burning zone of kiln	chrome magnesite
Preheating zone (charging end)	fire bricks

burner end	fire bricks
Acid pickling tanks	acid resistant bricks
Rotary kilns for cement	fire clay, high alumina bricks and magnesite bricks.
Soaking Pits:	
upper part of the wall	fireclay brick
hearth and lower part of the wall	chrome magnesite
pit cover	fireclay/chrome magnesite
ceramic recuperator	silicon carbide.
Reheating furnaces:	
chimney, flue channel, roof, recuperator	fireclay bricks
bottom hearth and bottom falls	chrome magnesite bricks
monolithic hearth (near discharge end where temperature is higher)	fireclay bricks followed by corrundum bricks with chrome magnesite lining
Glass melting furnaces	Sillimanite bricks.
Slide gate of steel teeming ladle	High alumina bricks/magnesite bricks/zirconia bricks.
Tunnel kilns:	
Preheating and cooling zone	firebricks
firing zone	silica bricks/high alumina bricks.
Boiler Furnaces	firebricks/high alumina bricks.
Laboratory Furnaces	High alumina refractory backlined with insulating bricks.

Properties & Testing of Refractory

An understanding of the important properties of refractories is required for the development, improvement, quality control and selection of refractory linings for high temperature processes.

Though the properties of refractories alone cannot be used to predict refractory lining performance, comparisons among properties of different refractories are used as screening tools to select and upgrade refractory linings. Matching refractory lining requirements to service demands proceeds with a review of the hot physical and chemical properties of the lining materials under consideration.

Important properties of refractories which can be determined most readily are chemical composition, bulk density, apparent porosity, apparent specific gravity and strength at ambient temperatures. These properties are used as controls in the manufacturing and quality control process. The chemical composition serves as a basis for classification of refractories. The density, porosity & strengths of fired products are influenced by factors such as type & quality of the raw materials, the size & "fit" of the particles, moisture content at the time of brick pressing, pressure of brick pressing, temperature & duration of firing, kiln atmosphere and the rate of cooling.

The mechanical strength of refractory brick is governed largely by the ground mass material between the larger grains. At ambient temperatures, the compressive strength is typically much higher than the tensile strength. For example, a fireclay or 50% alumina product may contain glassy material in the bonding phase. This glassy material is difficult to compress but may be broken quite readily in tension. A product's room temperature strength is an important indicator of its ability to withstand abrasion & impact in low temperature applications and to withstand handling & shipping. Other important properties determined at ambient temperatures are porosity, permeability and pore size distribution.

The strength of brick at ambient temperatures may provide little or no indication of their strength at furnace operating temperatures. For example, a fireclay or 50% alumina material that has good strength at ambient temperatures will exhibit significantly reduced strength at elevated temperatures where the glassy phase may become quite soft or even fluid.

Important properties determined at elevated temperatures are hot modulus of rupture, hot crushing strength, creep behavior, refractoriness under load, spalling resistance, dimensional changes and thermal conductivity.

Evaluation of refractory properties is also vital to the manufacture of superior refractory products that meet defined standards. Refractory properties are equally important as product design criteria in the development of new and improved refractories.

Post-mortem analyses of the properties of refractories taken out from service are useful in improving both refractory and furnace operating performance. Comparison of properties of used refractories against original properties helps to identify wear mechanisms and actual operating conditions, including operational upsets.

The most significant properties of refractories are those which enable them to withstand the conditions to which they are exposed in service at elevated temperatures. These conditions may be classified broadly as thermal, mechanical, chemical and various combinations of the three.

Refractories must be sufficiently refractory to withstand the maximum temperatures to which they will be exposed in service. Frequently, they must be resistant to the stress effects of rapid temperature change within critical temperature intervals. They may be required to withstand varying loads, abrasive action & penetration and corrosion by solids, liquids & gases.

High temperature properties of refractories which are relevant to service conditions include the following:

- Refractoriness
- Melting behavior
- Mechanical strength and load bearing capacity
- Changes in dimensions when heated
- Resistance to abrasion, mechanical erosion and impact of solids
- Resistance to corrosion and erosion at high temperatures by solids, liquids, fumes & gases
- Resistance to spalling
- Thermal conductivity

The one condition common to all furnace operations is high temperature. A closely related combination of other thermal effects results from high temperatures and from the rate of temperature change. Heat energy flows into the refractory structure when the furnace is heated and a temperature differential develops between the inside and outside surfaces. Part of the heat energy is stored in the refractory structure & part flows through the walls and other parts of the furnace structure and is lost to the outside air through radiation & convection. When refractories are hot their linear dimensions and volumes are greater than when they are cold due to "reversible thermal expansion." With some refractories, additional volume change, usually not reversible, results from mineral inversions & transformations and sometimes from other causes.

Within a single brick in a furnace wall or arch, there is commonly a temperature drop of many hundreds of degrees centigrade between the hot face and the cold end. Consequently, differences in the amount of thermal expansion in various sections of the brick will cause internal stresses to develop within the brick. In a refractory structure, pressure inherent in the furnace construction, such as arch stresses, is increased by thermal expansion of the brickwork. Unequal volume changes resulting from rapid fluctuations in temperature may cause the development of higher stresses within the refractories than they are capable of withstanding.

When no single kind of refractory material is adequate to meet the various conditions prevailing in different parts of a furnace, two or more different kinds of refractories may be employed in the construction. However, refractories of dissimilar chemical compositions may react chemically with each other if in contact at high temperatures. Furnace fumes, dust, and slags can accelerate these reactions. In the high temperature portions of a furnace, it is good practice to separate reactive refractories by an intermediate course of refractory brick which will not react with either at the temperature of operation, or by a joint of non-reactive bonding mortar.

Correct selection of refractory materials can be made on the basis of full knowledge of service conditions and properties of the refractories. The properties of refractories are evaluated by testing it in the laboratory.

The following properties of refractory bricks are tested for assessing its suitability:

- Chemical composition
- Fusion/melting point (Also called refractoriness or softening point)
- Refractoriness under load (RUL)
- Permanent linear change on reheating (PLCR)
- Porosity
- Bulk density/True density
- Permeability
- Cold crushing strength (CCS)
- Modulus of rupture
- Spalling resistance
- Thermal conductivity and electrical resistivity/conductivity.
- Abrasion, erosion, slag and vapor/gas resistance
- Warpage

The testing methods of the above properties of refractory materials are described below.

Chemical Composition

Chemical composition serves as a basis for classification of refractories and as a guide to their chemical properties. Another important application of chemical analysis is in the control of quality of both raw materials & finished products. Minor components once thought to be unimportant are now recognized as controlling factors in the performance of many refractories.

Among the applications of chemical analysis to control of quality are: the determination of iron oxide & alkali contents of some clays; the alumina content of bauxites; the alumina & alkali content of silica refractories; the chromic oxide content & accessory oxide levels in chrome ores; the carbon content of graphite; and the boron oxide level in magnesites.

The chemical composition of a refractory material may not be the most important selection criterion, as brick of almost identical chemical composition may differ widely in their behavior under the same furnace conditions. Chemical analysis of a refractory alone does not permit evaluation of such properties as volume stability at high temperatures or ability to withstand stresses, slagging or spalling.

In any given class of refractory brick, the various brands commercially available differ considerably in chemical composition. Relatively few consist of more than 90% of any single component and some, particularly in the basic category, are rather complex chemically.

Of the three main types of refractory (fireclay, high-alumina &silica), alkalies constitute the most serious impurity and, generally, the less the better. However, other oxides, such as TiO, FeO, CaO, and MgO, should also be low to obtain best performance characteristics. The lime (CaO) content of silica brick is an exception. It is deliberately added during manufacture and actually improves service performance.

For the basic refractories, MgO, Cr_2O_3 & Al_2O_3 are considered the primary refractory components, with SiO_2, FeO & CaO the accessory oxides. However, basic refractories often have a spinel mineral structure which is tolerant of considerable iron content.

The special refractory brick are essentially single component systems. Zircon is naturally occurring zirconium silicate and is refractory in that form. Silicon carbide will contain a small amount of SiO_2 as a result of oxidation during manufacture.

REFRACTORINESS OR FUSION POINT

Pure substances melt sharply at a definite temperature. Most refractory materials consist of high melting particles bonded together. At high temperature, glass fuses and as the temperature rises, the resulting slag increases in quantity by partial solution of the refractory particles. The temperature at which this action results in failure of a test pyramid (cone) to support its own weight is called, the melting point of the refractory. Table 4.1 shows the melting point of some pure compounds used as refractories.

Refractoriness of a refractory is the temperature at which is the material softens and hence is the measure of fusibility of the material. Addition of an impurity in the refractory material lowers its lowest fusing constituent and the capacity of the lowest fusing constituent to dissolve the material of higher fusion point at comparatively low temperature.

A ground sample of the material to be tested is molded into the form of test cones and mounted in a ceramic plaque with a series of standard pyrometric cones having known high temperature softening values. The plaque is heated at a fixed rate until the test cones soften and bend. The number of the standard cone whose tip touches the plaque at the same time as the tip of the test cone is reported as the PCE (Pyrometric Cone Equivalent) value of the test cone.

The PCE does not indicate a definite melting point or fusion point because the test is not a measurement, but merely a comparison of the thermal behavior of the sample to that of standard cones. The test is used in evaluating the refractory quality of clays &the softening temperature of slags, as well as in the manufacture & quality control of fireclay products

Acid oxides lower the fusion temperature of a basic material while basic oxides reduce the fusion point of an acidic material. Some impurities is deliberately allowed in refractory mixtures so that more fusible mixtures serve as a bond material for it. The fused material is called 'glass' as it solidifies to a slag like usually non-crystalline mass.

Table 4.1: Fusion/Melting Point of Pure Compounds

Pure Compound	Formula	Melting temperature, °C
Alumina	Al_2O_3	2050
Lime	CaO	2570
Chromite	$FeOCr_2O_3$	2180

Chromium Oxide	Cr_2O_3	2275
Magnesia	MgO	2800
Silica	SiO_2	1715
Titania	TiO_2	1850
Mullite	$3Al_2O_3.2SiO_2$	1810
Sillimanite	$Al_2O_3.SiO_2$	1816
Spinel	$MgO.Al_2O_3$	2130
Carborundum	SiC	Solid at 2240
Zirconia	ZrO_2	2677
Beryllia	BeO	2530
Thoria	ThO_2	Solid at 3000
Carbon	C	Solid at 3600

'Fusibility temperature range' is a more appropriate term for refractory materials as mixture of refractory materials do not have a sharp fusion/melting point. The heat resistance of refractory material is expressed in terms of Pyrometric Cone equivalents (PCE) which is determined by comparing the softening of a test pyramid cone with that of standard pyramid cones of known fusion temperature by heating them under similar conditions.

Pyrometric Cone Equivalent (PCE)

Temperature at which a refractory will deform under its own weight is known as its softening temperature which is indicated by PCE. Refractories, due to their chemical complexity, melt progressively over a range of temperature. Hence refractoriness or fusion point is ideally assessed by the cone fusion method. The equivalent standard cone which melts to the same extent as the test cone is known as the pyrometric cone equivalent. Thus in Fig. 4.1 refractoriness of Sample *A* is much higher than *B* and C.

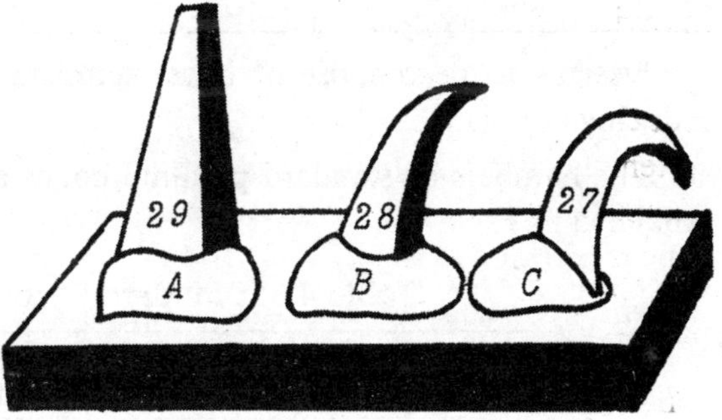

Fig.4.1: Pyrometric Cones.

The pyrometric cone equivalent indicates only the softening temperature. But, in service the refractory is subjected to load which would deform the refractory at a much lower temperature than that indicated by PCE. With change in the environmental conditions, such as reducing atmosphere, the P.C.E. value changes drastically.

At some temperature between that at which liquid first forms and that at which the whole brick is fluid, the proportion of liquid will be such that the brick cannot support its own weight and it will slowly deform. If the test piece is in the form of a pyramid, 1.5 inch high on a triangular base of 0.5 inch side (one edge of the pyramid being perpendicular to the base), and the specimen is heated at the rate of 50°C, per five minutes, the temperature at which the test piece has bent until the tip is level with the base is called the "refractoriness". The temperature at which this will occur increases with increasing alumina content.

A sillimanite or high alumina brick containing over 72% Al_2O_3 will not form liquid at 1545°C: no liquid will be produced (neglecting the fluxes present) until a temperature of 1810°C has been reached. At this temperature any mullite present decomposes into corundum and a liquid. The high refractoriness of sillimanite bricks is therefore readily explained. Most other refractories have a fusion point (refractoriness value) above that at which normal furnaces are operated.

The standard cones are called '*segar cones*' and '*orton cones*' and they are numbered according to their refractoriness/fusion points.

Average refractoriness values of some refractory bricks are given in Table 4.2.

Table 4.2: Average Refractoriness Values of Refractory Bricks

Brick	Segar Cone No.	Refractoriness, °C	Brick	Segar Cone No.	Refractoriness, °C
Silica	32-34	1710-1750	Sillimanite	35	1770
Siliceous	28-31	1630-1690	Magnesite	38	1850
Firebrick (23-30% Al_2O_3)	27-30	1610-1670	Bauxite	38	1850
Firebrick (30-35% Al_2O_3)	28-31	1630-1690	Chrome	38	1850
Firebrick (35-40% Al_2O_3)	29-33	1650-1730	Silicon carbide	38	1850
Aluminous firebrick	32-35	1710-1770	Carbon[1]	38	1850
			Zirconia	38	1850

Designs to decompose at lower temperatures. Decomposes unless atmosphere is strongly reducing.

The numbers of standard pyramid cones and their corresponding temperature of fusion are tabulated in Table 4.3.

Table 4.3: Standard Pyrometric Cone Equivalent (PCE)

Cone No.	Value of Orton Cone, °C (British Standard)	Value of Segar Cone, °C (German Standard)
12	1337	1375
13	1349	1395
14	1398	1410
15	1430	1440
16	1491	1470
17	1512	1490
18	1522	1520
19	1541	1530
20	1564	1540
23	1605	1560
26	1621	1585

27	1640	1605
28	1646	1635
29	1659	1655
30	1665	1680
31	1683	1695
31½	1699	—
32	1717	1710
32½	1724	—
33	1743	1730
34	1763	1755
35	1785	1780
36	1804	1805
37	1820	1830
38	1850	1855
39	1880	1875
40	1920	1900
41	1960	1940
42	2000	1980

Pyrometric cone equivalent (PCE) test to determine the refractoriness/softening points of refractory is described below.

Pyrometric Cone Equivalent (PCE) Test. The object of this test is to determine the softening points of refractory material by comparing the test cones prepared from a refractory under test with standard cones heated under a suitable furnace. The dried cones are utilized for comparative study of unknown material with the known behaviour of the used cones but with this we may get results within the temperature range of ± 15°C.

The furnace in which the neutral or oxidising atmosphere can be maintained is preferred for the test. The refractory material is ground after proper sampling and quartering to pass through IS sieve No. 20 in order to avoid excessive reduction of sizes. A particular quantity of sized material is taken and with the help of some bonding material (like dextrin, glue) and addition of water, the P.C.E. cones are prepared. Thc samplc thus prepared measures 8 mm on the sides of the base and 25 mm height. When dry, the test cones and standard pyrometric cones are mounted at a plaque of refractory material of such composition which will not affect the fusibility of cones. The test cones thus fitted are then put into the furnace and the furnace is electrically heated. Reducing conditions in the furnace is avoided during heating and also the flame should not strike directly against the cones on the plaque. The furnace is tested at intervals for uniformity of distribution of heat. The softening of the cone will be indicated by the top bending over at the

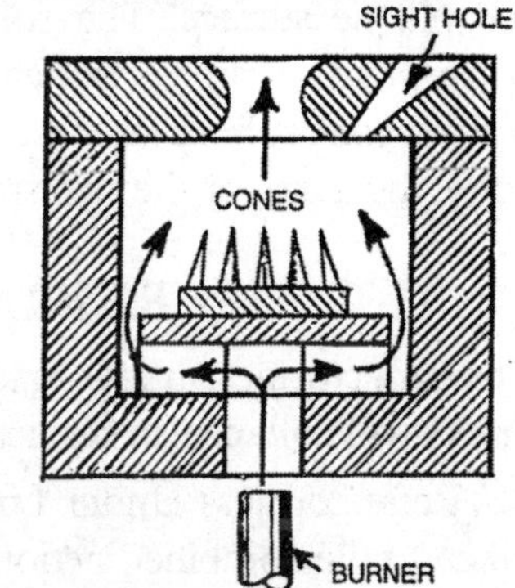

Fig.4.2: A gas-fired fusion-point furnace.

tip of the plaque surface. The softening point is reported in terms of the standard pyrometric cone which corresponds in time of softening with the test cone. A fusion point furnace (gas fired) is shown in Fig. 4.2.

This test does not involve actual melting of refractories. Rather, it gives a comparison of the softening characteristics of test cones with standard pyrometric cones when heated under similar conditions. Each cone is fixed on the plaque forming an angle of 82°with the horizontal. Standard cones are chosen keeping in view the anticipated fusion temperature of the test cones. After arranging the cones on the plaque, it is heated in the furnace at a standard heating rate as given in the table 4.4.

Table 4.4: Standard Heating Rate for PCE Determination

Room temperature ASTM cone number	Time interval, minutes	Cummulative time lapse, minutes	End point temperature, °C
20	45	45	1621
23	16	61	1640
26	7	68	1649
27	7	75	1659
29	8	83	1665
31	10	93	1683
31½	6	99	1717
32	7	106	1724
31½	3	109	1743
33	7	116	1763
34	9	125	1785
35	9	134	1804
36	7	147	1820
37	7	148	1835

Softening of cone is indicated by top bending over of the cone and the tip of the cone touching the plaque surface. The softening point of the standard pyrometric cone which most nearly corresponds in time of softening with the test cone is recorded as the P.C.E. of the test cone. In case the test cone softens later than one of the standard cones but earlier than the next higher pyrometric cone, the mean of the two standard cones is taken as the P.C.E. of the test cone.

REFRACTORINESS UNDER LOAD (RUL)

The refractoriness under load test (RUL test) gives an indication of the temperature at which the bricks will collapse, in service conditions with similar load.

Refractoriness Under Load (RUL) of a refractory material is a measure of its failure resistance to the combined action of heat and load. RUL is the softening temperature of a refractory brick (indicated by breaking of the test specimen) when a load of 2 kgf/cm^2 is applied as per the standard test procedure. Refractory bricks generally having more than two constituent compounds fail under load conditions at much lower temperature (*i.e.,* at RUL) than the fusion point (*i.e.*, refractoriness) of the highest melting and predominant constituent. That signifies that the temperature corresponding to the refractoriness under load of a refractory brick is much lower than its refractoriness (fusion/melting point). A refractory material composed of a single pure compound will have a closer value of its refractoriness and refractoriness under load.

In practical situations, some load is always acting on the refractory bricks/materials from the stock/charge of the furnace from air/gases passing through the furnace, from the weight of super incumbent brickwork *etc.* Moreover, the load may be unevenly distributed on various parts of a furnace. Refractory failure under load in absence of such a situation is comparatively rare. Cool outer parts including side thrusts of furnace structure have to take all the loads. Uneven expansion and brick movement results in failure of structure at these points due to increased local stresses.

Refractory failure is also influenced by its exposure time to load and heat. Longer exposure time results in failure of refractory under relatively low load & temperature conditions. Similarly refractories will fail at a lower temperature with increased load. Refractories containing small yet viscous fused glass also fails even at otherwise safe load and temperature conditions on prolonged exposure.

However, under actual service conditions where the bricks are heated only on one face, most of the load is carried by the relatively cooler rigid portion of the bricks. Hence the RUL test gives only an index of the refractory quality, rather than a figure which can be used in a refractory design. Under service conditions, where the refractory used is heated from all sides such as the checkers, partition walls *etc.* the RUL test data is quite significant.

Determination of Refractoriness Under Load (RUL). This *load test* determines the softening temperature of refractories under load indicated either by complete sloughning down *or* breaking of the test specimen.

The test furnace (electrically heated) consists of a heating tube of 100-120 mm inside diameter and about 500 mm length with a wall thickness of 10-15 mm. Heating tubes of corrundum, magnesite *etc.* are used. In India, magnesite tubes are widely used. The zone of approximately uniform maximum temperature shall have minimum length of 120 mm. The load application arrangement is such that a constant load of 2 kg/cm^2 can be applied vertically to the test piece. Provisions are there for recording changes in the height of the test specimen and to allow to permit it to be compressed by at least 20 mm.

A cylinder of 50 mm dia and 50 mm height obtained after boring out the central portion of a brick to be tested is used as test specimen. The load is applied by carbon or mullite rods. Carbon rods of 55-60 mm diameter between the test specimen is imposed. The temperature is raised at the rate of about 15°C/minute upto 1000°C and above 1000°C at a constant rate of 8 ^{0}C per minute. Temperature in the horizontal plaque may vary widely, the maximum difference of 20 to 30°C may be allowed in the hot zone. The change in height is plotted against the time on rectangular coordinates. As the temperature is raised at approximately constant rate and the change in height with respect to time is plotted, this chart will give the temperature deformation curve. The temperature is measured with an optical pyrometer sighted or adjusted upon the bottom of a refractory tube closed at its bottom and suspended in the furnace at the beginning of the test at about the middle of the test specimen. As such there should be a sensitive instrument to measure the deformation every moment.

Change in the height of the specimen on heating is plotted against time on a 10:1 scale to represent temperature deformation curve for the test. From the plotted curve, the temperature corresponding to the point at which the curve has dropped 3 mm below its highest point denoted by the bend of the curve downwards with respect to the horizontal tangent) is noted. The temperature (T_e) corresponding to the point at which the height of the test specimen has decreased by 20 mm from its original height is also noted. The temperature (T_b) at which premature breaking of the test specimen, without its actual softening, takes place denotes the breaking point of the test specimen. The temperature is reported correct to the nearest 10°C.

The refractoriness of a product is of less importance than the refractoriness-under load. In furnace construction, the brick-work is subjected to stress. This arises from the weight of the superincumbent brickwork, by the thrust of an arch, or by pinching due to inadequate expansion allowance. This stress may cause the bricks to deform at a temperature considerably below their fusion point. The effect is particularly marked in firebricks. Under a load of 2 kg/cm^2 (the standard adopted in setting this type of brick) 5% subsidence will generally occur at a temperature about 200°C below the fusion point.

In practice, only one face of a brick is at the highest temperature, so that frequently the colder part of a brick can support a load when the temperature of the working face is such that the brick would be expected to collapse.

The texture of a refractory involves a knowledge of the volume and nature of the pore system in relation to the solid material, the size & orientation of the grog particles, and the presence or absence of manufacturing faults. The texture has a bearing on the refractory properties, the resistance to spalling, abrasion & slagging and the thermal conductivity.

In fireclay products, the texture is controlled by the grain size of the clay, the nature & grading of the grog, and the temperature & duration of firing. Fineness of grain favour vitrification. A fine textured firebrick is expected to soften under load at a slightly lower temperature than a coarse product of similar composition. RUL value of some bricks are given in the Table 4.5.

Table 4.5: Average Refractoriness-Under-Load (RUL) Values

Refractory Bricks	Refractoriness-under load at 2kg/cm^2	
	Segar Cone No.	°C
Silica	29-32	1650-1710
Siliceous	18-27	1500-1610
Firebrick (25-30% Al_2O_3)	13-16	1380-1460
Firebrick (30-35% Al_2O_3)	16-19	1460-1520
Firebrick (35-38% Al_2O_3)	19-26	1520-1580
Aluminous firebrick (38-45% Al_2O_3)	26-29	1580-1650
Sillimanite	> 29	> 1650
Magnesite	18-28	1500-1630
Chrome	12-16	1350-1460
Chrome magnesite	18-32	1500-1710
Silicon carbide	>34	> 1750
Carbon	>34	> 1750
Zirconia	18-20	1500-1530

PERMANENT LINEAR CHANGE ON REHEATING (PLCR)

In the firing of refractory brick, permanent changes in dimensions generally occur, altering the individual brick from the mold size to the fired size. It is desirable that the changes be completed in the firing step to prevent expansion or shrinkage later in service. However, changes in dimensions and volume require time as well as temperature for completion. As firing proceeds, the rate of dimensional change gradually diminishes. Yet, it is rarely feasible to attain complete expansion or contraction during the firing process.

Therefore, in service, if the furnace temperature is high enough and maintained for a sufficient period of time, there can be an additional change in dimensions, generally slight, but permanent. Excessive dimensional changes in service are objectionable due to their potentially harmful effect on the stability of the furnace. A reheat test determines the amount of permanent change in dimensions of refractory brick which may occur at high temperatures. This permanent reheat or linear change is in addition to the reversible linear thermal expansion that all refractories exhibit when they are heated or cooled.

High temperature reheat tests is used to reveal:

- whether a refractory composition of a given quality has been fired long enough; or
- at a high enough temperature; or
- if not fired, whether it is made of volume stable materials; and
- whether a composition has adequate refractoriness and volume stability.

The tests may be used in these ways to estimate refractoriness. The amount of shrinkage or expansion depends not only on the refractory's characteristics, but also on the temperature and time of heating during the test. The results of a short term reheat test may not be a sufficient basis for predicting the long term volume changes which a refractory may undergo in service.

In the standard reheat test,, brick are placed in a furnace, gradually heated to a pre determined temperature and maintained at that temperature for five hours. After cooling, measurements are made to determine changes in linear dimensions and volume.

In the reheat test at 1400°C, high-duty fireclay brick generally show a slight (up to 1.5%) linear contraction. At 1600°C, superduty fireclay brick usually shrink 0 to 1 % in length. Some fireclay brick expand slightly in the reheat test, however, the expansion does not occur, if the brick are in compression. For high-alumina brick of the 50 to 80% alumina classes, the temperature of the reheat is usually 1600°C. Some brick show a linear shrinkage of as much as 2.0 to 2.5%; others, an expansion up to 3.5% or more. For brick of the 90% alumina class, corundum brick, and mullite brick, the temperature of the test may be 1725°C and the linear change is usually within the range of minus 0.5% to plus 1.5%.

The reheat test is not usually applied to silica brick, as the apparent specific gravity provides an adequate measure of the amount of permanent or residual expansion. In the reheat test at 1600°C, most basic brick will show a slight linear shrinkage, varying from 0.1 % or less for some types of brick, up to 2.5% for others. High fired chrome- magnesite brick usually have little or no shrinkage in the reheat test at 1600°C and a shrinkage of only 0.1 % at 1650°C.

Knowledge of the thermal expansion behavior of refractory materials is crucial in designing furnace linings. If the lining is constrained by the furnace structure, destructive forces may result as the lining expands during furnace heat-up. In such cases, the lining must be designed to allow some free expansion. The amount of expansion allowance, however, must not be so much as to cause lining instability. An understanding of thermal expansion characteristics is also important in assessing the thermal shock resistance of a material. Thermal shock occurs when severe temperature gradients in a material cause internal stresses due to differential thermal expansion.

The measurement of the linear thermal expansion of a refractory is usually made by determining the change in length of a bar-shaped specimen heated uniformly at a constant rate. The movement of the sample may be detected by connecting an electrical transducer, located above the furnace, with the top of the specimen by means of a sapphire rod. A device of this type is known as a dilatometer.

Thermal Expansion of Fired Refractories. If no changes of a permanent nature occur during heating, fired refractories return to their original dimensions when cooled. This is known as "reversible thermal expansion ". Typically, refractories that are heated to temperatures below their firing temperature show this behavior.

The expansions of different brands in each class of refractory differ somewhat among themselves, and the data therefore represent mean values. Magnesite & forsterite brick have relatively high rates of thermal expansion; fireclay, silicon-carbide, zircon brick, & most insulating firebrick have relatively low rates. The thermal expansions of other refractories have intermediate values. For most fired refractories, thermal expansion increases at a fairly uniform rate over the working range of temperature. Notable exceptions are silica brick which undergo the greater part of their expansion below 425°C.

Thermal Expansion of Unfired Refractories. In general, the thermal expansion behaviors of unfired refractories are more complex than those of fired refractories. During initial heating, dramatic expansions or contractions may occur in an unfired material as a result of changes in bonding structure, changes in minerology, and sintering effects.

The thermal expansion characteristics of cement-bonded refractories during initial heat-up show shrinkage over the temperature range 205 to 315°C which is associated with thermal decomposition of the cement. The amount of shrinkage is determined by the quality & amount of cement. At temperatures above 980 to 1090°C, additional shrinkage occurs as a result of sintering. The underlying thermal expansion is determined by the characteristics of the aggregate. The shrinkages which take place during initial heat-up to 1430°C are permanent in nature and commonly are of the order of 0.2 to 1.5%.

The thermal expansion curves for initial heat-up of several phosphate- bonded plastics and clay-bonded plastics materials show linear behavior below 1040°C. At about 1040°C, however, mineralogical changes among the clay constituents result in densification and the materials contract. Permanent shrinkages of 0.2 to 0.6% are common after initial heating to 1430°C

Changes in Dimensions Under Load. In service, refractory materials must support a load which, at its minimum, is equal to the weight of the lining above the reference point. The pressure which is exerted depends on the height of the lining and the density of the material. Therefore, for applications in which the entire lining component is at high temperatures, the load-bearing capabilities of refractory materials is important. Examples of such applications include blast furnace stoves and carbon bake furnaces in which the refractories are heated relatively uniformly to high temperatures.

The standard laboratory tests for refractories include one for the determination of the behavior of brick under load at high temperatures. In this test, two 9-inch straight brick are set on end in a furnace of specified design and a vertical load of 25 lb/in^2 is applied & the temperature is gradually raised in accordance with a prescribed schedule. For fireclay and high-alumina brick, the maximum temperature is held for 1½ hours, after which the furnace is allowed to cool by radiation to 1000°C or lower before the load is removed.

The amount of subsidence is reported as percentage of the original length. The normal temperature of the hold for high duty fireclay brick is 1350°C ; for superduty fireclay brick, 1450°C. For most classes of high-alumina brick the top temperature of testing is 1450°C ; for those of the 90% alumina class and mullite & corundum brick it is 1595 and 1760°C. Load testing provides data relative to maximum service temperatures.

Like most structural materials, refractories show creep behavior when exposed to high temperatures. Most refractories show two characteristic stages of creep. In the first stage, called primary creep, the rate of subsidence declines gradually with time. In the secondary stage, called steady state, the rate of subsidence is constant. At very high temperatures, steady state creep is sometimes followed by a tertiary creep region where the rate of subsidence accelerates and leads to catastrophic failure or creep rupture. Primary creep is generally short in duration, while secondary creep can occur over a long term. Hence, secondary creep usually provides a more meaningful comparison of refractories. Creep behaviors of brick cannot be predicted based only on their chemistry. Important variables which affect creep behavior are chemistry of the bonding phase and firing temperature. Formation of low viscosity glassy phases results in poor creep behavior, whereas well-formed crystalline bond phases give good creep behavior. Aggregate sizing and porosity also affect creep behavior with larger aggregate & lower porosity giving better creep resistance.

Volume stability, expansion and shrinkage at high temperatures. The contraction or expansion of the refractories can take place during service. Such permanent changes in dimensions may be due to:

- The changes in the allotropic forms which cause a change in specific gravity.
- A chemical reaction which produces a new material of altered specific gravity.
- The formation of liquid phase.
- Sintering reactions.
- It may also happen on account of fluxing with dust and slag or by the action of alkalies on fireclay refractories, to form alkali-alumina silicates, causing expansion and disruption. This is an example which is generally observed in blast furnaces.

While it is desirable that all these changes are affected during manufacturing itself, it is not possible due to economic reasons, as the process is time dependent. Permanent Linear Change (PLC) on reheating and cooling of the bricks give an indication of the volume stability of the product as well as the adequacy of the processing parameters during manufacture. It is particularly significant as a measure of the degree of conversion achieved in the manufacture of silica bricks.

Reversible thermal expansion. Any material when heated, expands, and contracts on cooling. The reversible thermal expansion is a reflection on the phase transformations that occur during heating and cooling. The PLC and the reversible thermal expansion are followed in the design of refractory linings for provision of expansion joints.

As a general rule, those with a lower thermal expansion coefficient are less susceptible to thermal spalling.

Permanent linear change on reheating (PLCR) Test. This test is carried out in the laboratory furnace by using refractory brick cut to specific size (as per IS specification) and heating the same at a particular temperature for a particular time (also specified in IS specification) and then the results can be reported in two ways: (a) linear change and (b) volumetric change. For the results to be reported in terms of linear change, the length of the original specimen is to be measured with the help of a vernier scale and after heating, the length is to be measured by the same apparatus. The PLCR is expressed as the percentage of increase/ decrease in length as under:

$$\text{PLCR \% (Linear)} = \frac{\text{Increase / decrease in length}}{\text{Original length}} \times 100$$

Similarly, in the volumetric method, the volume before & after firing is measured and the result is reported as:

$$\text{PLCR \% (Volume)} = \frac{\text{Increase / decrease in length}}{\text{Original length}} \times 100$$

Determination of Permanent Linear Change (PLC). A test specimen of refractory brick (of size 5 × 5 × 12.5 cm) cut by cutting off wheel from a given refractory shape is placed in a kiln having an oxidising atmosphere. In case of smaller refractory shapes, the specimen of largest possible size is cut in such a way that it retains maximum number of original faces of test refractory ensuring that its structure does not get damaged.

The largest face of the test specimen rests on a supporting refractory brick drawn from the same lot of refractories which are under test or which has equal refractoriness. It should be ensured that the flame does not impinge on the specimen directly from the burner. Fused alumina or kyanite of 85 micron (IS seive) is placed between the supporting refractory and the test specimen. When more than one test specimen is kept in the kiln, they should be placed 4 cms apart from each other or from the kiln wall.

Kiln heating is done according to a set time temperature schedule depending upon the type of refractory under test. After holding the test specimen at the maximum desired temperature for the stipulated period, it is cooled to room temperature for over 10 hours in the closed kiln itself. The test specimen volume is measured again after taking out from the kiln and removing small blisters, if any from the surface of the specimen. The linear change is then calculated on the basis of volume change by the following relationships:

$$\text{Volume change (\%)} = \frac{\text{final volume} - \text{Original volume}}{\text{Original length}} \times 100$$

$$\text{or, Linear change (\%)} = \frac{\text{final volume} - \text{Original volume}}{\text{Original length}} \times 100$$

Time-temperature heating schedule for some bricks are given in Table 4.6 (for PLCR test).

In PLCR test, the furnace hearth should have a uniform temperature which is measured by Pt - Pt/Rh thermocouple upto 1450°C and by optical pyrometer if temperature exceeds 1450°C. Maximum variation in temperature should be ± 10 - 15°C depending upon whether the temperature measured is below or above 1450°C. The kiln temperature is recorded every 15 minutes to maintain a constant temperature. The final reheating temperature of the test specimen depends upon the type of refractory and hence its refractoriness as shown in Table 4.6.

Table 4.6: Heating Schedule for Various Refractories

Time from start of heating	Fireclay refractories	Sillimanite refractories,	Silica refractories	General purpose basic refractories	Insulating refractories
1 hour	750-1000°C	900-1000°C	—	750-850°C	565-965°C
2 hours	1150-1325°C	1275il325°C	—	1275-1325°C	770-1300°C
3 hours	1250-1400°C maintained for 5 hours	1375-1400°C		1375-1400°C	850-1250°C maintained for 24 hours Maximum 1470°C
3.5 hours	Upto 1450°C maintained for further 2 hours	—	—	—	1500°C maintained for 24 hours
4 hours	1500°C Maintained for further 1 hour	1500°C maintained for further 5 hours	—	1475-1500°C	—
5 hours	1550°C maintained for further 1 hour	—	upto 800°C	1600°C maintained for further 2 hours	—
6 hours	—	—	1450°C maintained for further 4 hours	—	—

Dilatometric Method of PLC Test. A dilatometer assembly used for measurement of linear change of a refractory sample with temperature is shown in Fig. 4.3.

A cylindrical test specimen (after cleaning its surface of dust etc) cut from a given refractory is placed with its one end on a supporting tube so that other end is free to expand. This free end of the specimen is joined to the core of an electronic displacement transducer. The sample is heated and any change in its length makes the free end to move the ceramic rod which pushes the core of the transducer as a result of which deflection is recorded in millivolts which is proportional to the change in length of the specimen. Timely standardisation of the dilatometer is done using a standard specimen of known co-efficient of

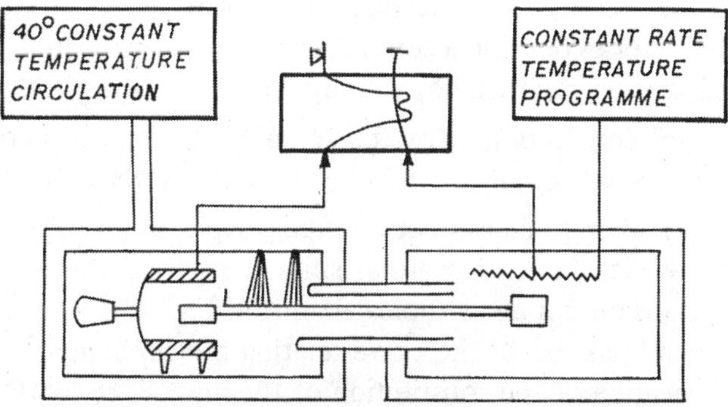

Fig.4.3: Dilatometer.

expansion where from correction factor is determined to account for the expansions that take place in the ceramic tube and the carrier refractory on heating simultaneously with the specimen.

Dilatometric analysis also helps in determination of structured changes and the percentage of different phases formed due to transformations on heating which involve change in total volume of the refractory specimen *e.g.* % of cristobalite, tridymite and quartz in a silica brick on heating to the desired temperature.

Effect of Re-firing on the Volume Stability

During the firing of refractory, the chemical and physical action/changes are stopped before they have gone to completion. During the firing of firebricks, for example, the actions taking place result in a certain amount of contraction. If these actions are not sufficiently completed, this contraction may continue during use. If such after-contraction is considerable, the masonry of the furnace will become displaced, open joints, cracks and distortions resulting. For fireclay products, an after contraction of 1% after retiring for two hours at 1400°C is considered high. The actual temperature at which a brick should be tested and the allowable after contraction depends on the conditions of service.

With silica bricks, the kiln treatment leads to expansion, but the reactions are generally arrested a little before completion. During use at high temperatures, the residual quartz may re converted into the forms of lower density, and growth may thus occur. This growth will take place even against high compressive forces, so that it is essential that the after-expansion of silica bricks should be low.

Magnesite shrinks during firing due to the crystallisation of periclase (MgO) from the originally amorphous magnesia. Chrome and chrome magnesite bricks may contract or expand. The kiln atmosphere plays a determining part with bricks containing chrome ore. Adequate firing is necessary always to take the reactions as near to completion as is practicable.

THERMAL EXPANSION

The ordinary or reversible thermal expansion of the materials of constructions is of particular significance as some allowance for its accommodation must generally be made in the furnace design. The thermal expansion of a refractory is also an important factor in determining its resistance to thermal shock.

In general, the thermal expansion is slightly above 1000°C. At high temperatures, most refractories develop a certain amount of "give", due to the softening of the bond, and some of the expansion can then be taken up within the brick itself.

Firebricks have a low, fairly uniform expansion up to 1000°C; this is generally of the order of $\leq$ 0.5% or less. Silica bricks behave quite differently.

Tridymite shows two comparatively small irregularities, while cristobalite and quartz each show one large. A normal silica brick will show each of these discontinuities in its thermal expansion curve, as indicated in the Fig. 4.4. Examination of a thermal expansion curve of a silica brick, or, of the curve relating the coefficient of expansion with temperature, gives a good indication of the constitution of the brick. The total thermal expansion of a silica brick between 0 and 1000°C is usually 1.2-1.3%.

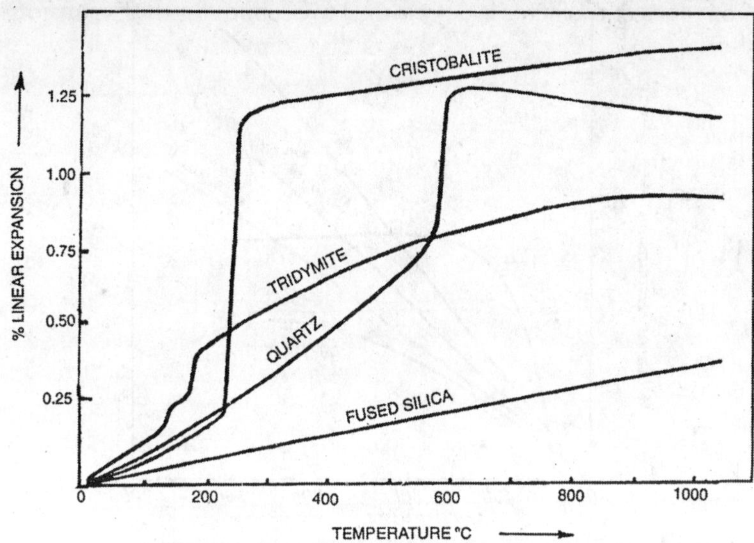

Fig.4.4: Expansion of the silica minerals.

The thermal expansion of magnesite bricks is comparatively high, 1.3-1.4% up to 1000°C that of a chrome or chrome magnesite brick is much lower unless the atmosphere is reducing

Sillimanite has a low thermal expansion, approximately 0.5% up to 1000°C. Sillicon carbide is of the same order but the thermal expansion of corundum brick is higher (about 1%). Typical variation of thermal expansion of some common bricks with temperature is shown in the Fig. 4.5.

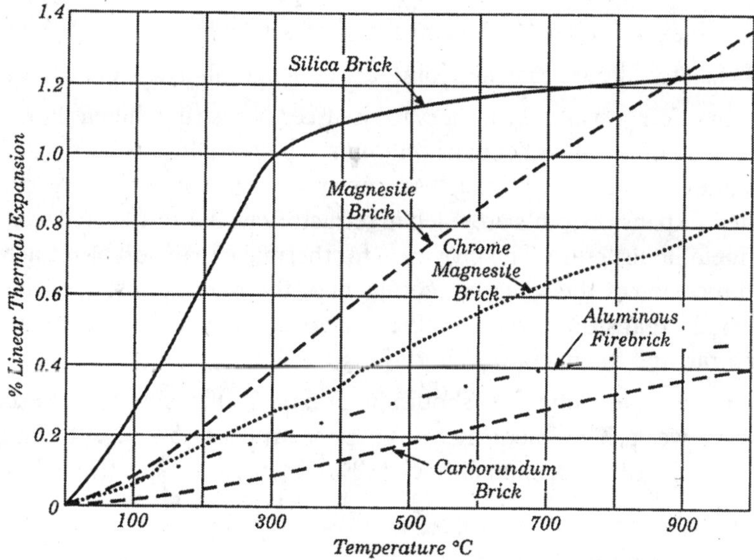

Fig.4.5: Reversible linear thermal expasion of refractories.

Typical linear expansion curve of some refractories subjected to higher temperature is shown in Fig. 4.6.

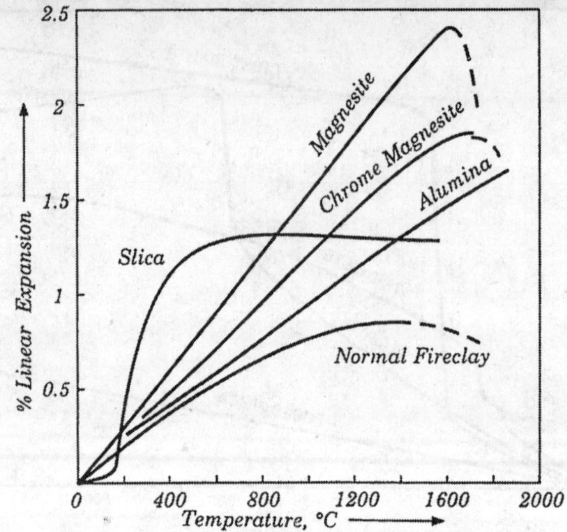

Fig.4.6: Typical expansion curve of some refractories.

Percentage linear thermal expansion of some more refractory bricks at various temperatures is tabulated in Table 4.7.

The thickness of the brick joints is influenced by the characteristics of both the brick & mortar, the kind of bond employed, and the provision made for thermal expansion of bricks. Sometimes, no mortar is required for special applications. The mortar joints, however, should be as thin as possible.

The usual thermal expansion and allowances for refractory bricks are tabulated in Table 4.8.

The materials used for filling expansion joints are either combustible or compressible materials.

The vertical expansion in walls allow them to rise freely upward, whereas horizontal expansion is taken care of by providing vertical expansion joints. The metallic structure used for holding the furnace gets heated up, although to a much lesser extent, due to comparatively lower temperatures than the refractories. Expansion joints provided in refractory walls are adjusted to accommodate the expansion of the metallic structure. The provision for thermal expansion of refractory bricks varies according to the dimension of the wall, the coefficient of thermal expansion of the refractories, the coefficient of thermal expansion of the metallic structure, the type of insulation used, the nature of the bonding mortar, and the operating conditions.

Although adequate expansion joints should be provided, they should not exceed a safe limit, making the structure weak. The linear thermal expansion (% at 540°C) of some constructional materials like cast iron, portland cement & steel (0.2% carbon) are respectively 0.82, 0.675 & 0.875.

Table 4.7: % Linear Thermal Expansion of Various Bricks at Different Temperature

Bricks	Temperatures,°C													
	100	200	300	400	500	600	700	800	900	1000	1100	1200	1300	1400
70% alumina brick	—	—	—	—	0.30	0.37	0.40	0.50	0.60	0.70	0.76	0.85	0.95	—
90% alumina brick	—	—	—	—	—	0.40	0.50	0.60	0.70	0.75	0.80	0.95	1.10	—
99% alumina brick	—	—	0.20	0.30	0.35	0.45	0.55	0.65	0.75	0.85	0.95	1.05	1.15	—
Burned Magnesia brick (93.5% MgO)	0.05	0.20	0.35	0.45	0.60	0.75	0.95	1.10	1.25	1.40	1.50	1.70	1.85	—
Magnesia chrome brick	—	—	—	—	—	—	0.85	0.95	1.10	1.25	1.40	1.55	1.70	—
Chrome magnesia brick	—	—	—	—	—	—	—	—	—	0.90	1.05	1.15	—	—
Chrome brick	—	—	0.20	0.30	0.40	0.50	0.60	0.70	0.75	0.85	0.95	1.10	—	—
Moulded synthetic graphite	0.02	0.05	0.075	0.125	0.175	0.200	0.250	0.300	0.350	0.375	0.450	—	—	—
Stabilised zirconia brick	—	—	0.25	0.35	0.45	0.50	0.60	0.70	0.75	0.80	0.85	1.00	1.10	—
Silicon carbide (clay bonded)	—	—	0.20	0.25	0.30	0.35	0.40	0.50	0.55	0.60	0.65	0.70	0.75	—
Zircon bricks	—	0.10	0.15	0.20	0.25	0.30	0.35	0.40	0.45	0.50	0.60	0.65	0.70	—
Electrocast alumina (α-β)	0.05	0.10	0.20	0.30	0.35	0.40	0.45	0.60	0.70	0.75	0.85	1.00	1.10	—
Electrocast Magnesia	0.05	0.10	0.25	0.40	0.60	0.70	0.85	1.00	1.15	1.30	1.45	1.60	1.80	—
Insulating bricks for temperature														—
(a) upto 1425°C	—	—	—	—	0.25	0.30	0.35	0.40	0.45	0.50	0.55	0.60	0.65	—
(6) upto 1650°C	—	—	—	—	—	0.35	0.45	0.50	0.55	0.65	0.70	0.80	0.85	—
(c) upto 1815°C	—	—	—	—	0.35	0.45	0.50	0.60	0.70	0.75	0.85	0.95	1.05	—

Table 4.8: Thermal Expansion Allowances for Refractory Bricks

Type of Bricks	Usual thermal expansion (mm / metre) at operating temperature	Expansion allowance at operating temperature, mm / metre
Fireclay	7.81	0.7
High-alumina with 50-80% alumina	7.81	0.7
High-alumina with more than 80% alumina	10.42	1.0
Silica	13.02-15.63	1.3
Magnesite and Mag-chrome	20.83	1.9
Chrome and Chrome-mag (fired)	10.42	1.0
Chrome-mag (chemically bonded)	18.23	1.7
Forsterite	15.63	1.5

APPARENT POROSITY

The apparent porosity is a measure of the volume of the open pores or interconnected pores in a refractory, into which a liquid can penetrate, as a percentage of the total volume. This is an important property in case where the refractory is in contact with molten charge and slags. A low apparent porosity is desirable since it would prevent easy penetration of the refractory size and continuity of pores will have important influences on refractory behaviour. A large number of small pores is generally preferable to an equivalent number of large pores. However, a measure of the true porosity which also takes into account the volume of closed pores, gives a reasonable idea of the texture of the material, as well as sintering characteristics. The porosity of a refractory has an effect upon its ability to resist penetration by metals, slags & fluxes and, in general, the higher the porosity, the greater the insulating effect of the refractory. In contrast to apparent porosity, true porosity represents a sample's total porosity, both open and closed pores. Closed porosity refers to porosity within the coarser grains, which cannot be readily determined. Hence, true porosity may be difficult to measure.

True porosity of a refractory brick is the percentage of the volume of the combined open and close pore space of its total volume. Density of a refractory brick is controlled by both open and close pores while its permeability is affected by only open (i.e., interconnected) pores.

With increase in the porosity of the refractory bricks, its:

- thermal conductivity decreases because the air which fills the pores acts as an insulator, that is why insulating bricks have higher porosity.
- resistance to the penetration of metal, molten slag and flue gases reduces due to increased permeability.
- temperature fluctuation sensitivity reduces as pores can accommodate the thermal expansion due to heating.
- strength reduces due to decreased compactness obviating their usage in taller structure demanding higher strength.

Two methods namely 'Boiling point method' and 'Evacuation method' are used for porosity determination.

Porosity determination. In raw/green refractory materials the pores are all open and fluids can pass through them. When these materials are fired some liquid formation due to fusion takes place and as a result some pores will be sealed. The apparent porosity as such takes the open pores and not the closed pores. True porosity is defined as:

$$\text{True porosity} = \frac{\text{Volume of open pores} + \text{Volumeof closed pores}}{\text{External volume}} \times 100$$

$$\text{And, Apparent porosity} = \frac{\text{Volume of open pores}}{\text{External volume}} \times 100$$

Boiling Point Method of apparent porosity determination

The apparent porosity of a clean small sample (measuring 6.5 × 6.5 × 4 cm) is measured by making three weighings. The first the dry specimen ; the second the saturated specimen immersed in water and third the saturated specimen suspended in air. To ensure that the specimen is thoroughly saturated, it should be boiled in water for two hours. The apparent porosity is determined from these weighings by the equation :

$$\text{A.P.} = \frac{W - D}{W - A} \times 100$$

where,

A.P. = % of apparent porosity

W = Weight of saturated specimen in air

D = Weight of dry specimen in air (dried at 110°C in an oven)

A = Weight of saturated specimen submerged in water.

The determination of total pores (including the closed ones) is done by the grinding of the specimen finely enough to open all pores and measure the true density (by pyconometer method). The total pores is given by

$$P = \frac{dt - db}{dt} \times 100$$

where,

dt = true density,

db = bulk density

p = % of total pores in relation to bulk volume

Apparent Porosity Determination by Evacuation or Vacuum Desicator Method

Clean test refractory specimen (of size 6.5 cm × 6.5 cm × 4 cm wheel cut from the core of a refractory shape to be tested) is dried at 110°C in an oven to a dry weight (D) maintaining an accuracy of 0.1 gm. The dry specimen is kept in an empty vacuum desicator which is then evacuated to a pressure of < 25 mm Hg. The test specimen is then covered by admitting immersion liquid (water or a fraction of liquid paraffin boiling above 200°C). The specimen is allowed to remain there under reduced pressure for 5-6 hours after which air is allowed to enter the desicator. Test specimen, held in sling thread, is weighed (S) while still suspended in immersion liquid.

The test specimen is then lifted up slowly from the immersion liquid by means of the sling thread. Liquid drops on the surface is removed by lightly contacting with a piece of blotting paper ensuring that it does not make a physical contact with the specimen surface itself. The soaked specimen is then weighed (S) while keeping it suspended in air. The apparent porosity (P) is then given by

$$P = \frac{W-D}{W-S} \times 100(\%).$$

It is difficult to get a typical value of porosity of various bricks as it depends upon many parameters. The porosity of firebricks may vary from 12 to 25% or more ; that of silica bricks from 20 to 30%.

BULK / TRUE DENSITY AND SPECIFIC GRAVITY

A useful property for refractories is bulk density, which defines the material present in a given volume. An increase in bulk density of a given refractory increases its volume stability, its heat capacity, as well as resistance to slag penetration.

The bulk density is a measure of the ratio of the weight of a refractory to the volume it occupies. Bulk density is expressed in kg/m^3. The density of refractories is an indirect measurement of their heat capacity or ability to store heat. This is particularly important in applications such as regenerator installations.

This property is important for both the lightweight (insulating) and dense refractory bricks (recuperator refractories and regenerator checkers).

True density is the weight per unit volume of the refractory including the volume of its open and close pore space. Density factor signifies the overall weight coming upon the foundation of a refractory structure and thus limits the size of a furnace. Foundation of the structure of less porous refractory materials should be stronger than those built from comparatively more porous refractory material in order to successfully bear the heavier load coming on it.

Determination of bulk density. The bulk density of any refractory is measured by the equation.

$$\text{Bulk Density} = \frac{\text{Total weight}}{\text{Total volume}} \text{ of refractory}$$

Bulk density is measured using either direct measurement method or direct volume determination method.

Direct Measurement Method. The rectangular refractory test specimen brick is accurately weighed (to nearest 50 gms) and its dimension is noted using a hook rule for calculating its volume. The bulk density is calculated using the formula:

$$B = \frac{W}{V}$$

where,

B = Bulk density of brick, gm/cc

W = Weight of the test specimen, gms

V = Volume of the test specimen, c.c.

Direct Volume Measurement Method. The test specimen of refractory measuring 6.5 x 6.5 x 4 cm is cleaned of its adhering dust. Its dry weight (D) is recorded (after drying it at 110°C in an oven to a dry weight maintaining an accuracy of 0.1 gm). The dry specimen is then placed in an empty vacuum desicator which is then evacuated to a pressure of < 25 mm Hg. The specimen is then covered with immersion liquid (water or liquid paraffin boiling above 200°C) and allowed to remain there under reduced pressure for 5-6 hours after which air is allowed to enter the desicator. Suspended weight (S) of the test specimen is found by holding it in a sling thread while still suspended in immersion liquid. The test specimen is then lifted up slowly from the immersion liquid by means of the sling thread. Saturated weight (W) of the test specimen is found by weighing the soaked specimen (after wiping off the liquid droplets from the specimen surface by a blotting paper) while suspended in air. The bulk density is then calculated by

$$B = \frac{D}{W - S}$$

The direct volume determination method (generally used for deeply branded and irregular refractories) gives more accurate results than the direct measurement method (used for rectangular shapes of refractory only).

This is one of the main criterias for assessing the quality aspect of the particular brick.

The true specific gravity of the material composing a brick as distinct from the bulk density of the brick as a whole (including the pore spaces) is of special significance only for silica bricks and magnesite bricks. This is due to almost constant composition of other bricks.

Various forms of silica have different specific gravity viz. quartz has a specific gravity of 2.65, cristobalite 2.33 and tridymite 2.28. Hence, specific gravity of silica brick is a good guide to the degree of conversion. Thus a silica brick of specific gravity 2.45 evidently contains a considerable amount of unconverted quartz which may cause permanent growth during use. Specific gravity of most of the silica bricks lies between 2.32 to 2.36 indicating adequate quartz conversion to cristobalite and tridymite.

The fact that density of magnesia increases with the heat treatment, has been used as a method of assessing the adequacy of the kiln firing of magnesite bricks. A magnesite brick in which periclase (magnesia) crystals are fully developed attains the maximum specific gravity of 3.65. However, this figure for magnesite bricks is somewhat high and should be modified according to the chemical composition of the brick tested; a high iron oxide content increasing and a high silicon content decreasing the actual value.

The specific gravity of firebrick is of little significance owing to the complex nature of the material.

Determination of True Specific Gravity. 1 gm of powdered (< 100 mesh) and dry refractory material is weighed into previously calibrated and weighed pycnometer bottle. The bottle is half filled with freshly boiled distilled water and is kept at 60°C for half an hour and the contents are agitated to wet the particles thoroughly and remove entrapped air. The bottle is then placed in mild vacuum to ensure that all air has escaped.

The bottle is then filled to the mark with distilled water and placed in a constant temperature bath for one hour. The bottle is capped ; the capillary & outside of the bottle is wiped clean and the whole is weighed. The specific gravity is then calculated from :

$$\text{Specific gravity} = \frac{W - P}{(W_1 - P)(W_2 - W)}$$

Where,

P = the weight of pycnometer and stopper

W = the weight of pycnometer, stopper and sample

W_1 = the weight of pycnometer, and stopper full of water

W_2 = the weight of pycnometer, stopper, sample and water

Tetrachlorethane (because of its low viscosity and high wetting properties) is useful in the case of non plastics in place of water, as the boiling and deairing steps are eliminated.

Some typical data of the specific gravity and bulk density of refractory is tabulated in Table 4.9.

Table 4.9: True Specific Gravity and Bulk Density of Refractory Bricks

Type of Refractory Brick	True Specific Gravity	Bulk Density gm/c.c.
Silica	2.3-2.4	1.7-1.9
Firebrick	2.6-2.7	1.9-2.1
Bauxite	3.1-3.4	1.8-2.1
Magnesite	3.5-3.7	2.7-3.0
Chrome	3.7-4.2	2.8-3.5
Zirconia	4.8-5.9	4.0-4.6
Carbon	2.7-3.1	1.9-2
Sillimanite	3.0-3.2	2.0-2.2
Silicon carbide	3.1	2.2

PERMEABILITY

Permeability is a measure of the rate of passage of fluids through a porous refractory material. Impervious (unpermeable) refractories eliminate the leakage of gases and penetration of liquids through the refractory walls of a furnace.

The flow of gases through refractory wall is important in many furnaces, especially where large pressure differences exist, as in recirculating types. Also the permeability of such parts as thermocouple tubes, muffles, and recuperators is of interest in efficient design.

Permeability is used as a check on the manufacture, as laminations, voids, and soft spots are shown up in the permeability test. In the case of heat insulators, this property is useful in indicating the extent of inter connecting pores.

Porosity of refractory bricks bears little relation to permeability which is a measure of the rate of diffusion of liquids and gases through the refractory. Permeability is governed by the size and number of connected pores/channels which are continuous from one side of the refractory to the other. Permeability to liquids will increase with increasing temperature as the viscosities of liquids will decrease while the condition is just opposite in case of gases whose viscosities increase at higher temperatures. Permeability of refractories is important in such applications as blast furnace linings where disintegration of refractories can occur as a result of penetration by carbon monoxide and subsequent carbon deposition ($2CO = C + O_2$) within the brick.

Measurement of Permeability. Absolute permeability is the property of the refractory and does not in any way depend on the fluid. It is expressed as

$$\lambda = \eta \frac{Q}{A} + \frac{l}{(P_1 - P_2)t}$$

where,

λ = absolute permeability

η = viscosity of fluid, poise (gm/cm sec)

Q = volume of gas flowing at mean pressure $(P_1 + P_2)/2$, c.c.

A = area, cm^2

l = length of path, cm

P_1 = exit pressure, dynes/cm^2

P_2 = entrance pressure, dynes/cm^2

t = time, sec

The dimension of this equation is L^2 = area. Measured permeability *'K'* (usually measured with air at room temperature) is expressed by:

$$\frac{\text{Volume of gas X length of path}}{\text{Area of X time X pressure difference}} \text{ which has the dimension } \frac{L_2}{ML^{-1}T^{-1}} = \frac{\text{area}}{\text{viscosity}}$$

Permeability 'K' is expressed as cm^3/(cm) • (cm^2) (sec) (cm of water).

Determination of Permeability. Permeability is a measure of the rate at which a fluid will pass through a porous materials and is denoted by volume (c.c.) of a gas or air that passes through one cm. refractory cube at a unit pressure difference (gm/cm^2) per minute. Refractories subjected to direct contact of gases/liquids should be impervious to eliminate the leakage of gases and penetration of liquids through the refractory walls of a furnace.

In permeability determination apparatus (shown in Fig. 4.7), purified air is passed through a cubic refractory sample of known surface area and thickness, fixed in a holder with mercury seal. The volume of air passed in a given time is noted. The pressure difference is measured by a manometer.

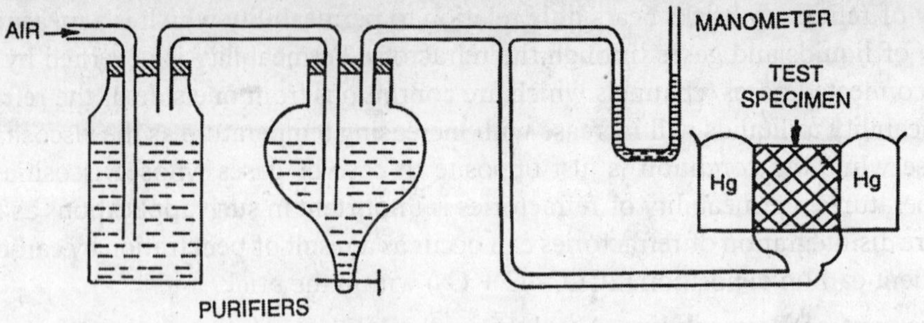

Fig.4.7: Permeability Determination Apparatus.

The permeability of the refractory sample is calculated using the equation

$$f = \frac{V.T.}{A.T.\,\Delta p}$$

where, f = permeability of refractory sample

V = volume of air passed through the sample c.c.

T = thickness of the sample, cm

A = cross-sectional area of the sample, cm^2

t = time for which air was passed through the sample, minutes

Δp = pressure difference across the two surfaces of the sample, gm/cm^2

Prevention of air escaping from along the side is a problem in permeability measurement of a specimen. Test equipments contain the specimen (a cube, cylinder or brick) in an inflatable rubber sleeve. Sufficient pressure is used to seal the edges completely. Gas flow volume is measured by a standard gas meter (or by water displacement) while the pressure difference is measured by a simple manometer. For low permeability, the testing time should be of considerable duration to obtain good precision.

Table 4.10 gives the permeability values for a number of bricks. The values have been measured for flow along the three axes of the brick but are given here only for edgewise flow.

Laminations or other structural discontinuities greatly affect the permeability values. All dry pressed fireclay bricks have substantially the same permeability but silica bricks have higher values.

Table 4.10: Edgewise Permeability of Some Refractory Bricks

Brick type	Average porosity, %	Average Permeability, f*
High alumina, 60%	28.0	0.117
Super duty fireclay	19.6	0.183
High duty fireclay	19.1	0.101
Mullite	20.8	0.032
Chrome magnesite	15.0	0.005
Magnesite chrome	13.5	0.001
Magnesite	24.2	0.143
Silica, standard	30.3	0.461

$$f^* = \frac{(cm^3)(cm)}{(sec)(cm^2)(gm/cm^2)}$$

Bricks made with more careful grog size, higher moulding pressures and higher firing temperature would have somewhat lower permeabilities.

COLD CRUSHING STRENGTH (C.C.S.)

Strength is one of the most widely used parameters for evaluating refractories. Strength can be measured at room temperature or at any temperature for which suitable test equipment exists.

Room temperature (cold) strength measurements cannot be used directly to predict service performance, but gives idea about the degree of bond formation during production. Room temperature testing also indicates the ability of brick to withstand handling & shipping without damage and to withstand abrasion & impact in relatively low-temperature applications.

Strength testing at elevated temperatures is valuable in assessing the ability of a material to survive stresses caused by restrained thermal expansion, thermal shock and mechanical loading. Impact and abrasion resistance depend on material strength, as well. It is also common to use high temperature strength numbers to predict resistance to erosion & corrosion by metals & slags, although other properties, such as mineral composition and porosity, may be of equal or greater importance.

Two types of strength tests are common. The modulus of rupture (MOR) test measures the flexural (transverse) breaking strength and the crushing strength test measures the compressive strength. Both tests have a version for cold (ambient) temperatures and high (service) temperatures. In the cold crushing strength test, a sample is loaded at a standard rate using a suitable mechanical testing machine. The load is applied vertically to the face of the sample until failure. The crushing strength is calculated by dividing the maximum load supported by the sample over the surface area of the face which receives the load.

The equation for calculating C.C.S is: $\mathbf{S = W/A}$

where,

S = cold crushing strength, N/mm^2
W = maximum load, N
A = cross-sectional area, mm^2

In the cold modulus of rupture test , the test specimen is placed on two bearing cylinders with the top & bottom faces oriented horizontally. The specimen is broken at mid-span in flexure at a standard loading rate. The modulus of rupture is calculated using the following equation:

$$\mathbf{MOR = 3PL/2bd^2}$$

where,

MOR = modulus of rupture, N/mm^2
P = load at rupture, N
L = span between supports, mm
b = breadth or width of sample, mm
d = depth or thickness of sample, mm

Strength values vary widely among classes of refractories and even among refractories of the same classification. Variables which significantly affect cold strength of refractories are bond chemistry, firing temperature, porosity, and the strength & sizing of the aggregate.

The cold crushing strength (C.C.S.) of fired refractory is not of high importance but it serves as a guide to other characteristics such as vitrification, burning temperature and purity of material. For insulating bricks, the strength determines the ability to withstand handling and shipping.

The C.C.S. reveals the ability of the brick to withstand the rigours of transport and can be used as a useful indicator to the adequacy of firing & abrasion in consonance with other properties such as bulk density & porosity.

The mechanical properties (such as cold crushing strength) of a refractory determined at room temperature should be used with caution as a guide to the probable behaviour of the refractory at higher temperature.

Since all the refractory bricks used are subjected to load bearing services ; as such C.C.S. is very vital to consider whether the brick can be actually used for a particular lining job or not.

Determination of Cold Crushing Strength (C.C.S.). It is measured in a compression testing machine. A particular size of brick (say 9" × 4½") is taken and in the hydraulic machine, this specimen is placed in between two rams which move by hydraulic pressure when the machine is switched on. Upto the point, the brick can take load, the machine indicates the total pressure in terms of tonnage, every moment and when the brick fails the indicator returns back to 0 (zero) for a particular specimen under test. The C.C.S. value is indicated by taking the reading of the machine.

Cold crushing strength of the refractory bricks is found out using a standard mechanical or hydraulic compression testing machine. The load at which cracks appear in the test specimen represents the C.C.S. of the specimen. The size of the plane bearing surface should be equal to or more than the size of the test specimen surface. The usual size of the test specimen is 230 mm standard brick except in case of smaller or other special refractory shapes where the test specimen are of smaller sizes. 77 mm cube refractory specimen may also be used. The test specimens are kept in between 5 mm thick asbestos card boards/fibre boards while conducting the test.

C.C.S. of refractory bricks is particularly important in shaft type furnaces or such parts of a furnace where charge load is concentrated as the bricks must be able to withstand the structural load coming over them. Porous coarse grained refractories have poor crushing strength, whereas dense and fine grained refractories generally have good crushing strength.

The cold crushing strength (C.C.S.) of some typical bricks of given composition and porosity has been given in Table 4.11.

Table 4.11: C.C.S. of Refractory Bricks

Bricks type	Approximate composition, %	Apparent porosity, %	Cold crushing strength (C.C.S.), kg/cm^2
Magnesia	MgO-93.6 SiO_2-4.1	19.7	620
Forseterite	MgO-50.3 SiO_2-29.5	22.1	337
Chrome-magnesite	Cr_2O_3-25.9 MgO-34.2	22.4	275
Chrome	Cr_2O_3-29.7 Al_2O_3-34.1	15.8	992

99% alumina	Al_2O_3-99.4 SiO_2-0.3	23.4	361
90% alumina	Al_2O_3-91.2 SiO_2-7.3	21.3	706
70% alumina	Al_2O_3-69.2 SiO_2-24.8	22.9	623
60% alumina	Al_2O_3-59.8 SiO_2-33.2	19.5	508
Super duty silica	SiO_2-96.3 Al_2O_3-0.2	19.6	340
Conventional silica	SiO_2-95.8 Al_2O_3-0.5	22.5	282
Coke oven silica	SiO_2-94.9 Al_2O_3-1.0	25.6	289
Semi-silica	SiO_2-77 Al_2O_3-20.4	25.5	157
Super duty fireclay	SiO_2-52.9 Al_2O_3-42.0	9.8	813
Stabilised zirconia	$(ZrO_2 + HfO_2)$-93.8 Ca0-4.0	23.2	409
Clay bonded silicon carbide	SiC-89.2 SiO_2-7.3	9.0	642
Silica insulating firebrick	SiO_2-92.5 Al_2O_3-0.9	—	12

HOT CRUSHING STRENGTH (H.C.S.)

Hot crushing strength (H.C.S.) of refractory bricks are also tested in various apparatus which vary in heating and loading arrangements.

In one apparatus for testing the H.C.S. of the refractory bricks, the specimen is put under a hydraulic ram (operated by a hydraulic pump) at a maximum temperature of 1400°C. The pressure on the ram is increased at a "given rate (indicated by pressure gage attached on it) till the specimen fails.

In another apparatus for testing the H.C.S , the load is applied on the specimen by lever arrangement while the heating element is of molybdenum disilicide. The refractory sample/ specimen deformation is read directly on a dial gauge.

In yet another apparatus, the refractory samples are loaded horizontally and the deformation is measured by two telemicroscopes sighting on the specimen ends.

SPALLING RESISTANCE

Spalling is the loss of fragments or spalls from the face of a refractory brick or structure through cracking and rupture. In various types of furnace operation, service conditions necessarily expose the brickwork to spalling influences. The resistance of refractory brick to spalling is enhanced by the addition of spall inhibitors, proper design of the brick shapes, optimum sizing of grains, proper choice of grains and close manufacturing control. For a given refractory, spalling can also be minimized by proper design, construction, and operation of the furnace.

Spalling of refractory bodies can be divided into three general types: thermal, mechanical and structural spalling.

Thermal Spalling. This type of spalling is caused by stresses resulting from unequal rates of expansion or contraction between different parts of the brick and is usually associated with rapid changes in temperature. Brick with the greatest resistance to thermal spalling are those having the lowest average rate of thermal expansion, high tensile strength and a texture conducive to flexibility & relief of stress. Other factors being equal, temperature gradients and stresses which cause spalling are least destructive in brick having the highest thermal conductivities. A refractory of a given mineral composition will usually have maximum resistance to spalling when the ratio of strength to the modulus of elasticity is a maximum.

The spalling characteristics of fireclay brick are generally dependent upon the amount of free silica present in the clays, the composition of the glassy bond and the size & "fit" of the particles. The amount & character of the glass are fixed by the quantity & kind of accessory minerals in the clays and by the time & temperature of firing. Within the temperature range at which the glass is rigid, a high glass content is conducive to spalling. However, at temperatures high enough to cause the glassy bond to become somewhat viscous, fireclay brick can be highly resistant to spalling.

Relatively porous, light-duty fireclay brick are usually more resistant to spalling than denser, hard-burned brick. Superduty fireclay bricks are of the spall resistant variety, while dense & hard burned, have high resistance to spalling and retain this resistance upon exposure to high temperatures.

Some high-alumina brick also have excellent resistance to spalling. In many cases, brick are designed to generate a system of stress-relieving microcracks to enhance their spalling resistance. With other conditions being equal, higher density, lower porosity compositions tend to have poorer spall resistance than average density, average porosity compositions.

Most silica brick are sensitive to rapid temperature changes at low temperatures due to the abrupt volume changes associated with the crystalline inversion of the mineral cristobalite. However, silica brick may be heated or cooled quite rapidly as long as they remain above the inversion temperature of about 650°C.

Burned basic brick, in general, do not have as high a resistance to thermal spalling as fireclay and high-alumina brick. However, with proper choice of raw materials, basic brick with improved thermal shock resistance can be manufactured.

Mechanical Spalling. The spalling of refractory brick caused by stresses resulting from impact or pressure is known as mechanical spalling. Shattering or spalling of brickwork may result from such influences as the rapid drying of wet brick work, inadequate provision for thermal expansion, or pinching of hot ends of brick, especially in furnace arches. Brick which are strongest and toughest at operating temperatures have the greatest resistance to mechanical spalling. "Pinch spalling" is observed in sprung brick arches because the hot ends of the brick expand more than the cold ends. This condition is aggravated by rapid heating because the furnace structure cannot rapidly adjust to the expansion forces. The differential expansion may cause a concentration of arch stresses upon relatively small bearing surfaces at the inner ends of the brick. Insulation decreases the temperature gradient through a roof arch and tends to somewhat reduce pinch spalling.

Structural Spalling. Structural spalling of a refractory unit is caused by stresses resulting from differential changes in the structure of the unit. The word "structural" does not refer to the furnace or the furnace lining assembly, but to the texture or structure of the individual brick units in the lining. Contributing to structural spalling are changes which occur in service to the texture & composition of the hotter portions of the brick through the action of heat, the absorption of slags or fluxes and their reactions with the refractory.

These alterations may result in zones in the brick which differ in mineral content and chemical and physical properties from the unaltered portions of the brick. Following the development of a zoned structure, hot ends of the brick may spall or fall off for various reasons, including differences in thermal expansion or contraction between adjacent zones, increased sensitivity to rapid temperature changes or the presence of shrinkage or expansion cracks.

Monolithic refractories, not having been fired prior to installation, are more resistant to thermal spalling than are their fired brick counterparts, at least until they have become vitrified or slagged in service.

Being lower in strength than fired brick, however, monolithic refractories may be more subject to mechanical spalling.

Several tests are commonly used to determine the relative resistance of refractory compositions to spalling. The most common are the panel spalling test, the prism spalling test and the loss of strength test. In nut shell , spalling is defined as the fracture of the refractory brick or block due to any of the following causes:

- A temperature gradient in the brick due to uneven heating or cooling that sets up stresses causing failure. (*Thermal Spalling)*
- Compression in a *structure of refractories* due to expansion of the whole from a rise of temperature causing shear failure.
- Variation in co-efficient of thermal expansion between the surface layer and the body of the brick due to surface slag penetration or to a structural changes in service resulting in shearing off the surface layer.

The first type is usually considered the true spalling but that is incorrect. It is undoubtedly the predominating effect in refractories such as silica, magnesite and chrome. The first type of spalling is caused by the inherent properties of the refractories.

The second type of spalling is due to poor furnace design and the refractory itself.

The third type of spalling can be minimised by using bricks burnt at higher temperature than the subsequent operating temperature and by preventing slag penetration as far as possible.

Mechanical strength and elasticity have a controlling influence on the spalling resistance. Thermal spalling (as distinct from *physical and mechanical spalling*) is caused by the development of stresses within a body of brick due to changes in temperature gradient. Calculation of these stresses and their comparison with ultimate strength of the material at the temperature in question leads to following conclusions :

On sudden heating:

$$\text{Spalling tendency } \alpha \; \frac{\text{co-efficient of expansion}}{\text{maximum shearing strain} \times \sqrt{\text{diffusivity}}}$$

On sudden cooling:

$$\text{Spalling tendency } \alpha \; \frac{\text{co-efficient of expansion}}{\text{maximum tensile strength} \times \sqrt{\text{diffusivity}}}$$

It shows that, to resist spalling, a brick should have a low co-efficient of expansion, high strength and high thermal diffusivity. These requirements are fulfilled by silicon carbide, sillimanite and most firebricks. Silica bricks have a low spalling resistance below 600°C owing to the high coefficient of expansion up to this temperature; above 600°C the expansion becomes small & uniform and the spalling resistance increases. Magnesite bricks have a fairly high diffusivity and strength, but also have a high coefficient of expansion. Therefore, magnesite bricks fall between silica bricks and firebricks as regards their resistance to thermal shock.

Spalling is most liable to occur at comparatively low temperatures. At such temperatures the bond is glassy and will not yield to the expansion stresses ; as the temperature rises, the bond becomes more deformable and will take up some of the strain. Spalling chiefly occurs during the warming up & cooling down of furnaces ; particular care must be taken during the warming & cooling of any furnace lined with silica or magnesite bricks.

Dolomite bricks should not be exposed to the furnace atmosphere, since with these products a peculiar type of spalling occurs on exposure to high temperatures. This applies to very high temperature iron melting furnaces, but not to reheating furnace hearths.

If a furnace is to be used intermittently, firebricks, sillimanite or silicon carbide bricks should be used wherever possible, the ultimate choice depending on the presence or absence of slag & its nature and on the cost.

Thermal fatigue resistance (i.e. thermal spalling resistance) signifies the fact that refractory bricks subjected to frequent/alternate cycles of heating & cooling should not crack *(i.e.* spall) and crumble to pieces due to sudden changes in temperature.

During heating of the bricks, refractory may fail by shearing due to expansion of refractory grains causing compression in its structure. That is why, refractory materials with coarse texture and low thermal co-efficients of expansion have more thermal spalling resistance. Refractories made from non-uniform mixtures of raw materials having wide variation in co-efficients of thermal expansion are liable to fracture and fail. Failure/fracture of refractories during service due to structural changes during service can be controlled by proper mixing of raw materials, firing/burning refractories at a temperature higher than its subsequent operating temperature and preventing/minimising slag penetration.

Failure of refractories by uneven/sudden cooling & heating due to thermal stress is predominant in silica, magnesite and chrome refractories.

A refractory material heated/burnt at high temperature for longer duration has a higher thermal spalling resistance due to complete mineral inversion.

Refractory bricks having higher strength resists spalling better because of its improved property to yield under stress without rupture.

Refractory bricks having greater diffusivity have lesser tendency to spall due to its greater tendency for temperature equalisation within it. Diffusivity increases with thermal conductivity hence, bricks with poor conductivity are more liable to spall due to internal stresses development caused by temperature difference between different parts of the refractory. Also, the diffusivity is less, if the specific heat of the refractory material is high. Hence the refractory having high specific heat are more liable to thermal spalling.

Spalling also occurs due to 'carbon transfer' (causing carbon deposition) and due to slag penetration as both of them cause differential in the composition of the refractory material.

This property of refractory is very important for the construction of intermittant kiln furnaces/ ovens (which are fired & cooled cyclicly/frequently) and also for the use in such position as regenerators (in which heating & cooling is done alternatively).

Determination of Spalling Resistance. It is determined by two standard methods namely :

(a) Panel Test *and* *(b)* Small Prism Test

Panel Test for Spalling Resistance. In this test, 12 to 14 test specimens of refractory (of size 65-70 mm thick, 230 mm long and 113 mm wide) after drying (at 150 ^{0}C and weighing each one accurately are laid in the panel frame by Kaolin or any other suitable material (having a fusion point not less than that of the test specimens and which does not react with it). The test specimens are labelled with ceramic paint on the inner face which is not exposed to heating. Joint thickness of test specimens in the panel should be < 2 mm. The test panels are backed with suitable insulation to allow heat loss of about 605 kcal/m^2/hr(°C/m) under operating conditions and its mean temperature should be about 700°C.

The panel assembly is then preheated (in a gas/oil fired preheating furnace) to maintain an average face temperature of 1600°C for 24 hours. The heating rate is so maintained as to achieve 1600°C in 5 to 8 hours. After completion of preheating, the condition of the test specimen is recorded and photographs of panel surface are also taken for record.

The preheated panels after the removal of insulation is transferred (by track or trolley) into another furnace (gas/oil fired) called spalling furnace where the test specimens are heated to attain a temperature of 1400°C in 3 hours. During heating, the panel position is changed at 1000,1200 and 1300°C to ensure uniform heating. Heating at 1400°C is allowed for 10 minutes and the panel is then cooled in another 10 minutes using a blast of air-water mist delivered through a vertical manifold (having an aperture of 90 mm x 600 mm). Air is admitted at the rate of 40 m^3/minute whereas water enters at the rate of 11 litres during first 8 minutes of cooling and no water is used during the last 2 minutes.

The cooling unit has means for reciprocating motion to facilitate its to and fro movement (125 times in 10 minutes) over the entire surface of the panel. The test specimens are subjected to 12 such heating & cooling cycles and the furnace is then shut off. The panel is subjected to two additional cooling cycles without using any water mist. Subsequent cooling is allowed to take place overnight after spacing the panels 25 mm face to face.

The refractory test specimens are then removed from the panel and each of them are reweighed. They are then rearranged in the order, they were placed in the test panel and photographed for comparison with the first photograph giving the original condition of the refractories. Average loss of the refractories for the whole panel is reported. The extent of the loss of weight of refractories gives an indication of the spalling resistance of the refractories.

Small Prism Test. In this test, three refractory test pieces are cut from the refractory sample and ground to prism shapes with a square base of 50 mm side & 75 mm height in case of standard shapes and 50 mm high rings are cut from other shapes (like nozzles, sleeves etc.)

After drying the three test pieces, they are placed in a cold muffle or semi-muffle type furnace and then uniformly heated to achieve 450°C in 3 hours in case of silica refractories and 1000°C in 3 hours in case of fireclay, siliceous or basic refractories. The test pieces (which are maintained at this temperature for 30 minutes) are removed from the furnace (using a pre warmed pair of tongs) and are cooled for 10 minutes on a brick floor in a position free from draughts. The test pieces are examined for presence of cracks, if any, recharged in the furnace (maintained at test temperature) for 10 minutes and then cooled similarly again for 10 minutes. Such cycles of heating and cooling are repeated till fracture is promoted in the test pieces. The cycle during which the crack first appear is also noted. The number of cycles that the test pieces withstand before fracture is taken as a measure of the spalling resistance of the refractory.

Spalling Characteristics of Various Bricks

Silica bricks are not at all resistant to spalling in the low temperature range. But above the inversion temperature, they are very resistant to wide changes in temperature because of the flatness of the expansion curve in this region.

Table 4.12 gives the spalling resistance of various refractories.

Table 4.12: Spalling Resistance of Various Refractory Bricks

Type of brick	Resistance to spalling
High duty fireclay, regular	Fair
High duty fireclay, spall resistant	Good
Superduty, regular	Very good
Superduty, spall resistant	Excellent
Semisilica	Fair to good
High alumina	Good to excellent
Silica	Poor below 650°C, excellent above
Chrome magnesite	Fair to good
Magnesite chrome	Excellent
Forsterite	Fair
Silicon carbide	Excellent
Carbon and graphite	Excellent

Chrome and magnesite brick, because of their high coefficient of thermal expansion, are not so resistant to spalling as fireclay brick. Silicon carbide bricks have an excellent resistance to spalling.

THERMAL CONDUCTIVITY

When a furnace is heated, thermal energy flows into the refractory structure, causing a temperature difference to develop between inside & outside surfaces of walls and roofs. Part of this thermal energy is stored in the refractory structure & its foundation, and part flows through the walls, roofs, & hearths, and is lost to the outside air by radiation & convection. The amount of heat which escapes in this manner is of considerable importance in the economy of the process. In estimating the quantities of heat flowing through the parts of a furnace, use is made of a co-efficent known as the thermal conductivity, or K-value, of each material involved in the construction. The thermal conductivity value differs not only for different materials, but usually also for the same material at different temperatures. The thermal conductivity is reported at the mean temperature of the test brick.

Major factors which affect the thermal conductivity of a refractory material are mineral composition, the amount of amorphous material (glass or liquid) which it contains, its porosity and its temperature. For materials which have similar mineralogical compositions, the proportion of pore space is the most important factor affecting the amount of heat which will flow through it at a given temperature. Within the temperature range seen in most applications, thermal conductivity decreases with increasing porosity.

At ambient temperatures, the thermal conductivity of glass is considerably lower than that of crystalline material of the same composition. With rising temperatures, the conductivity of glasses tends to increase, while that of crystalline material tends to decrease. However, in refractory bodies consisting of crystal aggregates with limited amount of glass, the temperature effects are difficult to predict. The conductivity of a refractory in service at high temperature is often changed somewhat, either by an increase in the amount of glass or liquid it contains, or by devitrification of any glass it may contain.

High thermal conductivity is desirable for refractories used in constructions requiring efficent transfer of heat through brickwork, as in retorts, muffles, byproduct coke oven walls and recuperators. In most types of vessels, however, low thermal conductivity is desirable for heat conservation, but is usually less important than other properties of the refractories.

The rate of heat flow in refractory constructions can be calculated only approximately for several reasons:

- The accuracy of the methods used to measure thermal conductivity is unconfirmed.
- The temperatures of the hotter & cooler faces of the refractory wall or roof are rarely known with exactness.
- The conductivity of a refratory may be altered in service by mineral changes in the refractory, by vitrification of the refractory, or by absorption of slags, metals or other materials.
- The thickness of a furnace lining may not remain constant during operation. The lining may become thinner because of wear & transmit more heat; or it may become thicker & transmit less heat because of the formation of a solid coating (as in rotary cement kilns).
- The rate of heat flow is influenced by the pressure of the furnace gases and by the permeability of the refractory. A positive pressure tends to force hot gases out through the walls, while a negative pressure tends to draw cold air from the surroundings into the furnace.
- The rate of heat flow may be influenced by the thickness of the joints & gaps between lining components and by the character of the bonding mortar, if used.

Other important factors which affect the amount of heat flowing through refractory linings include the emissivity of the refractory or metal shell; the kind of gases within the furnace; and the external convection currents.

Effects of Gases. In refractory structures built of insulating firebrick or other very porous refractories, the type of furnace atmosphere can have a very appreciable effect on heat loss through the refractory walls. This is especially true of atmospheres with a high content of hydrogen. Most protective atmospheres, including dissociated ammonia, exothermic gases, and endothermic gases, contain appreciable percentages of hydrogen. The conductivity of the hydrogen must be taken into consideration in calculating the heat losses through the refractory walls of controlled atmosphere, heat- treating furnaces.

The thermal conductivity of hydrogen is approximately seven times as great as that of air. Consequently, the presence of hydrogen gas in the pores of a porous refractory product increases the rate of heat flow through the refractory. With an atmosphere of 100 percent hydrogen, the heat flow through an insulating firebrick or other refractory with similar porosity would be 2-2.5 times as great as in an atmosphere of air. At lower concentrations of hydrogen, the increase will be less and the relationship is almost directly proportional to the percentage of hydrogen present. As compared with the conductivity of air, that of nitrogen is 100 percent, carbon dioxide about 59 percent, and carbon monoxide about 97 percent.

Thermal conductivity depends upon the chemical mineralogical compositions, glassy phase contained in the refractory and the application temperature. Although it is one of the least important properties as far as performance is concerned, it evidently determines the thickness of brick work.

The conductivity usually changes with rise in temperature. In cases where heat transfer is required through the brick work, for example in recuperators, regenerators, muffles etc, the refractory should have high conductivity. Low thermal conductivity is desirable for conservation of heat by providing insulation.

The provisions for back up insulation, conserves heat but at the same time it increases the hot face temperature and hence the demand on the refractory quality increases. Depending on the characteristic of the refractory used in the hot face, such as the high temperature load bearing capacity, it may be required that the quality of the brick be increased to the rise in temperature caused by over insulation.

Light weight refractories of low thermal conductivity find wider applications in the moderately low temperature heat treatment furnaces, where its primary function is usually conservation of energy. It is more so in case of batch type furnaces where the low heat capacity of the refractory structure would minimise the heat storage during the intermittent heating and cooling cycles.

The thermal conductivity of both firebricks and silica bricks increases with temperature. The precise value of their thermal conductivities depends upon the texture, degree of firing and other factors. Since the rate of increase of thermal conductivity with temperature is greater for silica bricks, the conductivity of this class of bricks at higher temperature is greater than that of firebricks. Increasing the firing temperature increases the conductivity of both silica bricks and firebricks.

A more porous brick has, in general, lower thermal conductivity. However, the size & number of the pores is also important as heat transfer can take place across die pores by radiation and is prominent at high temperature.

Refractories used in melting furnace should have low thermal conductivity to ensure least heat losses and maximum heat efficiencies whereas recuperators require refractory with high thermal conductivity to ensure maximum heat transfer.

Mean value of thermal conductivity of some more typical bricks over the temperature range 25°C to *t*°C in kcal/m.hr.°C is given in the Table 4.13.

Table 4.13: Thermal Conductivity of Refractory Bricks

Brick type	Thermal conductivity, kcal/m.hr.°C		
	t = 100°C	500°C	1000°C
Corundum (fused alumina)	—	3.01 (between 650 and 1250°C)	—
Silimanite brick	1.36	1.48	1.48
Silicon carbide brick	—	—	6.88 + 8.61
Chrome brick	1.48 ± 0.37	1.48 ± 0.37	1.48 ± 0.37
Fireclay brick	0.56 + 0.12	0.74 ±0.12	1 ± 0.12
Magnesite brick	5.58 ± 0.62	4.46 ± 0.62	3.47 + 0.62
Silica brick	0.74	0.99	1.24
Insulating firebrick	—	0.24	—

Because of the air entrapped in the pores (which acts as a bad heat conductor), thermal conductivity of a refractory brick decreases with the increase in porosity as also the amount of entrapped air increases with the porosity. Fig. 4.8 gives the thermal conductivity of some refractories at various temperatures.

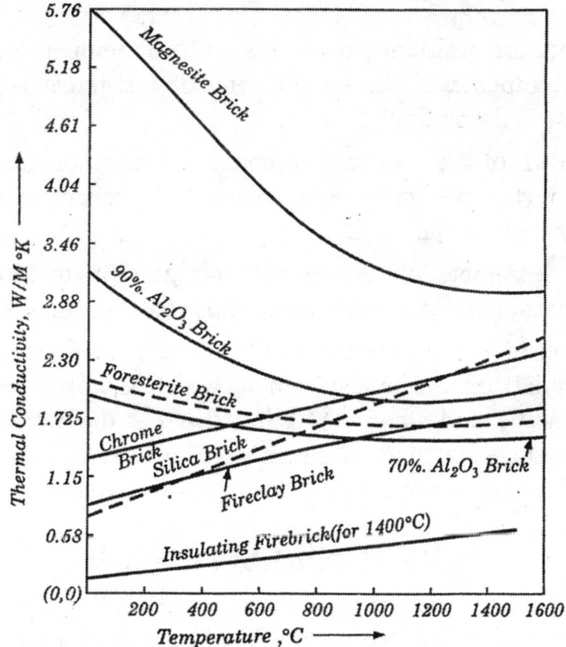

Fig.4.8: Temperature *vs.* thermal conductivity of refractories.

ABRASION RESISTANCE

In many types of service, refractories are subjected to impact by heavy pieces of material charged into the furnace; abrasion by metallic or nonmetallic solids; or direct impingement by abrasive dusts & high velocity gases. For resistance to these actions, brick should be mechanically strong and well-bonded. The strongest brick generally show the highest resistance to abrasion. As compared with other properties, the modulus of rupture or crushing test gives the best indication of resistance to abrasion.

Refractories used in the hearth of the furnaces or other places where materials/objects are slid along them (as in blast furnace) must have a good resistance to abrasion. Also, refractories are worn away by particles in rapidly moving gas. Abrasion test at room temperature gives no idea of its behaviour at high working temperature.

Descending solid charge material in a blast furnace is an example which results in abrasion/ wearing away of the refractory lining material. Use of a dense, fine grained and wear resistant refractory lining ensures good abrasion resistance.

In one of the abrasion tests, samples of refractory are rubbed together inside a furnace and the loss in weight is determined for a certain time of test. A typical result is given in Table 4.14.

Table 4.14: Result of Abrasion Test of refractory

Temperature of test,°C	Loss in weight, gm
20	0.115
1050	0.255

It shows that the abrasion resistance decreases with the temperature. A fine grained structure and a high burning temperature increase the abrasion resistance. A brick which shows good resistance to load will also resist abrasion.

The abrasion resistance of an insulating firebrick is less than that for a heavy refractory. Abrasion test is also conducted by circulating fine particles by means of a high velocity air stream through a duct lined with a particular brick.

Although there are a number of testing methods, the BS method for Marshall-Morgan abradibility index is important and this apparatus is illustrated in Fig. 4.9.

The important features of this test are angle of impingement of the abrasive grit (fused brown alumina or similar material) on the test specimen, which is normally 45°, the hardness, size (generally -10 + 22 BS mesh), and amount of grit, as well as the nature of the standard carbon block against which the test specimen is to be compared. All these parameters are specified in the BS method, though the result obtained is only a comparative value.

EROSION RESISTANCE

Erosion means 'Chipping off of the particles from the refractory surface due to impingement/ striking by molten metal or dust and slag particle laden gas against the refractory lining causing damage due to mechanical actions. Proper refractory should be selected to resist the action of erosive media.

Arches and bends in the refractory structure are liable to be eroded by slag particles carried along by flame or gases while splashing of slag erodes the refractory walls.

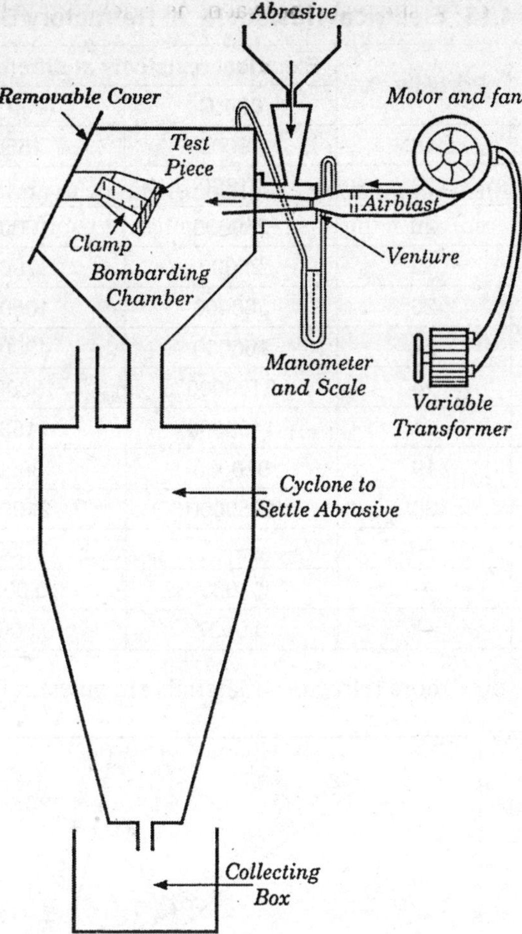

Fig.4.9: Schematic diagram for testing Marshall-Morgan abradibility index.

ELECTRICAL RESISTIVITY/CONDUCTIVITY

In electric furnaces, the electrical resistance of refractories is of considerable importance while in many other cases, a high temperature electrical insulator is desirable.

Graphite and metals are good electrical conductor and others are all electrical insulators among refractories. The refractory bricks for lining electric furnaces should normally have very low electrical conductivity. Graphite (an electrically conductive refractory material) is used for electrodes either in the prebaked form or as lining in high temperature electric furnaces. Porous refractories have low electrical conductivity.

Resistivity measurement at high temperature is beset by difficulty in reducing the contact resistance.

Electrical resistivity of some commercial refractories are given in Table 4.15.

Table 4.15: Electrical Resistivity of Refractory Bricks

Brick type	Porosity, %	Electrical resistivity at different temperature, Ohm.cm.		
		800°C	1200° C	1400°C
Fireclay, super-duty	18	19000	1550	720
High alumina (70%)	23	11800	900	460
Mullite (38 PCE)	26	210000	16000	7200
Mullite, fusion cast	15	25000	17000	760
Silica, superduty 97% SiO_2	26	360000	10500	3300
Silica, standard 90% SiO_2	22	150000	43000	1650
Chrome magnesite	14	2100000	130000	2400
Forsterite	21	1450000	11500	680
Magnesite Chrome	19	370000	3900	400
Zircon, 65% ZrC_2	30	1250000	21000	3600
Zirconia	—	—	1250	300
Carbon	—	0..003	0.003	—
Graphite	—	0.0007	0.0008	0.0009

Electrical resistivity of some more refractory materials are given in Fig. 4.10.

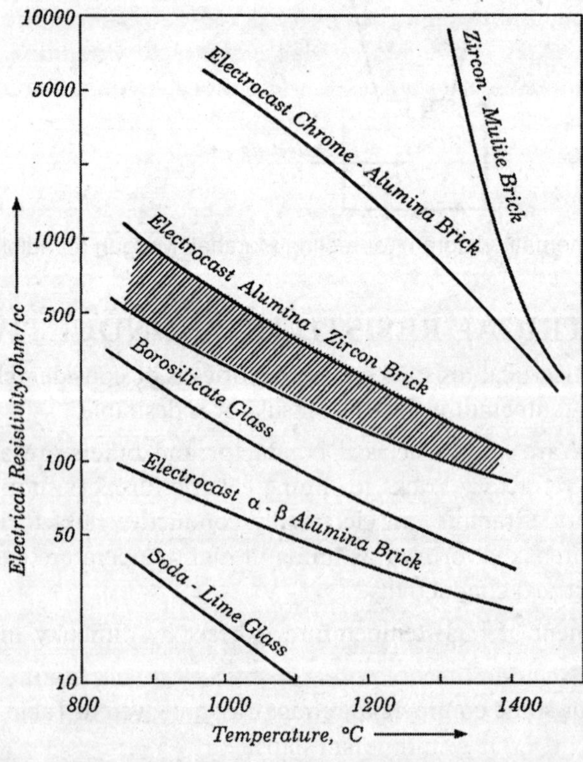

Fig.4.10: Temperature (°C) Vs. Electrical resistivity of refractories.

At low temperatures, refractory conducts very slightly. As the temperatures is raised, the conductivity increases rapidly until the melting point is reached where electrolytic conduction occurs. For pure materials, the conduction will be low upto the melting point. But with the average refractories, liquid phases are formed by the impurities at relatively lower temperatures so that there is an early falling off of the resistance occurs.

In general, a pure high melting point material will give the best high temperature electrical resistivity.

SLAG AND VAPOUR/GAS RESISTANCE OF REFRACTORY

Attack on refractory takes place by chemical solution of brick work, plus erosion due to moving fluids, and spalling, following changes in temperature. Eutectics are formed by acid-base reaction or by fluxes. A eutectic is a compound of two or more substances melting at a lower temperature than any single component of the mixture.

Coal ash is the most common cause of attack of refractories by slagging. It contains SiO_2, Al_2O_3, Fe_2O_3, FeO, CaO, MgO, Na_2O, FeS_2 and FeS. Of these CaO, Na_2O, FeO, FeS_2 and FeS cause the greatest degree of chemical attack by slagging.

Under reducing conditions, ferrous oxide, ferrous sulphide, and iron are formed. These reduce the melting point of the refractory walls and dissolve the outer layers.

The comparative resistance to slag attack of various refractories changes with temperature, change in the slag composition and whether the slag is moving/stationary or is carried by gases/ flame. Nature of the slag and the refractory which come in contact with each other dictate the resistance to the action of slag on refractory. A basic refractory addition to a basic slag and acid refractory addition to an acidic slag makes the slag more viscous (due to increase in its melting point) and inactive. This follows from the fact that melting point of an acid slag gets reduced by addition of a basic refractory and that of basic slag gets lowered by addition of an acidic refractory. That is why an acidic slag attacks a basic refractory and a basic slag attacks the acidic refractory very rapidly.

Slag attacks the refractory invariably at high temperature and also if the slag contains such constituents which may form mixture or new compounds with the refractory. Slag film in contact dissolves some refractory thereby its composition and melting point gets altered (either raised or lowered). If its (slag's) melting point gets increased, the slag film protects the refractory walls from further damage because the slag film becomes highly viscous, sticky and inactive. As long as thus formed slag film does not get detached from the refractory surface and gets dissolved in the main slag, it continues to protect the refractory from further damage due to slag attack/action. However in practice, this viscous slag film gets diffused in the main slag and is diluted by the fresh slag thereby reducing its viscosity and increasing its fluidity enough to have further damaging effect on the refractory. On the other side, if the melting point of the slag film gets lowered by the dissolved refractory, its diffusion rate into the slag body increases (due to increased fluidity) and hence its chance to get replaced by fresh slag increases. It, then, continues to rapidly dissolve more fresh refractory till the chemical composition of the slag film gets altered thereby raising its melting point again. Slag attack action is thus far intense in this case (compared to the case when melting point of the slag film gets raised due to dissolution of refractory) due increased fluidity and thus increased penetration power of the slag film.

Dense/well vitrified refractory reduces the attacking effect of moving slag as it provides a smoother surface to the slag movement. While rough and porous refractory surfaces (providing increased surface area to the action of slag) accelerate the action of moving slag which will detach the slag film from the refractory wall and replace it with a film of fresh slag. Fusible glass of refractories is more prone to be dissolved in the slag than the refractory grains.

Slag attack rate is also accelerated by the presence of defective joints and cracks in the refractory structure.

Many gases/vapours present in furnace atmosphere have destructive action on the brick-work. Decomposition of carbon monoxide ($2CO \rightarrow C + CO_2$) even at low temperatures of 450 - 550°C (catalysed by presence of free iron oxide) results in carbon deposition at the joints of brick thereby disrupting the structure of the brick. Such bricks are greyish and cracked or friable. Dilution of carbon monoxide by carbon dioxide (as is usual in flue gases) considerably reduces the disintegrating tendency of the bricks. Water vapour has a damaging effect on basic refractories and reduces the fusion temperature of fireclay and silica bricks. Alkali vapour in the gases penetrate the brickwork and react to form compounds of low melting point.

Slag attack refers to chemical reactions and solutions which corrode the surface of a refractory lining in service & reactions which can take place between the molten slag, refractory & fluxing agents which have been absorbed. Erosion of the refractory often follows corrosion. An example is the washing away of refractory grains after the bond between grains has been dissolved by fluxing agents in the slag.

The composition of industrial slags, of the refractories which they contact in service, and of the reaction products derived from them are exceedingly complex. The most advantageous selection of refractories is governed by the chemical nature of fluxing agents present in the furnace.

Alkaline materials are highly basic; most metallurgical slags and metallic oxides are basic; and fluxes high in silica are acidic. To determine whether a particular slag is acidic or basic, the ratio of lime-to-silica in the slag is examined. As a general rule, if the CaO/SiO_2 ratio (or $CaO + MgO/SiO_2 + Al_2O_3$ ratio) is greater than, the slag is usually considered to be basic. If the ratio is less than 1, the slag is considered to be acidic.

Fireclay, high-alumina or silica brick are generally used where the corrosive agent is acidic. High-alumina refractories may be preferred if the flux is only slightly basic. For highly basic slags, magnesia, chrome or a blend of these two materials typically give the best service. Exceptions to these general principles are based on operating and reaction temperatures, reaction rates and formation of protective glazes & coatings on the refractory surfaces.

In many industrial applications, the refractory is in contact with a slag or metal during service. A chemical reaction results between the refractory and the slag or metal. Some reaction products can be extremely detrimental to brick service life while other reactions result in little or no change in service life.

In order to determine the relative resistance of a refractory to the slag or metal present in an industrial application, various slag tests have been developed. Among the most common are the cup slag test, drip slag test, gradient slag test, rotary slag test, dip & spin tests and aluminum boat test. The slag used in the tests may be the actual material or a synthetic material prepared to be chemically similar to the slag or metal chemistry.

Resistance to Carbon monoxide gas attack. For this test, two refractory specimens are cut. One of the test specimens is cut from the centre of a refractory while the second one is cut from the exterior of another refractory shape. The test specimens are normally cylindrical (of 50 mm length and not less than 30 mm diameter) in shape, however they may be cut to rectangular or prismatic shapes also.

To determine the resistance to carbon monoxide attack on the refractory, the two test specimens are kept in a suitable size wire wound furnace which is purged with pure nitrogen. The furnace is then heated to 450°C and then pure carbon monoxide (which is produced by the reaction of CO_2 with charcoal at 1000°C or generated from formic acid and sulphuric acid) is passed through the furnace at 2 litres/hr. The test is continued for 100 hours or until the test specimens disintegrate, if it happens earlier. The test specimens are regularly examined for decolourisation, carbon deposition & disintegration occurring during the test period and the observations are recorded for the two specimens. The temperature (450 - 500°C) and flow rate of carbon monoxide is maintained constant throughout the test. The time after which carbon deposition and disintegration takes place is taken as a measure of resistance of the refractory to carbon monoxide attack. The disintegrating effect of carbon monoxide depends particularly upon the presence of ferric oxide in the refractories.

Carbon monoxide and nitrogen used in this test is purified (to remove carbon dioxide, oxygen and water vapor) prior to use and the flow rate of carbon monoxide is measured by a manometer & flowmeter in the gas line before it enters the furnace.

Measurement of Slag Resistance. Slag test evaluates the resistance of refractory to slag penetration. Testing methods measure the penetration of slag into the refractory, the fusion point of powdered slag & refractory when mixed together and petrographic study of mixtures of slag & refractory after definite heat treatments.

Running Slag Test. This consists of heating a single brick or a column of bricks to a uniform temperature & regularly dropping powdered slag on top and allowing it to run down the face in a groove, as shown in Fig. 4.11. The depth to which this groove has been cut at the end of the test gives a very good indication of the slag resistance of the brick, because the test is carried on under conditions closely approximating those of service. Although this test gives excellent relative values, it is not a quantitative test and does not maintain a proper temperature gradient from the slag surface to the other face.

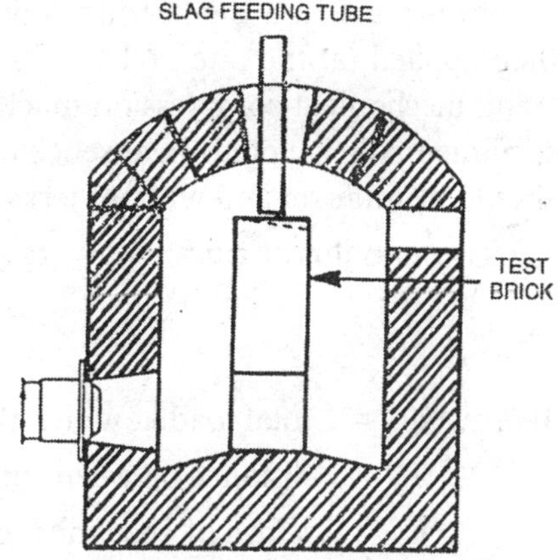

Fig.4.11: Slag Test

Spray Slag Tests. Another more satisfactory test consists in spraying powdered slag against the heated refractory. Here the powdered slag is fed through a revolving burner impinging on the test bricks set on the inside of the cylindrical furnace. Thus, all the bricks get the same treatment and the temperature gradient through the bricks is similar to that found in service. If the test is run for several days, truly quantitative values can be obtained.

Table 4.16: Maximum Temperature (°C) at which Two Refractories can be in Contact without Reaction

Brick Type	Chrome Magne-site	Chrome	90% alumina	70% alumina	Super duty fire clay	Clay bonded silicon carbide	Silica	Zircon
Magnesite	1705	1650	1705	1595	1315	1485	1485	—
Chrome-magnesite	—	1705	1595	1485	1495	1485	1495	1485
Chrome		—	1595	1485	1485	1485	1595	—
90% Alumina	—	—	—	—	—	1595	1595	—
70% alumina	—	—	—	—	—	1595	1485	—
Superduty fireclay	—	—	—	—	—	1595	—	-
Clay bonded Silicon carbide	—	—	—	—	—	—	1595	
Silica	—	—	—	—	—	—	—	1650
Zircon	—	—	—	—	—	—	—	—

Reaction between Refractories in Contact. It is desirable to know the maximum safe temperature at which one refractory can be heated when set on another. Table 4.16 gives some data on commercial refractories.

MODULUS OF RUPTURE TEST

Failure of brick in practice rarely takes the form of cold crush. In general bending moments are involved and the bricks fail by breaking in two rather than by crushing.

In modulus of rupture (MOR) test (which is performed on a standard mechanical or hydraulic compression machine), the test specimen (of size 2 × 11.3 × 6.5 cm cut from the desired refractory shape) is placed on the bearing edges of the machine positioned 18 cms apart. Uniform load is then applied (at the rate of 10 kgf/minute in case of hydraulic machine) and 1.25 kgf/minute if using mechanical compression machine at the middle of the specimen and the load at which the specimen fails is noted. In absence of mechanical or hydraulic machines, the application of load should be uniform and without jerks.

The modulus of rupture, R, kg/cm^2 is given by,

$$R = \frac{3wl}{2bd^2}$$

Where, w = total load at which the specimen failed, kg

l = distance between support, *i.e.* bearing edges, cm

b = width of specimen, cm

d = depth of specimen, *i.e.* thickness, cm

In the same compression test machine used for cold crushing strength determination, with slight modifications, the test for modulus of rupture can be carried out.

WARPAGE OF REFRACTORIES

Differential structural changes in the refractory during use causes its "distortion" which is called warpage.

To measure the warpage of a 'warped refractory (with a graduated wedge and a straight edge) it is kept on a horizontal base (with its concave surface upside) and the straight edge is placed along the diagonal of the refractory as shown in Fig. 4.12. The gap between the straight edge and the refractory surface at the point of warpage (*i.e..* W_r) is measured by inserting the graduated wedge. The length of the diagonal (*i.e.* L_d) of the warped refractory is also measured.

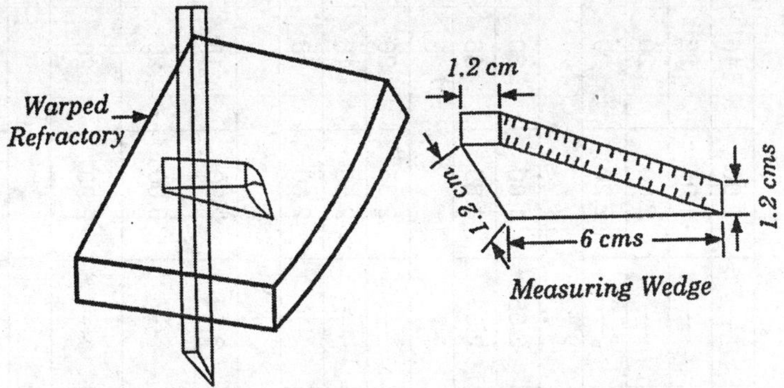

Fig.4.12: Warpage measurement method of refractories.

This measurement experiment is repeated by placing the straight edge along the other diagonal of the refractory surface. The % warpage of the refractory along each diagonal of the refractory is calculated from the following relation:

$$\% \text{ Warpage} = \frac{Wr}{Ld} \times 100$$

The higher value of the two warpages so calculated is taken as the final warpage of the distorted refractory under test.

Table 4.17: Typical properties of various commercial refractories

Type of refractories	Major constituent, %	Refractoriness		Refractoriness under load (RUL), °C	Density, gmlc.c.		Apparent Porosity, %	Cold Crushing strength (CCS), kgf/cm^2	PLCR % at °C for hour
		PCE (orton cone no.)	°C		True	Bulk			
1. Firebricks % Al_2O_3 =									
(a) Bloating Bricks	20	18	1522	—	—	2.00	16	300	+ 3.0,1400,3
(b) MHD firebrick	30	30	1655	1300	—	1.95	22	200	1.0,1350,5
(c) HHD firebrick	40	32	1717	1400	—	2.00	20	250	1.0,1400,5
(d) SD firebrick	45	34	1763	1500	—	2.20	20	300	0.5,1500,2
2. Silica Bricks %$Si0_2$ =									
(a) Standard brick	95	31	1683	1670	2.35	1.80	30	300	+ 1.0,1450,4
(b) Super duty brick	96	33	1743	1700	2.33	1.85	35	350	+ 0.4,1450,3
3. High alumina bricks %$A1_20_3$ =									
(a) Sillimanite bricks	60	36	1804	1600	—	2.35	20	400	1.0,1600,2
(b) Mullite bricks	72	38	1850	1650	—	2.50	18	450	1.0,1600,2
(c) Corundum bricks	90	40	1920	1700	—	2.80	18	500	0.5,1600,2
4. Basic (burned) bricks									
(a) Magnesite bricks	MgO = 85%	42	2000	1650	3.53	2.90	18	400	—
(b) Chrome-mag bricks	MgO = 45% & Cr_2O_3 = 25%	42	2000	1600	—	3.00	20	250	0.516,00,2
(c) Mag-chrome bricks	MgO = 70% & Cr_2O_3 = 12%	—	—	1550	—	2.90	20	250	0.5,1600,2
5. Special refractories									
(a) Zircon bricks	ZrO_2 = 65%	42	2000	1600	—	3.75	18	500	0.5,1550,2
(b) Carbon bricks	C = 90% & Ash = 6%	—	—	1750	—	1.55	16	400	—
(c) Silicon carbide bricks	SiC = 50%	38	1850	—	—	2.60	15	800	—

CHARACTERISATION OF REFRACTORIES

The characteristics of a refractory material depends on the type and purity of raw materials used, its microstructure and the type of bond developed. Quality refractory with uniform characteristics along any section is produced by the raw materials having uniformity in its composition.

Various physical and chemical tests are conducted for better quality control of refractory raw materials and their products. Analysis methods used for refractories include :

(a) Instrumental analysis by spectrophotometry, spectrography, polorography, thermal and thermo-gravimetric analysis, colorimetry etc.

(b) Decomposition by acid and fusion with fluxes.

(c) Wet chemical analysis by gravimetric and EDTA method.

Typical properties of some important commercial refractories are tabulated in Table 4.17.

REFRACTORY SHAPES

Various standard refractory shapes are shown below in Fig.4.13.

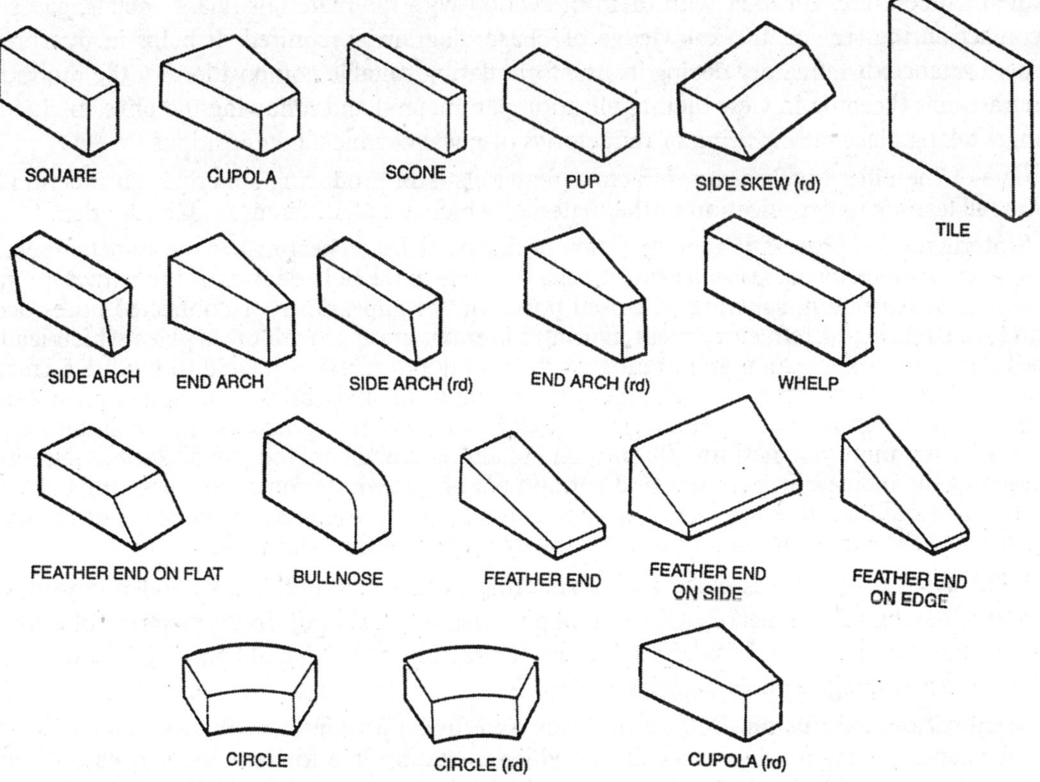

Fig.4.13: Standard shapes for refractory bricks.

Phase Equilibrium in Refractory Materials

Phase equilibrium in refractory materials indicates chemical stability and limits of its service temperature & pressure. Phase diagrams are graphical representations of the equilibrium system as defined by the phase rule.

To know the physico-chemical characteristics of refractory materials and their reaction at elevated temperatures together with their interaction with the materials (glass, metals, gases *etc.*) in contact during service, the knowledge of phase diagram is required. It helps in interpreting the performance of refractory during its use, formulating suitable composition for the production of refractories (keeping in view their application parameters) and following the physico-chemical changes taking place during firing of refractories of given chemical compositions.

One of the effects of heat on refractory materials is in producing sintering, vitrification and fusion (all leading to densification of the material) which are all different from each other.

Sintering. It is a process of agglomeration of fine particles on heating to a requisite temperature and is accompanied by decrease in porosity and increase in the bulk density of the refractory mass. Adhesion, densification, sphering of closed pores and disappearance of connected pores occurs during the sintering of refractory mass. During sintering, grain growth takes place which leads to removal of pores. Nucleation and grain growth from liquid phase is related to mullite formation in the alumino-silicate bricks, development of tridymite in silica bricks and spinel formation in chrome-mag bricks. Another type of grain growth from the fine grains occurs in sintering pure oxides of zirconium, magnesium, aluminium etc and is known as 'recrystallisation'. Sintering is retarded by the increase in pore size and entrapment of gas within the pores. Sintering behaviour of refractory materials in the presence of viscous liquid is influenced by the viscosity of the liquid, pore size and the surface tension of the liquid with respect to the solid grain.

Solid state reaction takes place during sintering of zirconia and direct bonded chrome-mag refractory [having solid spinel (Cr_2O_3) to solid periclase (MgO) bond]. In the sintering of alumino-silicate refractory liquid diffusion effect follows. Interaction of MgO, CaO and SiO_2 yields eutectic liquids in case of magnesite refractories.

Vitrification and Fusion. Both vitrification and fusion of refractory materials are associated with formation of liquid phase on heating which is responsible for reduction in its porosity & adhesion between individual grains and increase in density of the material (i.e., densification). Fusion and vitrification differ only in degree of melting. While fusion almost approaches melting ; vitrification is somewhat less than fusion in which proper shape is retained without any deformation.

Melting characteristics of conventional refractory materials (which do not have a sharp melting point but have a 'fusibility range') is indicated by fusion. Since presence of any appreciable amount of liquid phase in the refractory material is undesirable for its good performances, the desired characteristics in it should be achieved by sintering as far as possible in the absence of liquid. If sintering is not possible, then the refractory products are made by partial vitrification to achieve the specified characteristics.

Though fusion of any ingradient or mixture of ingradients is undesirable in any type of refractories; fusion as a process is utilised in producing fusion cast refractories viz. mullite, pure alumina, alumina-zirconia mixture, magnesite, mag-chrome *etc.* which are all very useful in their specific applications where bonded refractories of similar composition have failed or proved unsuitable.

Table 5.1 gives the fusion point and specific gravity of some pure refractory oxides.

Table 5.1: Fusion Point and Specific Gravity of Some Pure Refractory Oxides

Pure Refractory oxides	Fusion point, °C	Specific gravity
SiO_2	1715	2.32
MnO	1780	5.40
TiO_2	1850	4.24
BaO	1917	5.72
NiO	1950	6.80
V_2O_3	1977	4.87
Al_2O_3	2050	3.97
Cr_2O_3	2075	—
BeO	250	3.03
CaO	2570	3.23
ZrO_2	2677	5.56
MgO	2800	3.58
HfO_2	2810	9.68
UO_2	2875	10.96
ThO_2	3070	9.69

Phase Rule. Phase rule relates the number of phases (P) in co-existence under equilibrium at fixed temperature and pressure at a definite number of degree of freedom (F) in a system of fixed number of components (C) as given below :

$$F=C-P+2$$

The phase diagrams may be of various components viz. single component, double components, triple components or systems and so on. Phase equilibrium in refractory materials is highly complex as these are generally poly component and therefore poly phase systems. Complete phase equilibrium in refractory systems is rarely achieved.

Degrees of freedom (F) means the total number of independent variables that define the state of the system i.e. the number of variables that can be changed independently without changing the number of phases present. **Number of components of the system (C)** means the independent chemical species that comprise the system and these may be elements, compounds or ions depending on the type of the system. In refractory systems, components generally are compounds like Al_2O_3, Cr_2O_3, MgO, FeO, SiO_2 *etc.* ***Number of phases*** (P) means physically distinct, chemically homogeneous and mechanically separable parts in a system. Gases, liquids and solids form three different phases. Oil and water (being immiscible) separate into two distinct phases. Similarly in case of solids where different chemical compositions and crystal structures are formed, each is represented by a different phase.

The melting points (fusion temperatures) of some important refractory materials are given in the table 5.2 which is needed for clear understanding and following the phase diagrams.

Table 5.2: Formulae and Fusion Points of Refractory Minerals

Refractory Materials	Chemical Formulae	Fusion Points, °C
Andalusite	Al_2SiO_5	1816
Bauxite	$Al_2O_3.nSiO_2.nH_2O$	3722
Beryllia	BeO	2570
Calcite	CaCOs	2570
Carbon	C	3500
Chalcedony	SiO_2	1715
Chromic oxide	Cr_2O_3	1900
Chromite	$FeCr_2O_4$	2180
Corundum	Al_2O_3	2050
Cristobalite	SiO_2	1715
Diaspore	AlO(OH)	2035
Diatomite	SiO_2	1715
Dolomite	$CaMg(CO_3)_2$	2570-2800
Fireclay (high grade)	$Al_2O_3.2SiO_2.2H_2O$	1785
Foresterite	Mg_2SiO_4	1910
Gibbsite	$Al(OH)_3$	2035
Kadinite	$Al_2Si_2O_5(OH)_4$	1785
Kyanite	Al_2SiO_5	1816
Magnesite	$MgCO_3$	2800
Mullite	$3Al_2O_3.2SiO_2$	1810
Periclase	MgO	2800
Pyrophyllite	$Al_2Si_4Oio(OH)_2$	1621
Quartz	SiO_2	1715
Silicon carbide	SiC	2700
Silicon nitride*	Si_3N_4	1900
Sillimanite	Al_2SiO_5	1816
Spinel	$MgAl_2O_4$	2135
Thoria	ThO_2	3200

Tridymite	SiO_2	1715
Zircon	$ZrSiO_4$	250
Zirconia	ZrO_2	2700
*Decomposes and/or sublimes		

From phase rule, it is clear that a single phase system has the maximum number of independent variables and as the number of phases increases, the number of these variables decreases. The maximum number of phases that can exist in an equilibrium system (when degree of freedom is zero) is given by

$$P = C + 2$$

Hence, for single, double and triple component systems the maximum number of phases to exist in equilibrium will be 3, 4 and 5 respectively.

The important phase diagrams that are relevant for the manufacture and understanding the behaviour of refractories during use are discussed below.

SINGLE COMPONENT SYSTEM

Pure Silica (SiO_2) is an example of single component system. Temperature and pressure are the only independent variables in such a system as there is no possibility of change in composition and at atmospheric pressure, the only variable will be the temperature.

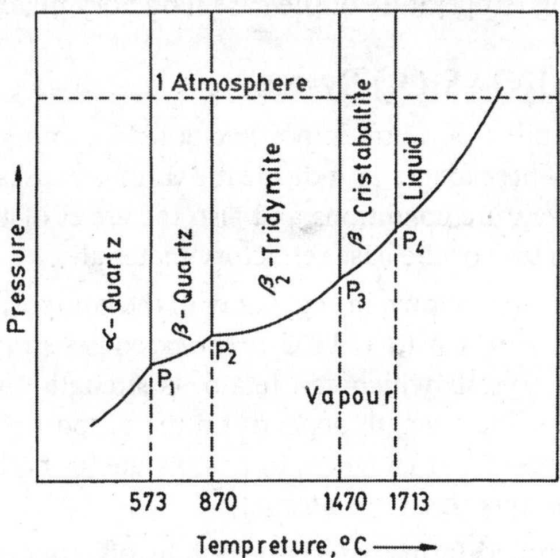

Fig.5.1: Temperature vs. pressure in a single component silica system.

Fig.5.1 shows the relationship between temperature and pressure for the pure silica (SiO_2) system.

In the above diagram each of the point P_1, P_2, P_3 and P_4 denotes a 'triple point' (where three phases coexist) and neither change of pressure nor that of temperature is possible. With change in either pressure or temperature, one or two of its phases will disappear and equilibrium of the system will shift to conditions shown by solid lines or the areas of predominance respectively.

DOUBLE COMPONENT SYSTEMS

Independent variables in this system are temperature, pressure and the composition of each of the phases that exist in equilibrium in the system. A single phase double component system has the least number of independent variables and increase in the number of phases would decrease the number of these independent variables. In double component systems, equilibrium diagrams are generally drawn with temperature and composition as variables at a constant pressure because most of the system work at/near atmospheric pressure and the effect of pressure on condensed phase system is very small.

The maximum number of independent variables would be two and zero respectively for single and three phase systems which indicates that composition of each phase and the temperature of the system are fixed. The two components in a system may have complete, partial or no solid solubility in each other. In cases where the two components have complete solid solubility in each other, and in all proportions, addition of one component to other decreases or increases its melting temperature depending on whether the base component has a higher or a lower melting point compared to the additive component. If the two components have partial or no solid solubility in each other, addition of one to another will decrease its melting temperature irrespective of whether it has higher or lower melting point and will form 'eutectic' alloys or compounds.

The phase equilibrium diagrams of some important double component refractory systems are shown and discussed below. These diagrams indicate the various phases that may be present in each system under different temperature conditions and also the effect of the presence of an impurity oxide on the melting temperature of the base refractory material.

The Alumina-silica (Al_2O_3-SiO_2) System

The phase equilibrium diagrams of some important double component refractory systems are shown and discussed below. These diagrams indicate the various phases that may be present in each system under different temperature conditions and also the effect of the presence of an impurity oxide on the melting temperature of the base refractory material.

The Al_2O_3-SiO_2 equilibrium is shown in Fig. 5.2 with the composition of the various alumino silicate refractories marked on it. Up to 1545°C firebrick consists of mullite and silica. Mullite is present as needle shaped crystals which can retain its strength and rigidity up to 1810°C at which it melts incongruently. The strength depends on the proportion of mullite present (i.e. on the Al_2O_3 content). Silica is present as tridymite or cristobalite (or both), but the effect of fluxes is to transform some of it to viscous glassy silicates.

Above 1545°C, viscous liquid forms but viscosity falls off rapidly with temperature and the structure becomes a matrix of mullite needles with liquid in the interstices. In siliceous firebrick (25% Al_2O_3), the proportion of liquid is about 70% and hence the remaining 30% in the form of mullite will not be able to maintain the shape of the brick for very long. The refractoriness of the brick has been exceeded at 1545°C (probably at lower temperature in the presence of fluxes). A brick containing 45% Al_2O_3 on the other hand contain only 40% liquid at 1545°C, the remaining 60% being mullite, which remains solid. Temperature increases beyond 1545°C cause the amount of liquid to rise slowly. At 1700°C the high alumina brick has still only 45% of liquid (although the other has 80% and has slumped). Clearly, the refractoriness more than 1700°C is possible with firebrick of suitable quality but Refractoriness Under Load (RUL) values better than 1545°C will be difficult to attain. Considering even higher Al_2O_3 content (say 72%), at the mullite composition the proportion of liquid at 1545°C has fallen to zero and remains negligible upto 1810°C. Hence, the advantages of high alumina are obvious.

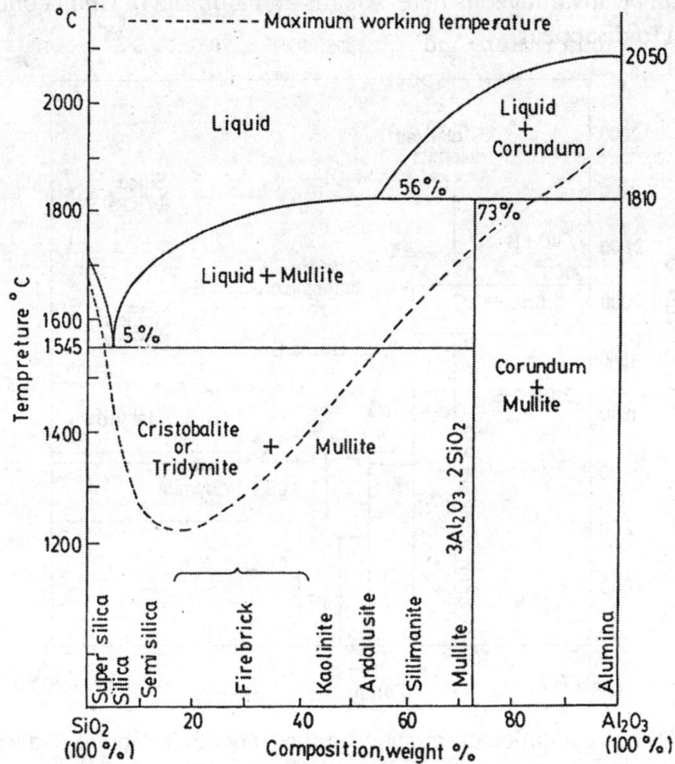

Fig.5.2: Al_2O_3-SiO_2 Equilibrium Diagram.

One of the modifications of this equilibrium diagram is that melting of mullite has been reported as being congruent at 1850°C with respect to the range of composition occupied by a mullite solid solution which is expected from theoretical considerations. It has also been reported that the eutectic should be at 10% SiO_2 rather than 5%.

SiO_2-CaO System

The equilibrium diagram of SiO_2-CaO system has been shown in Fig. 5.3. This system along with SiO_2-Fe0 system is quite important in the production of silica bricks.

Bonding in the silica brick is partly formation of calcium silicate glass and partly through an interlocking action between tridymite needles. Very high R.U.L. of silica brick is due to this tridymite bond. Rigidity can be maintained even under load very close to the melting points. 1.7% CaO is rather a powerful flux in such a brick. The above SiO_2-CaO equilibrium diagram shows, however, that there is a monotectic reaction at 1690°C below which a very small, proportion of liquid can form about 5% with 1.7% CaO down to 1440°C. This has little weakening effect on the strong tridymite skelton. FeO, MgO and MnO behave in a similar manner with silica but

Al_2O_3 and TiO_2 both form eutectics with silica at 1545°C with 6% and 10% additions respectively. These and alkalis should be kept low in general but TiO_2 itself can be tolerated in some of the high duty bricks of the silcrete type because the monotectic platform extends deep into the CaO-SiO_2-TiO_2 equilibrium diagram despite the eutectic in the SiO_2-TiO_2 system. A rather high CaO content can be advantageous here. Alkalis and alumina in small concentration cause the monotectic platform to disappear.

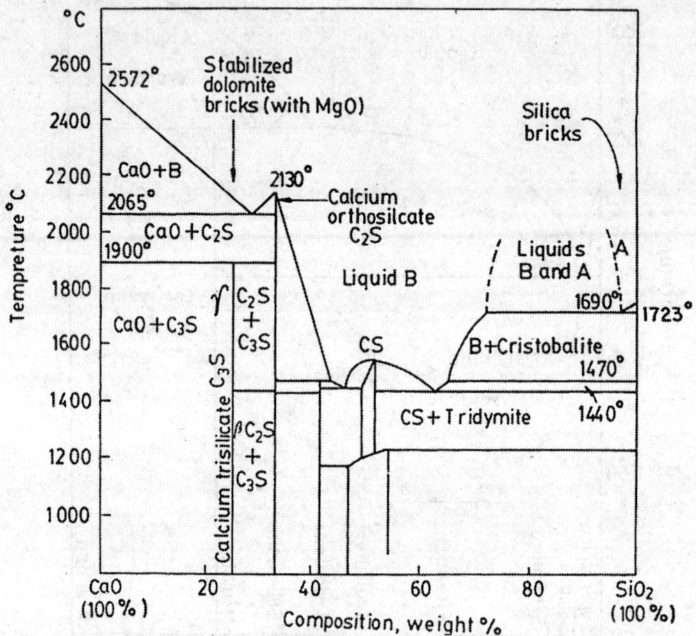

Fig.5.3: The CaO-SiO_2 equilibrium diagram showing the "monotectic" point and wide immiscibility gap which control the liquid content at a low level even in the presence of some CaO. The trisillicate and orthosilicate of calcium used in the stabilization of dolomite also appear in this diagram.
Notation : C = CaO, S = SiO2, C_3S = 3CaO.SiO_2.

FeO-SiO_2 System

The equilibrium diagram of FeO-SiO_2 system has been shown in Fig. 5.4.

This equilibrium diagram is also important in the manufacture of silica bricks. The behaviour of silica bricks versus slag can be predicted from the diagram. For example, it can be deduced from the above equilibrium diagram that iron oxide would rapidly form a slag with silica brick. However, in practice, silica bricks stand up remarkably well to iron oxide attack because the slag which forms at the face of the brick

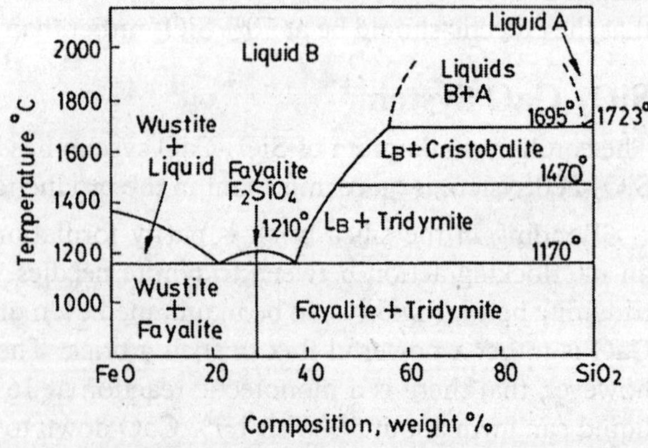

Fig.5.4: The FeO-SiO_2 Binary system

is very viscous and reduces the rate of attack to a very low value. Value .In most cases, a ternery (three components) or higher order phase diagram is more suitable to determine the probable slagging behaviour. FeO like CaO CaO also forms two immiscible liquids when added to silica, thus greatly increasing the tolerance of silica brick for FeO. Moreover, as with CaO, small amount of $A1_2O_3$ can eliminate this immiscibility.

MgO-SiO_2 System

The Mg0-SiO_2 equilibrium diagram which is useful in the manufacture and use of magnesite and foresterite refractories using magnesium silicate rocks like talc, serpentine, olivine *etc.* is shown in Fig.5.5.

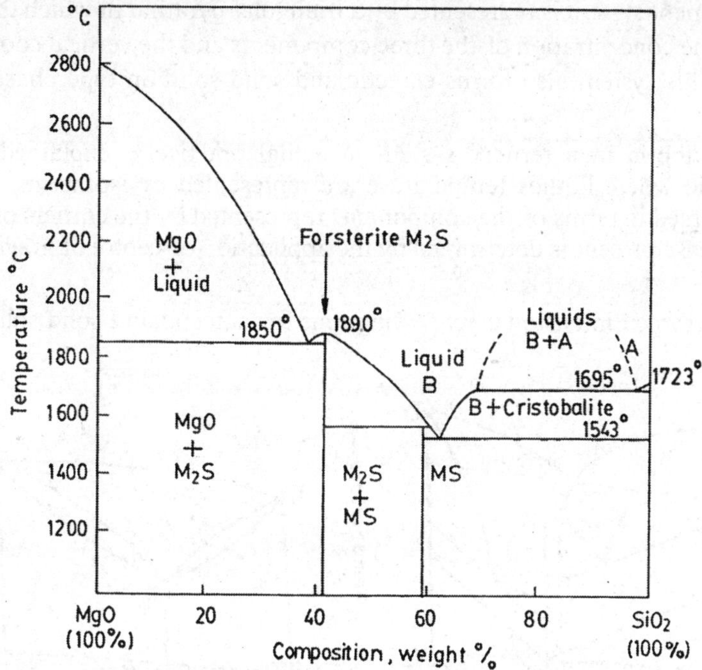

Fig.5.5: The MgO-SiO_2 equilibrium diagram showing the "monotectic" point on the right-hand side and also the very refractory nature of foresterite M2S compared with MS.

Notations: M-MgO, A-Al_2O_3, S-SiO_2, F-FeO, M_2S-Mg_2SiO_4, MS-$MgSiO_3$.

In magnesite brick, the bond that develops is a cement of small crystals of silicates and other compounds. Silica is an inevitable impurity and the desirable compound is foresterite (2MgOSi- O_2) which melts at 1900°C. Additions of pure MgO in the finest grade bricks is done to increase the MgO-SiO2 ratio in the bond and hence improve the refractoriness as is evident from above equilibrium diagram.

Silicates of magnesium have good refractory properties as is clear from the MgO-SiO_2 equilibrium diagram. It is one of the olivines and occurs in dunite in which it is strongly contaminated with varying amounts of FeO replacing MgO. Olivines decompose to minerals like serpentine ($3MgO.2SiO_2.2H_2O$) and talc & steatite (both $3MgO.4SiO.2H_2O$). All these are used to make foresterite bricks. Very fine magnesia is added to increase the MgO in bond. Refractoriness and RUL for foresterite bricks are about 1750 and 150°C respectively. Its resistance to acid and basic slag is moderately good-between silica on one hand and the chrome magnesite on the other. Resistance to iron oxide is also good.

TRIPLE COMPONENT SYSTEMS

Independent variables in a triple components system are temperature, pressure and composition of the phases which is governed by the concentrations of the components. Knowing the concentrations of two components, fixes the concentration of the third and as such this system has four independent variables. However, pressure is generally kept constant.

A triple component system is represented by a triangular pyramid in which three equal sides of its base represent the concentration of the three components and the vertical coordinates represent the temperature. This system also forms eutectic and solid solution type phases like the double component system.

The phase equilibria in a ternery system in actual practice is explained on paper by an equilibrium triangle where liquids temperature are represented by isotherms. The overall compositions are expressed in terms of the components represented by the corners of the triangle. The amount of each phase present is determined by the application of centre of gravity principle to the system.

Fig. 5.6 shows three dimensional space diagrams for eutectic and solid solution type ternery systems.

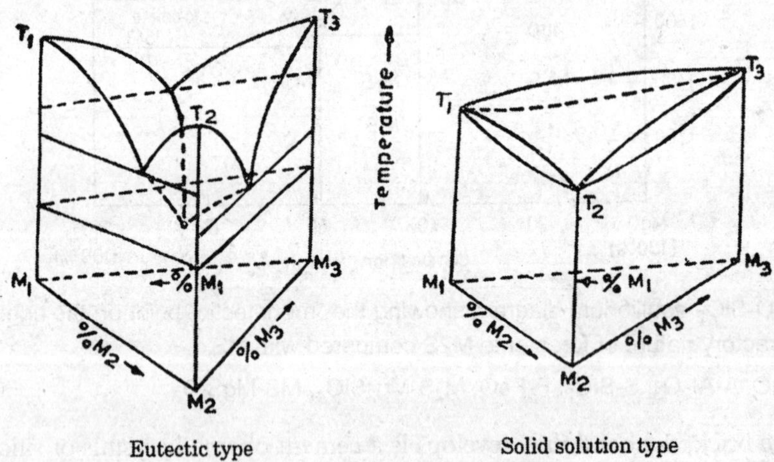

Eutectic type

Solid solution type

Fig.5.6: Ternery space diagrams

A ternery system in actual practice (where the sides of the triangle represent the concentrations of the three components) depicting schematic representation of the crystallisation of a given composition is shown in Fig. 5.7.

It is seen that cooling a liquid of composition C allows crystallisation of M_2 from the melt and its composition shifts along A-N as the crystallisation progresses. Lowering the temperature to T, the systems attains an equilibrium between the liquid and two solid phases Mj and M_2 and this follows by the crystallisation of M_1 The composition of the liquid then changes along N-O. The % of three phases M_1 , M_2 and liquid at any point M along N-0 is represented in (b).

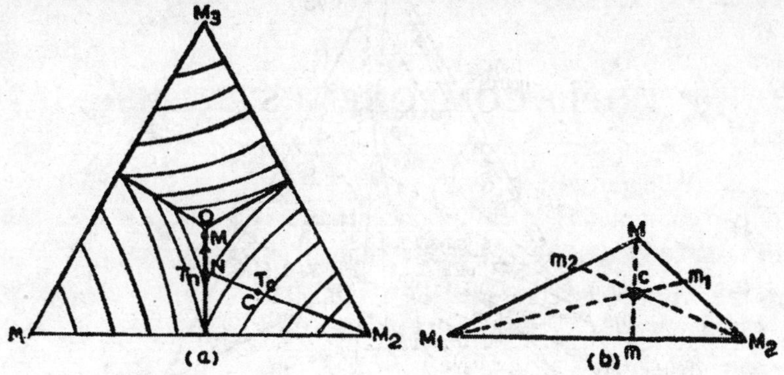

Fig.5.7: Schematic representation of the crystallisation of a given composition in a ternery system.

where,

$$M_1 = \frac{m_1 c}{m_1 M_1} \times 100\% \qquad M_2 = \frac{m_2 c \times 100}{m_2 M_2}\% \qquad \text{and, liquid} = \frac{mc}{mM} \times 100\%$$

On further crystallisation, the composition of the liquid finally shifts to the invariant point O-the ternery point where the three phases M_1 , M_2 and M_3 will be in equilibrium.

Some three components refractory systems are discussed below.

FeO-Al_2O_3-SiO_2 System. FeO-Al_2O_3-SiO_2 system applies to silica, alumina and high alumina refractory bricks and the oxide impurities like FeO, CaO, MgO, Na_2O, K_2O are largely responsible for reducing the refractoriness of these bricks. The FeO-Al_2O_3-SiO_2 system is shown in Fig. 5.8.

It is seen that with unlimited amount of FeO available for reaction with end members in the diagram (i.e. $A1_2O_3$ and SiO_2), the formation of some liquid in the alumino-silicate refractories occur even below 1100°C. At higher service temperature, even more damaging amounts will be formed. This is particularly true, as FeO bearing liquids are very fluid.

CaO-Al_2O_3-SiO_2 System. This system is most applicable to reactions of fireclay refractories with blast furnace slags. It also gives information on other reactions involving refractories such as those occurring with more complex open hearth slags or in the bond formed in magnesite bricks. This system is also useful in predicting the behaviour of silica bricks.

The CaO-Al_2O_3-SiO_2 system is shown in Fig. 5.9.

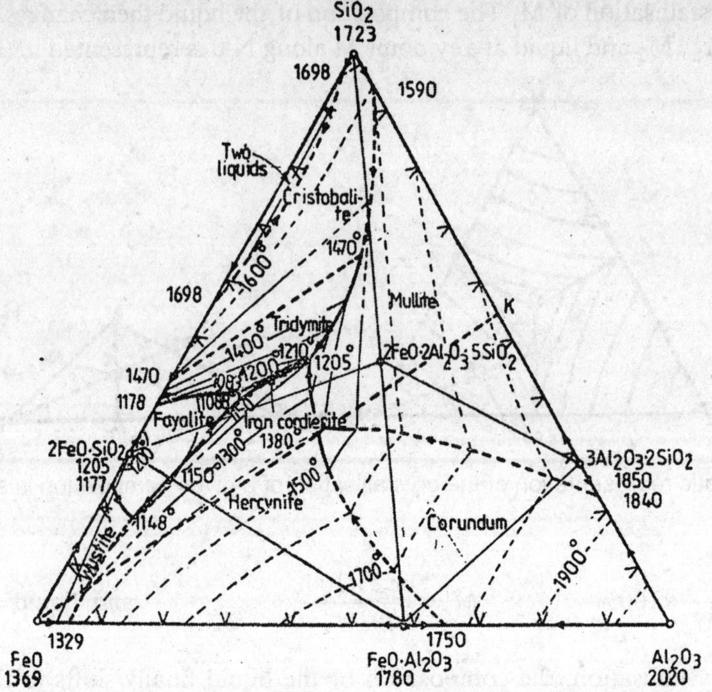

Fig.5.8: System $FeO-Al_2O_3-SiO_2$.

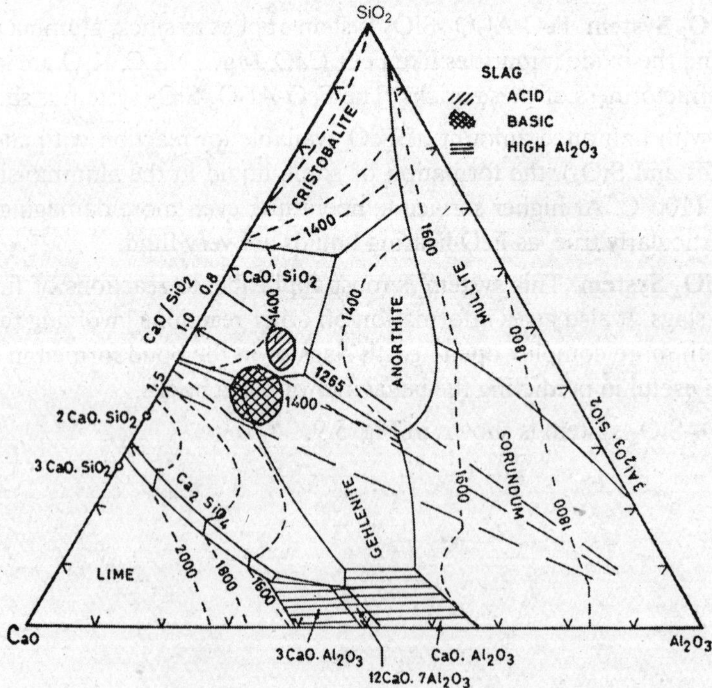

Fig.5.9: $CaO-Al_2O_3-SiO_2$ System.

The case of attack on alumino-silicate refractories by blast furnace slags has been investigated using the above diagram.

Composition corresponding to 20% slag and 80% firebrick all lie in the field in which the primary phase is mullite and the slope of the liquids surface down to the eutectic trough between the mullite and anorthite fields is so steep that at 20% slag and 1500°C (even ignoring the contribution of fluxes from the brick itself), about 90% liquid can be expected. Beyond 70% alumina and 20% slag could be absorbed without formation of much liquid at 1500°C but mullite bricks are very much expensive than firebricks.

Na_2O-Al_2O_3-SiO_2 and K_2O- Al_2O_3-SiO_2 Systems. The reactions of alkalis with alumina-silica refractories is clear from the Na_2O-Al_2O_3-SiO_2 and K_2O-Al_2O_3-SiO_2 system as shown in Figs. 5.10 and 5.11.

It is evident that refractoriness of alumina-silica refractories will be very seriously affected by very small amounts of Na_2O, less than 1% being sufficient to lower the temperature of initial liquid formation to less than 1100°C, while 10% is sufficient to completely liquify the more siliceous alumina-silica composition at about 1200°C.

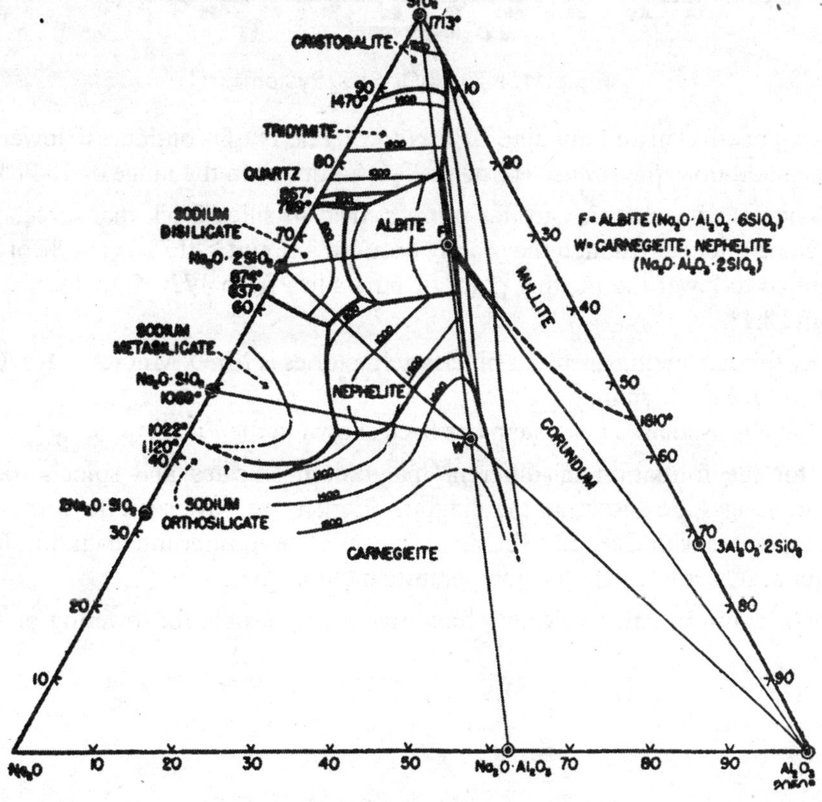

Fig.5.10: The Na_2O-Al_2O_3-SiO_2 System.

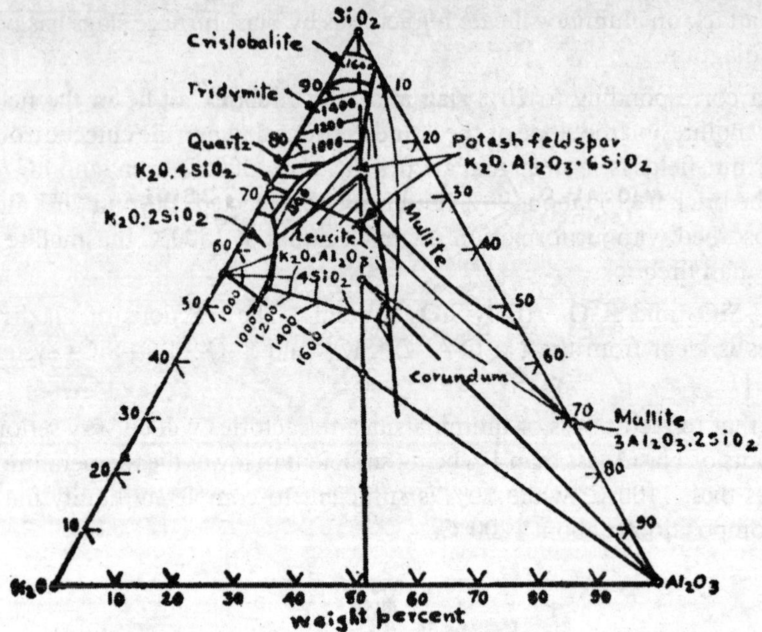

Fig.5.11: K_2O-Al_2O_3-SiO_2 System

K_2O has a similar effect in amount upto 10%, but, whereas Na_2O continues to lower the melting point with greater additions the reverse is true of K_2O additions in the range of 10-20%.

Small percentages of impurities are far more critical in silica brick than fireclay brick and alkalis are the worst offender (though they rarely occur in amount > 0.3%). The % of Na_2O, K_2O and Al_2O_3 required to lower the melting point of pure silica from 1726°C to 1676°C are respectively 1.4, 1.9 and 3.1%.

Another very serious fluxing agent for fireclay refractories is MnO. MnO-Al_2O_3-SiO_2 system is similar to FeO-Al_2O_3-SiO_2 system.

MgO-Al_2O_3-SiO_2 System. This system has been shown in the Fig. 5.12.

The areas for the formation of different magnesium silicates and spinels together with their temperature ranges are shown in the diagram. Earlier, we have seen the formation of low temperature compounds with CaO and FeO in their respective equilibrium diagrams, however, the compounds with MgO are not such low temperature melting ones.

Presence of MgO in alumino-silicate refractories is responsible for lowering of their refractoriness.

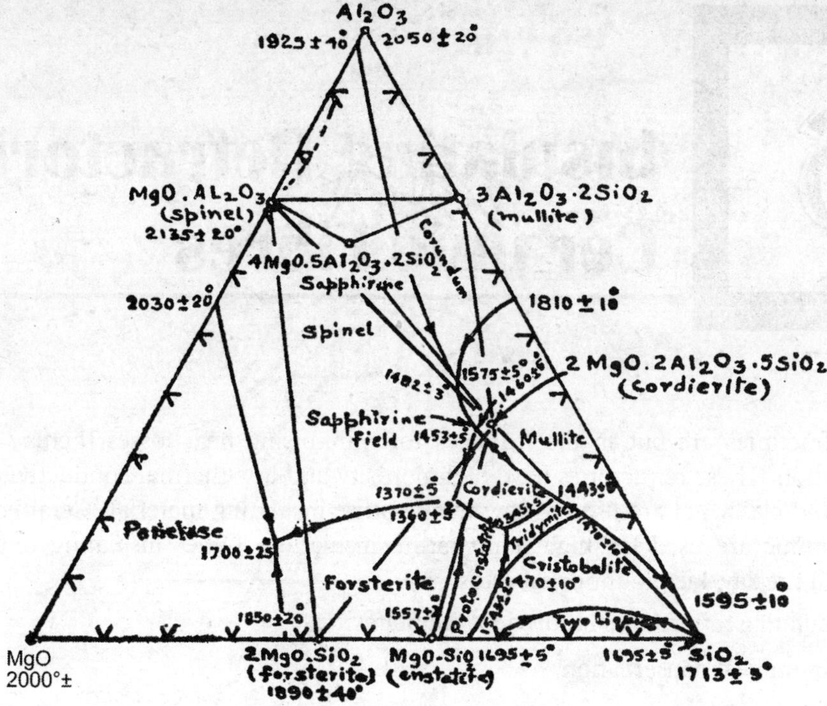

Fig.5.12: $MgO-Al_2O_3-SiO_2$ System

Insulating Refractories/ Ceramic Fibres

Insulating refractories are put in the furnaces for minimising heat losses thereby maximising heat conservation. These refractories have high porosity and low thermal conductivity. Slagwool, vermicullite and glasswool are used as low temperature insulating material. Ceramic fibre, wool and foam ceramic are used for high temperature insulation. Other insulating refractories are produced from fireclay, kieselguhr or asbestos.

Use of insulating refractories in furnaces facilitates:

- Maximum heat conservation
- Reduction in heat losses
- Maximum thermal efficiency
- Minimum fuel consumption
- Higher working temperature
- Better working condition for shop floor personnel
- Less fuel handling problems

The desired properties in a good insulating refractory are:

- Uniformly high porosity
- Low thermal conductivity
- Reasonably good cold crushing strength
- High PCE values to withstand the working temperatures.

Highly porous structure with air entrapped therein minimises the conductive capacity of the refractories and increases their insulating characteristics as still air is a good insulator. Insulators used as a backing material in the furnace prevent heat conduction through the heating zone through the sidewalls or bottom without unduly enlarging the furnace structure. Uniform distribution of pore spaces ensures better thermal insulating characteristics. Insulating characteristics are developed by adjusting the ratio of size and number of solid particles & pore spaces and also selection of a less conductive materials. To induce porosity in the insulating refractories, some carbonaceous combustible material like saw dust is put in the refractory mix which burns on firing and leaves open space in the insulating refractory products. Porosity is also induced by adding volatile matter like powdered naphthalene or chemicals like calcium phosphate, ammonium chloride/sulphate etc. Use of hardwood chips also results in large sized pores.

The production of insulating refractory bricks and other insulating materials require a different treatment. The insulating bricks are classified into two categories — one used for low temperatures i.e. < 1000°C and the other used for any temperature above 1000°C. The other insulating materials are the ceramic fibres of different compositions with corresponding application temperatures.

The low temperature insulating bricks are manufactured from granules of exfoliated vermiculite,, exfoliated mica or raw diatomite. In using any of the above mentioned raw materials, a good quantity of carbonaceous grain is used in the batch composition, which will burn out during firing, leaving voids inside the texture of the brick, which ultimately contribute to their insulating properties. Exfoliated vermicuiite granules may, however, be utilised for making bricks and blocks without firing by using hydraulic setting cement for their production.

The high temperature insulating bricks are produced from mixtures of grains of calcined clay or calcined kyanite or raw kyanite together with combustible carbonaceous materials in the batch. The use of combustible material may be eliminated by adopting the foaming technique during forming the shapes. Presently bubble alumina and bubble silica grains are produced for making both "shaped" & "unshaped" refractory insulating products for high temperature application. The rice husk ash is an important insulating material suitable for fairly high temperature up to 1500°C.

A small proportion of plastic clay as bond is used in both low temperature and high temperature insulating bricks. The bricks are formed at a low pressure or even by hand moulding. The firing of the insulating bricks is carried out at a temperature depending on the raw material used as well as on the temperature of their application. The firing temperature should preferably be higher than the temperature of their use.

Insulating refractories are laid with insulating cements of similar composition as the insulating refractories themselves ; thereby ensuring very thin joints.

Suitable grade fireclay is the most widely used insulating refractory material which is made by mixing fireclay with the saw dust (a combustible material used for inducing porosity) and a binder followed by moulding of the mixture into desired shapes (by hand, extrusion or dry pressing), drying and firing (at a temperature 100°C below the PCE value of the fireclay used). Flue gas generated due to combustion of saw dust is expelled from the refractory bricks result in a light product having the desired porous structure. Insulating bricks produced from china clay has higher refractoriness while those produced from asbestos and kieselguhr though less conductive have lower refractoriness than fireclay bricks. Fireclay is used upto 1500°C while asbestos or kieselguhr is used upto 900°C. Expansion on its heating reduces its density (by 16%) and the disintegration of the material because of the structural change due to expansion is taken care of by using a combustible material (e.g., saw dust) which burns off on firing and leaves open space in the product. For insulating acid furnaces, fireclay quartzite mixture with saw dust is used for making insulating refractory. Being highly porous & comparatively of lower strength, insulating refractories require careful handling, packing in the kilns and firing to avoid their disintegration.

Anorthite light weight refractories (artifacts) produced from kaolin and gypsum (by melting and casting) have high heat insulating properties, high strength & low apparent density (0.5 gm/c.c.) and suitable for use at 1100 to 1250°C. These are specifically used in carbon containing reducing media. High strength 'artifacts' (with open porosity of upto 85%) can be produced by adding polythene, polystyrene, polyvinyl chloride etc. (i.e. high molecular synthetic compounds).

Slagwool, glasswool, vermicullite, asbestos (in the form of wooly fibre) etc. are used for low temperature insulation. Zirconia and alumina powders are used for insulating laboratory type furnaces where loose packing of powder results in a porous structure. Uniform size powder ensures even distribution of open space and therefore improved insulating properties.

Hot top inserts (density = 0.85-1.5 gm/c.c.) produced from refractory mixture containing 64—80% SiO_2 and 8-20% each of lignin & saw dust are used as heat insulating inserts (i.e. hot top) for steel ingots to effect lateral heat insulation of steel in riser part of the ingot which reduces the rejections as top trimming.

Sufficiently strong & porous refractories are used in checkers of regenerative furnaces which helps in better heat transfer and regeneration. Hot gases passing through the pores rest the refractory grains and this heat is recovered when the cold blast (air) passes through the checker in the blast cycle. The more uniform is the distribution of pores, the better is the heat generation efficiency of such refractories.

High alumina wool slabs are used as insulating material in furnaces for heat processing steel stampings, melting and casting units producing aluminium alloys and in furnaces used for homogenizing aluminium semi finished products. It has a high refractoriness (> 1750°C), low thermal conductivity (< 29 W/m. °K at 800°C) and can be safely used upto a working temperature of 1150°C. However, this material has the disadvantages of loss of electricity during heating and it undergoes particle deformation after prolonged soaking at > 1000°C. Moreover, its low plasticity prevents the use of tiles for parts having large curvatures. These drawbacks of insulating slabs are overcome by replacing them with insulating fibre {e.g. ceramic fibre), matting or felt.

Insulating refractory brick (IRB) is a class of brick, which consists of highly porous fireclay or kaolin. IRBs are lightweight, low in thermal conductivity, and yet sufficiently resistant to temperature to be used successfully on the hot side of the furnace wall, thus permitting thin walls of low thermal conductivity & low heat content. The low heat content is particularly important in saving fuel & time on heating up, allows rapid changes in temperature to be made, and permits rapid cooling. IRB is characterized by the presence of large amount of porosity in it. The pores are mostly closed pores. The presence of porosity decreases the thermal conductivity of the insulating bricks.

There are two types of bricks namely (i) bricks based on clay and gypsum using the burnout of sawdust to create high porosity (and thereby provide better insulating value), and (ii) bricks based on lightweight aggregate and clays.

IRBs are made in a variety of ways, such as mixing of organic matter with clay and later burning it out to form pores; or a bubble structure incorporated in the clay-water mixture which is later preserved in the fired brick.

IRBs are characterized by the presence of large amount of porosity (45 % -90 %) in it. The pores are mostly closed pores. These bricks due to a highly porous structure exhibit low thermal conductivity values. The thermal conductivity not only depends on the total porosity, but also on the size and shape of the pores as well as chemical & mineralogical composition. The bricks also have low heat capacity. Due to the presence of high porosity in these bricks, the bulk density and the strength of the bricks are low. The application temperature of these bricks depends on the constituents. For example, kyanite based insulation bricks can be used at a temperature greater than 1250°C. Whereas fireclay based insulation refractories are usually used at low temperatures.

IRBs are shaped light weight refractory products with a total porosity of greater than 45 % and an application temperature of 800°C minimum. Important properties of insulation refractories are given below in Figure.

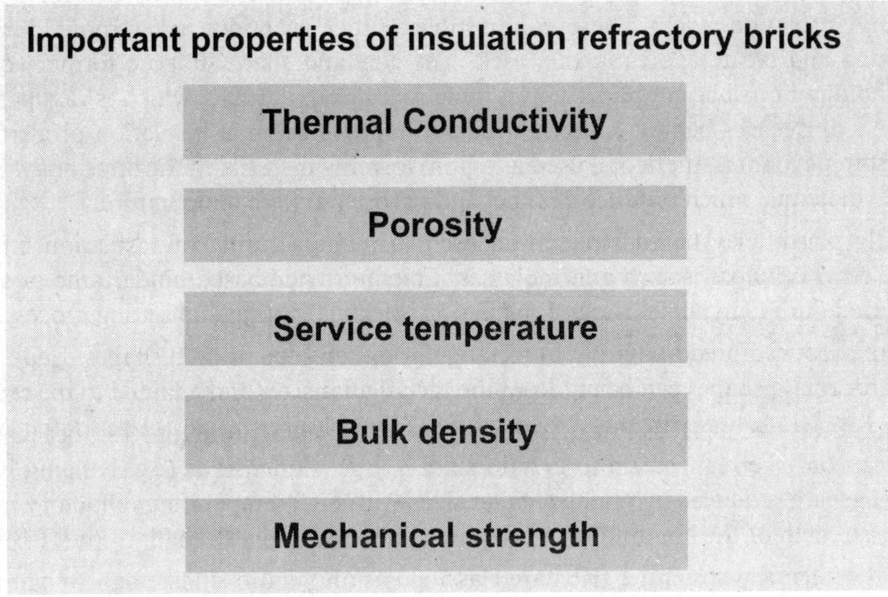

Fig.6.1: Important properties of insulation refractories

Different types of IRBs are mainly manufactured by using the raw materials such as diatomite, perlite, expanded vermiculite, calcium silicate, fireclay, kaolin, quartz, alumina and light weight refractory aggregates by conventional method. Different types of pore formers such as sawdust, foam polystyrene, fine coke, binders and organic foams or granular materials such as hollow microspheres and bubble alumina are commonly used to obtain decreased density or to produce porous bodies in the IRBs.

The raw material of diatomaceous earth products are microscopically small shells derived from diatoms. The heat insulation results from the large number of small pores within the shells of various shapes with sizes from 5 to 500 micro meters. Diatomaceous bricks are produced by extrusion presses. Bonding clay, burning out materials and fibres can be added. The fine pore structure of these bricks has a better heat insulation than vermiculite products with the same bulk density.

Vermiculite is a 3-layer mineral which expands like an accordion when heated quickly above 700 ^{0}C. Through the pressure of the evaporating water in the intermediate layer, the original volume enlarges by 20 to 30 times. This so-called exfoliated vermiculite (density 60 to 200 kg/m^3) serves as raw material for the production of vermiculite bricks, boards and shaped parts. Concrete, water glass and phosphates are used as binder.

Perlite is water containing volcanic rock, solidified to a glass shape. Water was absorbed by the magma under high pressure during litho genesis and can evaporate when heated quickly. Expanded perlite is formed with a settled density of 35-150 kg/m^3. The production process of bricks, boards and shaped parts is almost the same as that for vermiculite. The application limit of this product range is from 750 to 1000°C due to intensive shrinkage at higher temperatures.

Rice husk/sawdust straw or low cost biomaterial are used as combustibles or pore former in manufacturing of insulation brick. However naphthalene, starch are also used for pore former in high duty insulation brick. These combustibles either evaporate or burnout during initial stage of firing and creates pores in the brick. The size and shape of pore former controls the pore morphology in insulation brick. Polystyrene foam each particle, which is dissipated during firing process and leaves behind a cavity that can improve thermal insulation properties of the brick. Polystyrene foam is, therefore, used as a pore forming material in the brick body. Sawdust is combustible material, which produce channel and porosity at high temperature.

Normally plastic clay based binders are used in manufacturing of insulation brick. Other binders are ethyl cellulose, starch and molasses. Calcium oxide based binder lime or sometimes gypsum is used during manufacturing of insulation brick via casting and setting process.

IRBs are produced predominantly in rectangular dimensions or arch shapes. End arches and side arches for round shapes can be cut from the rectangular sizes and adapted to the radius.

The maximum size of IRBs bricks is limited by the manufacturing process, because uniform properties can only be realized up to certain sizes. Larger dimensions (e.g. hanging blocks and roller bushings) are produced by gluing smaller sizes with high temperature refractory mortars. In general, the strength of the bonding joint is higher compared to the strength of the IRB itself.

As per the chemical structure, IRBs are classified as aluminum silicate lightweight refractory bricks, silica lightweight refractory bricks, zircon lightweight refractory bricks and corundum lightweight bricks. The group of aluminum silicate IRBs (fireclay and mullite bricks) is the most important and common group. Raw materials based on Al_2O_3, SiO_2 and in some cases CaO are used for the production of these bricks. Raw materials such as clay, kaolin, fireclay, sillimanite, andalusite, kyanite, mullite, alumina, alumina hydrate and corundum are used as alumina carriers. In addition to the fine grained raw materials, coarse grained and porous raw materials are also applied. These include lightweight fireclay and hollow spheres (balls) consisting of corundum or mullite.

The thermal decomposition process is best known and most applied technique to produce pores in the insulating firebricks. Fine saw dust, petroleum coke, lignite abrasion, styropor balls, fine waste products of the cellulose & paperboard (carton) industry are used as burning out materials.

Decomposition materials with low ash content are required in order to prevent negative effects on the hot properties (e.g. alkalines). The so called foaming process is another further method for the production of pores in insulating fire bricks. Special soaps, saponins and sulfonates are used to produce stable foams. The slurry for the ceramic body is often made separately from a foam emulsion. Foam and slurry are homogenized in an intensive mixer. The required bulk density is adjusted by controlled mixing of foam and slurry. In practice the gas impellent process is used less frequently. Gas producing substances are mixed into the compound.

Lightweight refractory bricks which are produced by mixing in evaporating substances (naphthalene) have distinctive differences in their properties compared to other brick qualities. It is possible to produce bricks with low density and high strength. Very fine pores guarantee low thermal conductivity values.

Shaping of the lightweight refractory bricks is done by casting, slinger method or pressing. During casting, the perforated metal moulds (forms) are lined with filter paper before being filled. Sulphite liquor, gypsum or concrete can be added to improve the mixing of the raw material and to speed the setting.

The slinger method is very efficient due to the continuous shaping of large blocks.

Plastic, semi-dry and dry mixes are shaped with the corresponding presses (extrusion presses, hydraulic presses or mechanical presses).

Bricks, unfinished cylindrical pieces or blocks are fired in chamber furnaces, bogie hearth furnaces or tunnel kilns. The firing temperature corresponds approximately to the classification temperature indicated by the producers. Cutting or grinding is necessary for most brick qualities in order to obtain the standard shapes due to high drying and firing shrinkage.

IRBs with complicated shapes are produced by hand forming, vibration or moulding processes.

For long term exposures near the rated service temperature, shrinkage may occur sufficiently to allow joint opening. For this reason, it is important to use IRBs with a higher duty rating than required by the process temperature alone.

IRBs are susceptible to alkali attack. The products made using gypsum and sawdust burnout material (indicated by high CaO contents) may undergo an expansion reaction in a service environment containing alkali. In some cases, this reaction has been deleterious and reduces service life. By contrast, the IRB made with lightweight aggregate may experience surface glazing in an alkali environment.

The uses of IRBs ranges from laboratory furnaces to foundry furnaces and large tunnel kilns. The requirements on IRBs are diverse and in some cases even contradictory. On the one hand, high thermal insulating capability and low bulk density are the desirable properties of IRBs, on the other hand the bricks need to have sufficient mechanical strength but also good machining. Additionally the properties needed in these bricks are high thermal resistance under a multitude of atmospheric conditions as well as resistance to temperature shocks and changes.

The operation time of the industrial furnace is a decisive criterion for the IRBs. For furnaces operated on a continuous basis the mass of the constructed refractory lining is less important concerning energy efficiency. The thermal conductivity (k) is significant for efficient operation, therefore bricks with low k-values are preferred. The basic rule is 'The lower the bulk density, the lower the thermal conductivity'. The bulk density needs to be adapted to the service temperature of the kiln. The reason is the existence of a minimum thermal conductivity which is shifted to higher bulk densities at higher operating temperatures.

The cold compressive strength is less important compared to the hot properties of the brick. In general the mechanical loads are not high. The strength of insulating fire bricks is fully sufficient. Approximately 0.5 N / mm^2 is required as a minimum strength. That value ensures safe transport, handling and installation work. Often a compromise between strength, bulk density and thermal conductivity has to be done.

Constructions with higher mechanical loads require bricks with higher cold crushing strength. It is important that the higher cold compressive strength is not the result of a higher fluxing agent content in the brick, otherwise the creep under compression increases at high temperatures.

Thermal shock resistance is an important property for periodically operated furnaces.

Depending on the heating up and cooling down conditions of the furnace, it is important to know the stress limits of the material. The material will crack, if temperature differences between cold and hot side are higher than 130 to 250 K. Such temperature differences are often exceeded repeatedly in wall linings. Each temperature change affects the structure of the brick.

The critical heating up speed for lightweight fireclay bricks (depending on the shape) is 5-10 K/min.

The different IRBs vary in their thermal shock behaviour due to their specific composition and porosity. High cristobalite contents (more than 10 %) have a negative effect on thermal shock resistance. In case of less than 10 % other criteria are more essential.

Micro cracks in the structure are an advantage since they absorb stresses without further crack development.

IRBs remain a choice for construction and repair when castable or fiber products may not be preferred. IRBs constitute the working lining of many furnace structures. Construction is usually convenient since most IRBs can be cut with a hacksaw. Caution should be used in selecting IRBs when abrasion resistance or impact resistance is required.

Mortar joints and adhesive joints are the weak points in the brickwork. Due to high porosity and therefore fast removal of water, there are the following two important factors that are to be considered.

- The mortar must be 'plastic' and must have a high capability to bond water. If not, the bricks cannot be laid precisely according to measures and the bond of the individual bricks is lost when corrected later.
- The mortar should have low alkali content. The insulation firebricks are infiltrated with the binder liquid resulting in a higher alkaline concentration in the border areas. This causes lower melting points/increased sintering of the material and increased creep at high temperatures. Also thermal shock resistance is lowered by alkaline impurities. Hence mortars without water glass are preferred for high temperature applications.

CERAMIC FIBRE

Cermic fiber, a low thermal mass insulation material, which is used in furnace interior lining is manufactured by blending and melting alumina & silica at a temperature of 1800 - 2000°C, and breaking the molten stream by flowing compressed air or dropping the molten stream on a spinning disc to form loose or bulk ceramic fiber. The bulk fiber is used to produce various insulation products including blankets, strips, veneering and anchored modules, paper, vacuum formed boards and shapes, ropes, wet felt, mastic cement etc. Fibers are usually produced in two temperature grades based on A1203 content. A new product is Zr02 added alumino silicate fiber, which helps to reduce shrinkage levels and thereby making the fiber suitable for higher temperatures. Continuous recommended operating temperature for fibers are given below in the Table.

Table: Continuous recommended operating temperature for ceramic fibers

Operating temperature °C	Al_2O_3 %	SiO_2 %	ZrO_2 %
1150	43-47	53-57	—
1250	52-56	44-48	—
1325	33-35	47-50	17-20

Cermic fibers are generally produced in bulk wool form and needled into a blanket mass of various densities ranging from 65 to 190 kg/m^3. Converted products and over 40 different forms are made from blankets to suit various requirements.

The characteristics of ceramic fibers are a remarkable combination of the properties of refractories and traditional insulation material which is given below :

Lower thermal conductivity

Because of the low thermal conductivity (0.1 kcal/m/hr/°C at 600°C for a blanket with 128 kg/m^3 density) it is possible to construct thinner linings with the same thermal efficiency as conventional refractories. As a result of thinner lining, the furnace volume is higher. It is 40% more effective than good quality insulation brick and 2.5 times better than aabestos. Ceramic fiber is a better insulator than calcium silicate.

Light weight

The average density of ceramic fiber is 96 kg/m^3. It is one tenth of the weight of insulating brick and one third of the weight of asbestos/calcium silicate boards. For new furnaces structural supports can be reduced by 40%.

Lower heat storage capacity

Ceramic fiber linings absorb less heat because of their lower density. Furnaces can therefore be heated and cooled at faster rates, typically the heat stored in a ceramic fiber lining system is in the range of 2700 - 4050 kcal/m^2 as compared to 54200-493900 kcal/m^2 for conventionally lined systems.

Thermal stock resistant

Ceramic fiber linings resist thermal shock due to their resilient matrix. This also allows for faster heat up and cool down cycles, thereby improving furnace availability and productivity.

Chemical resistance

Ceramic fiber resists most of the chemical attack and is unaffected by hydrocarbons, water and steam present in flue gases.

Mechanical resilience

The high mechanical resilence of ceramic fiber makes it possible to manufacture fiber lined furnaces off-site, transport them to the site in assembled form without the risk of damage.

Low installation cost

As the application of ceramic fibers is a standardized process, no special skills are required. Fiber linings require no dry out or curing times and there is no risk of cracking or spalling when they are heated after installation.

Ease of maintenance

In case of physical damage, the section of damaged ceramic fiber can be quickly removed and replaced with a new piece. Entire panel sections can be prefabricated for fast installation with minimal down time.

Ease of handling

All product forms are easily handled and most can be quickly cut with a knife or scissors. Vacuum formed products may require cutting with a band saw.

Thermal efficiency

Thermal efficiency of a furnace lined with ceramic fiber is improved in two ways. First, the low thermal conductivity of ceramic fiber allows the lining to be thinner and therefore the furnace can be smaller. Second, the fast response of ceramic fiber to temperature changes also allows for more accurate control and uniform temperature distribution within the furnace. Other advantages offered by ceramic fiber lining of the furnace are:

- Lightweight furnace
- Simple steel fabrication work
- Low down time
- Increased productivity
- Additional capacity
- Low maintenance cost
- Longer service life
- Higher thermal efficiency
- Faster response

It is produced from various molten materials such as alumina, silica, titania, zirconia, thoria, tantalum oxide etc. in the form of wool, short fibres and long fibres. Cloths and blankets are then moulded out of such fibres of varying length and width. Similarly, papers and felts are also formed from the wool for their final application as high temperature insulating material.

Being very light and highly porous, ceramic fibres has an excellent insulation efficiency for high temperature insulation coupled with the advantage of decrease in material consumption of insulators by 40-60%. Apart from reducing the overall weight of the structure, ceramic fibre lining in the furnaces reduces the fuel consumption markedly and increases their productivity. Kaolin fibre in particular is very effective insulating material.

Various types of felts based on ceramic fibre are made by roll compacting fibres produced by melting alumina & sand in an electric furnace and dispersion of the melt into fibre by blowing it with steam carrying the necessary amount of bond material (e.g., silicone) to be incorporated into the fibre for imparting strength. The compacted felt is cut by disc knives and made into rolls for final use. Alumino silicate fibre felt (density 0.048 to 0.128 gm/c.c.) is used as working layer of industrial furnace linings at service temperature of 800-850°C and flue gas speeds upto 9 m/second. Similarly, silicone bonded high alumina insulating felt can withstand temperature upto 1150°C even under prolonged use and can be used as working layer of industrial furnace lining at service temperature of 850°C & gas speeds upto 5 m/second.

Due to their excellent insulating values, simplicity of application, low heat storage, low weight & low thermal conductivity, ceramic fibres are now being increasingly used in aerospace, jet engine and steel plant furnace insulation. These are economic and efficient insulation materials for those furnaces where there is no direct contact with molten metal. A typical specification of Indian ceramic fibres of two grades are given in Table 6.1.

Table 6.1: Typical Specifications of Indian Ceramic Fibre

Parameters	Unit	Grade 1	Grade 2
Composition:			
Al_2O_3	%	43-47	46-48
SiO_2	%	53-57	47-49
Fe_2O_3	%	0.02-0.08	0.7-1.2
TiO_2	%	0.02-0.05	1.5-1.9
Properties:			
Bulk density	gm/c.c.	0.128	0.128
Fibre diameter	micro metre	2-3	3
Tensile strength	MPa	1400	960
Melting point	°C	1760	1760
Specific heat capacity at 980°C	kcal/kg. °K	1.07	—
Safe working temperature	°C	1260	1260

Most of the fibres used in refractory applications are blown alumina-silicate fibres which are white glassy and non crystalline. The fiberization process differs from manufacturer to manufacturer but generally consists of pouring a stream of molten alumina-silica glass at about 2300°K into spinning rotors or in front of a high pressure stream of air or steam. Both techniques break up the stream into droplets and impart a high acceleration in those droplets causing them to attenuate into fibres.

PROPERTIES OF CERAMIC FIBRE

Fibre diameter

Average diameter ranges between 2.2 and 3.3 microns. Finer fibre gives a lower thermal conductivity at light density and high mean temperature, but its effect at higher density is marginal.

Thermal Properties

Ceramic fibres have a very low thermal conductivity which decreases with increase in bulk density of fibres and increases exponentially at increasing temperature due to their 90% high air proportions. Thermal conductivity in the preferred fibre direction of fibre blanket is about 30% higher than perpendicular thereto. The fibres block radiation, prevent connective currents, yet still allow only a small amount of thermal conduction.

By comparison with conventional refractory materials, fibres exhibit a thermal diffusivity that increases with temperature. Fibres have an emission coefficient of about 0.95 as compared to less than 0.7 for conventional refractory.

Their low heat capacity gives faster heating and cooling cycles in periodic kilns. Fibres are also completely resistant to thermal shock.

On heating fibres to about 980°C. devitrification starts which causes shrinkage, loss of strength and crumbling. At 1260°C to 1315°C, it accelerates quickly. A maximum linear shrinkage of 2.5-3% after heating over 24 hours is a realistic value for a classification temperature. The amount of shrinkage directly affects the thermal stability.

Strength, erosion & corrosion resistance

Low density of fibres enables use of light weight structural support. They are relatively resistant to mechanical shock and vibration. Erosion resistance of fibre blankets against gases can be improved by modular construction. The surface saturation of fibre lining with hardening & sealing materials (such as silicasol) and the application of protective coatings (SiC or Zircon) improve on erosion resistance.

Type of Fibre Products

A variety of products are made from fibres e.g., felts, blankets, vacuum formed shapes, paper etc. Felts are mats that are organically bonded in roll or sheet form which has good cold strength. Blankets differ from felts in being completely inorganic and are made by a self binding process.

ADVANTAGES OF LIGHT-WEIGHT FIBRE CONSTRUCTION

The benefits derived from fibre-lined furnaces are as follows:

- Fibre linings are much lighter
- Fibre linings store less heat
- Lighter-weight steel structure may be used
- Reduced wall thickness increases furnace capacity
- Furnace control are improved
- Heating and cooling cycles are shorter.

Ceramic Fibre Installation Techniques

Generally, there are three methods for lining ceramic fibre in furnaces, namely,

- wall papering
- modular construction and
- vaneering

Wall Paper Construction

It involves welding metal studs or anchors to a metal backing plate and then impaling layers of fibre insulation over these studs until the desired thickness of insulation is achieved. The hot face layers should be either singled or overlapped to prevent penetration of hot gases through joints. Less expensive back up insulation may be used by this method.

Metallic anchors, for temperature upto 1230°C, are welded to furnace shell at regular intervals. After the fibre blankets are impaled, they are secured by anchor washers. Above 1230°C, ceramic anchors are used. In one system, alloy stud is welded to the furnace shell and a ceramic pin (Corundum) is threaded over the metal stud.

Modular Construction

Refractory blanket is prefabricated into 305 mm square modules which are quickly and easily installed on site. Various modular systems differing in manufacturing and assembly methods are available.

(i) Laminated block consisting of fibre blanket layers which are compacted into one element of 305 x 305 mm and layed in a mortars bed. Expanded metal is welded into the furnace casing after the cementing operation. The modules are generally laid in perquet fashion. This method is easy & cheap to apply and offers good possibilities for repairs. It can be adopted to complicated geometrical shapes. It gives reduced heat insulation as the fibres are perpendicular to the furnace interior.

(ii) Precompacted block with fibres extending perpendicular to furnace interior - it consists of a folded blanket and is mechanically secured on cold side. Each module can be separately replaced in this method.

(iii) Precompacted block with fibres extending parallel to the furnace interior - it also consists of a folded blanket. It gives optimal heat insulation as the orientation of fibres are parallel to furnace wall.

(iv) Vacuum formed modular box which is filled up with fibres and backed by heat insulating material. Denser surface on hot side on these modules provide higher erosion resistance.

Typical specifications of yet another different grades of ceramic fibre is tabulated below:

Ceramic fibre	HTZ Grade	STD Grade
Chemical, (wt%) Analysis:		
Al_2O_3	33-36	45-48
SiO_2	44-48	52-55
ZrO_2	16-19	—
Bulk Density, kg/m^3	128,96,64 (Three grades)	128,96,64 (Three grades)
Fibre Dia., Micron, maximum	3.0	3.0
Shot content, % maximum	15	15
Tensile strength, Kpa, minimum	60	40
Shrinkage, % max, after 20 hrs. firing	At 1400°C—3.0	At 1200°C—3.0

Typical Specifications of ceramic fibre module is given below:

Chemical analysis, wt %

Al_2O_3-33, (min), SiO_2-52(max), ZrO_z-14(min)

Physical Characteristics:

Size—300 × 300 × 300 mm, Bulk density-190 and 160 kg/m^3, Fibre Diameter (micron)- 3.0, Shot Content (%) - (max 10), Shrinkage, % (after firing at 1400°C/24 hrs)-3.5

The desirable properties of ceramic fibre in their various forms may be summarized as:

Very low density, Very low thermal conductivity, Very low thermal mass, High flexibility in use, High melting point, High resistance to chemical attack, High spalling resistance.

Wallpaper Vs. Modular Construction

- Less expensive back-up insulation can be used in wall papering whereas one material is used in modules.
- Joints are offsetted in wall papering whereas the joints are fitted by expansion of the fibres in modular construction.
- Installation time is less with modular construction.
- Replacement of modules is selective.
- Modules enable the anchors to be disposed in the colder part of a lining. Hence lower cost alloys can be used.

Veneering

Using this method, fibre can be installed over conventional refractory lining that is still in good shape. It involves using modules of thickness made from strips of fibre held together with an open mesh organic cloth. The four ridges running across the module are a hot melt glue which holds the cloth together. Both glue and cloth burn off when the furnace is first fired. These modules are installed using air setting mortar.

Advantages of veneering are as follows:

- It is cheaper than tearing out existing lining and completely rebuilding.
- It takes less time than installing a completely new lining.
- It saves fuel (10 to 30%) because of added insulation.
- It acts as a cushion against rapid temperature changes which protects brick from thermal shock during heating and cooling.
- It can help increase in available production time as furnaces can be heated up faster and less prone to thermal shock failure.

APPLICATION OF CERAMIC FIBRES

Ceramic fibres are an efficient and economic solution to insulation problems in furnaces where there is no direct contact with liquid metal. A few application of ceramic fibre are discussed below.

Ladle lids. Concentric strips of blankets of ceramic fibre are a parallel to lag the ladle lid, retained by glueing with special adhesive to the bottom plates of the cover and by hooking on to metal pins. Sheet metal is spot welded to the bottom of the cover to facilitate the glueing by providing a series of anchor points in concentric circles, spaced to accommodate the thickness of lining required.

Advantages of use of ceramic fibre in ladle lids are as follows:

- increased ladle life due to reduction in thermal shock,
- higher lining temperature with consequent reduction in preheating cost and higher ladle availability.
- lowering of primary furnace temperature.
- more regular thermal state of ladle lining.
- reduction in steel temperature dispersion.
- fewer teeming problems.

Soaking Pits. Ceramic fibre sealing system for soaking pits results in about 10% fuel saving. A fibre-lined cover for soaking pits offers many advantages such as weight reduction (more than 90%), immunity to thermal shock and the ability to withstand accidental knocks (other than on the fibre surface) in handling. In Japan, trials on veneering castable-liquid covers with 60 mm thick modules showed good life endurance with fuel savings of about 20%.

Reheating Furnaces. Modules are attached to the furnace casing by means of a simple metal fixing system. A layer of 10-20 mm of fibre blanket is usually placed between modules and furnace casing. Life can be extended by the application of special coatings which increase fibre resistance to erosion and corrosion.

For high temperature installation, veneer module based on high alumina are glued to brickwork of reheating furnaces resulting in 10-30% reduction in fuel consumption depending upon thickness of modules and brickwork lining.

Skid pipes are lined with two layer insulation using ceramic fibre felt adjacent to the pipe and refractory castable or prefabricated tiles. Another method is to use ceramic fibre block or veneer the skid pipe support with fibre blanket.

Annealing furnaces. Relining of bell annealing furnace with ceramic fibre, replacing the refractory lining can result in the minimum reduction in specific fuel consumption by 10%. Ceramic fibre lining of the bell furnace enables reduction in lining thickness and a greater distance between hot face of heating bell and protective cover. As a result of ceramic fibre lining, more uniform heating of protective cover is obtained. These factors contribute towards increasing life of the protective cover and reducing incidence of the premature buckling. Changing over from an arch type to flat roof construction could be done by means of cerainic fibre lining. This permits to increase the height of the protective cover.

Forging Furnaces. The main reasons for use of fibre lining in forge and batch furnaces are lower accumulated heat and shorter heating & cooling times. One forge furnace was veneered with 1½ inch of alumina fibre over insulating fire brick for reduced maintenance and fuel savings (12%).

Some useful data regarding insulation by refractory or ceramic fibres have been given in table 6.2 & 6.3 and in figures 6.2 to 6.7.

Table 6.2: Thermal Data for Walls (35 ems thick) with Hot Face Temperature of 1760°C

Type of refractory	Cold face temperature,°C	Approximate heat flow, Kcal/m².hr
Fireclay	238	4070
70% Alumina	249	150
90% Alumina	288	5830
Silica	268	4960
Chrome	279	5370
Magnesite	352	8900
Forsterite	260	4610

Table 6.3: Effect of Ceramic Fibre as Insulation

Original refractory lining	Operating temp. (°C)	Additional ceramic fibre Thickness, mm	Cold face temp. (°C)	Heat loss, kcal m².hr	Heat loss reduction (%)	Heat storage (kcal lm²)	Heat storage reduction, %
Dense fire-brick 229 mm	870	0	95	772	—	93023	—
		50	80	560	27.5	70777	23.9
		75	74	485	37.0	62741	32.6
Insulating fire-brick (1093°C) 114 mm	1090	0	114	1060	—	122573	—
		50	98	810	23.6	98273	19.8
		75	92	718	32.3	89018	21A
Insulating castabie (1204°C) 229 mm	1090	0	129	1320	—	36357	—
		50	108	960	27.2	29891	17.8
		75	100	832	36.9	27306	24.9

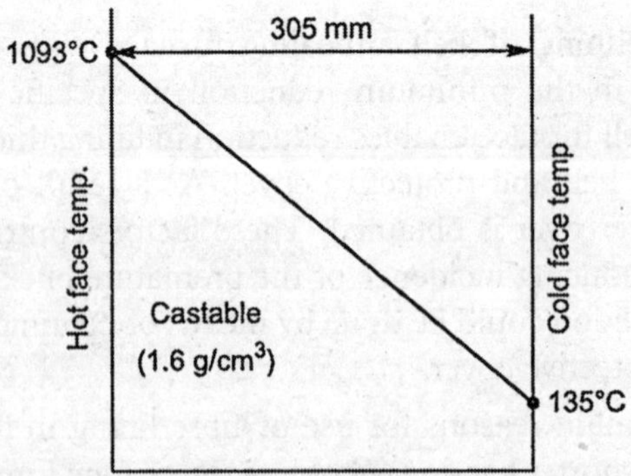

Fig.6.2: Cold face temperature vs. hot face

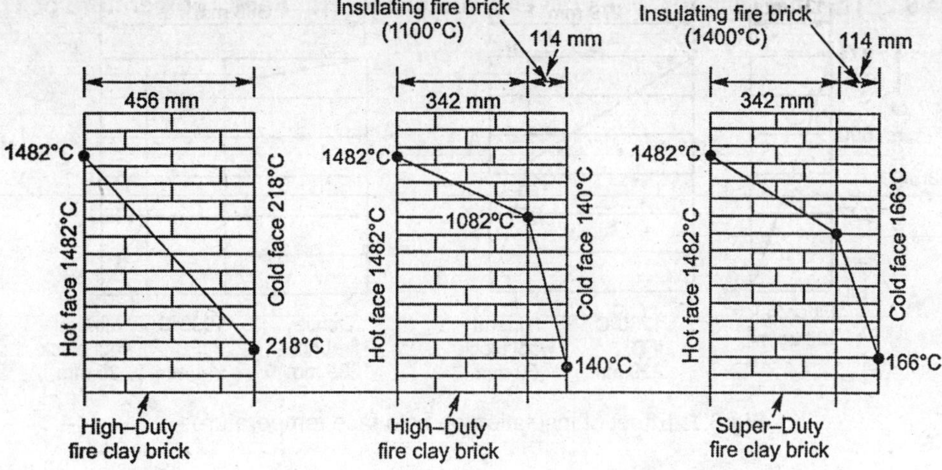

Fig.6.3: Effect of insulating firebricks on cold face temperature for firebricks.

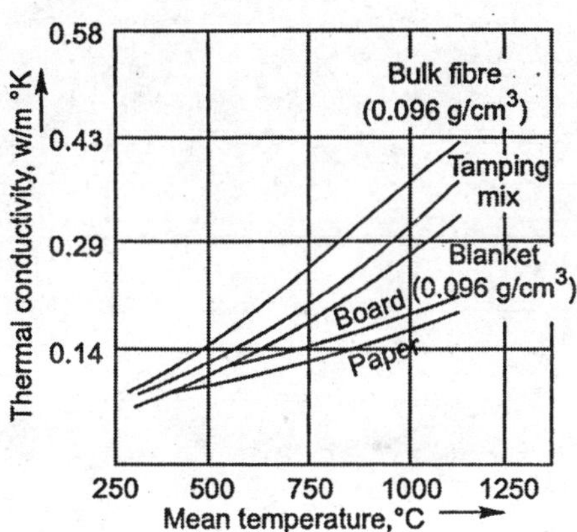

Fig.6.4: Temperature vs. thermal conductivity of ceramic fibres.

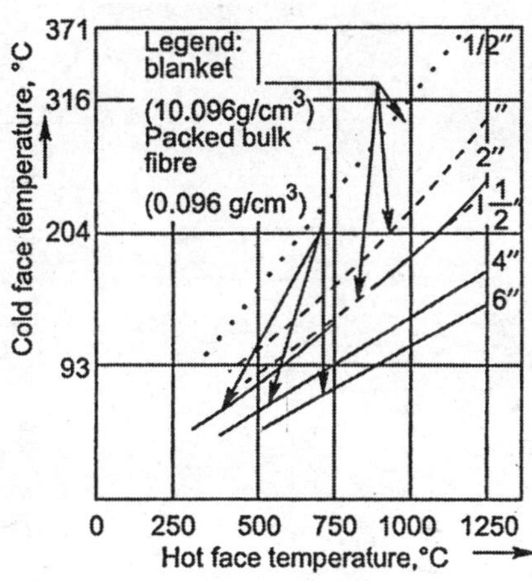

Fig.6.5: Cold face temperature vs. hot face temperature for ceramic fibres.

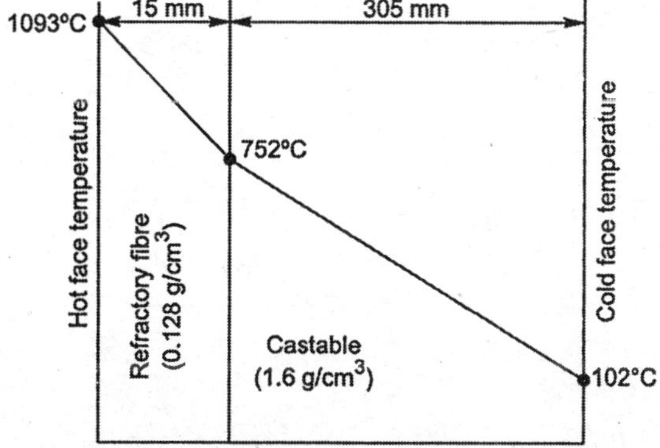

Fig.6.6: Effect of ceramic fibre on cold face temperature of castables.

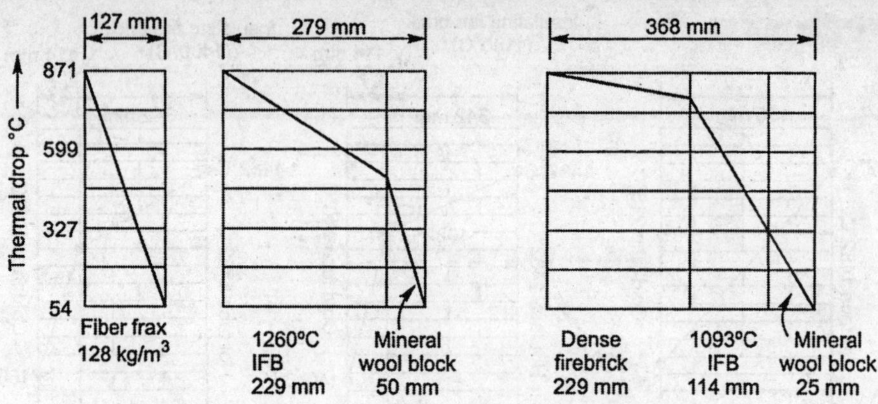

Fig.6.7: Effect of insulation on cold face temperature.

7 Special Refractories

Materials that surpass commonly used refractories in one or more of their essential properties are often required for industrial purposes. Carbon / graphite, silicon carbide, zircon, zirconia, fused cast, fused silica and insulating brick are some of the refractories with extraordinary properties for special applications and hence are termed as special refractories.

Expensive refractory materials used for the construction of crucible, experimental furnaces and other special purposes where their cost is no consideration are termed as special refractories and include high purity refractory oxides, borides, carbides, nitrides, silicides, alundum (mixture of fused alumina and clay), Sialons (Si-Al-O-N group of materials), electrocast refractorics. cermets, sillimanite etc. In fact, some of these materials have not become common due to their manufacturing limitations. Special refractories are mainly used in the furnaces employed for the production of rare & costly metals and also those operating at very high temperature where no contamination of the metal produced can be tolerated. Due to high production cost of special refractories (as they involve special manufacturing processes and purer raw materials), their use for common applications is prohibited. For example, nuclear plants, gas turbines, turbo prop jet engines involving high temperatures require the use of some highly pure oxides having high hardness, strength and thermal spalling resistance.

Zirconia(ZrO_2), Zircon ($ZrSO_4$), Silicon carbide(SiC), thoria (ThO_2), berellia (BeO), alumina (Al_2O_3), Magnesia (MgO), silica (SiO_2) and electrocast refractories of mullite, magnesite and mixture of magnesia, bauxite & chromite are some of the special refractories to mention.

Brief features of some of the special refractories are given below.

CARBON AND GRAPHITE

Carbon refractory is composed of the elemental carbon. Starting materials for the production of carbon refractories are typically the amorphous carbons, e.g., metallurgical coke, petroleum coke, heat treated coal tar pitches etc..

Naturally occurring flake graphite or artificial graphites are sometimes blended with amorphous carbons to achieve a desired thermal conductivity. These materials are combined with high carbon yielding resins or pitch and formed into blocks & slabs. Such shapes are well suited to places where high heat transfer is required, such as areas using water cooled panels.

Its use is limited to applications which are either strongly reducing or where the oxygen content of the atmosphere at a given operating temperature is low enough to prevent appreciable combustion of carbon.

Carbon blocks have been used to line the hearth and bosh of blast furnaces. Carbon blocks have also been used to line the hearth and sidewalls of aluminum reduction pots. Electrodes and anodes used in many industrial applications also are made from carbon.

Carbon is a desirabie element for refractory use because it is not wetted by most molten metals and slags; it has excellent thermal shock resistance; and its strength increases when it becomes heated. Because of its susceptibility to oxidation, however, this refractory should be used under reducing conditions; or efforts should be made to minimize reaction with gaseous oxygen by adding oxidation inhibitors to the shape, such as boron carbide, fine metals (Al, Si, Mg), or by coating the shape with a protective glaze. Metal melting crucibles made of clay-graphite have been used for a considerable number of years as have clay-graphite shapes, such as stopper rods and sleeves. Clay-graphite shapes have been replaced, by alumina- graphite shapes which provide longer service life.

SUPER REFRACTORIES

Highly pure oxides (having very high fusion point) which can be used in contact with molten metals in oxidising conditions are used for making super refractories. Purified oxides is sintered/fused in electric furnaces which is then cooled, crushed and ground to a fine powder. Since pure oxides are non plastic, they are shaped under high pressure and high temperature conditions. Shaping is also done by slip casting, extrusion, high pressure pressing or hot pressing.

Organic plasticisers are also sometimes added to impart plasticity to the mixture. During firing under controlled atmospheric & temperature conditions, the organic material burns off and recrystallisation taking place on heating imparts high pack density to the refractory product.

High purity oxides refractory bricks are very dense and resistant to the chemical action of slags. They are comparatively inert to H_2, N_2, cracked NH_3 and carbonaceous & sulphurous atmospheres besides having high mechanical strength and stability at high temperature. They offer high degree of resistance to corrosion, erosion & abrasion and also they show high electrical resistance at high temperature. Zirconia, thoria, berellia, and oxides of Al, Mg, Ce etc. fall under this category of refractory.

Thoria (which is radioactive) fuses at over 3000°C and is used for making crucible employed for melting some high purity metals. Berellia is used for making crucibles for melting zirconium and beryllium etc. It is also used in carbon resistance furnaces as radiation shields. Alumina is used as an abrasive, in rock cutting drills and in roller bearing used at high temperature. It is also used for making spark plugs, pyrometers, electric tubes, nose pieces for missiles etc. Pure silica is used for making dishes, crucibles, tubes, thermocouple sheaths etc. Magnesia is used for making crucibles for melting platinum and nickel. Zirconia is used for making crucibles (for melting glazes, enamels, special glasses etc.), thermocouple, ignition boats, burner tips etc. It is also used for lining high temperature ceramic kilns and for thermal barrier coatings for jet engines. Zircon refractories are used in aluminium remelting furnaces and boilers.

Monoliths of super refractories produce more corrosion & erosion resistant jointless linings thereby minimising leakage of air & heat and ensuring better controlled atmosphere. Super refractory mortars, refractory cements & patching mixes are used for laying super refractories and also for repair of such worn out linings. Use of monolith reduces overall installation cost by avoiding manufacture of and skilled workmanship for laying expensive special shapes.

SILICON CARBIDE (SiC) REFRACTORIES

Silicon carbide, a major component in this special refractory class, is produced by reacting silica sand and coke at temperatures above 2000°C. Various grades of silicon carbide bricks ranging from 90 to 98 % SiC are available. Silicon carbide by itself is extremely inert. Under normal atmospheric conditions, it will not self bond, even at high temperatures. Various schemes have been developed to bond silicon carbide using clay, silica, metallic powders and molten silicon.

Clay bonded SiC refractories are made by adding crude clay to silicon carbide grain and firing the shape to sufficient temperature to vitrify the clay & produce a glassy bond. These shapes are used with success in hot, abrasion prone applications with temperatures under 1427°C. Their usefulness is limited by the refractoriness of the clay bond.

Higher strength bonds can be achieved by bonding silicon carbide with a nitride phase or by self-bonding silicon carbide grains with secondary in situ formed silicon carbide. Nitride bonds are formed by adding fine silicon powder to SiC, forming a shape and heating the shape in a nitrogen atmosphere to above 1205°C. Gaseous nitrogen reacts with the dispersed silicon phase and forms silicon nitride crystals which readily bond to the surface of the silicon carbide aggregate. To produce a silicon oxynitride bond, an oxygen source (typically silica) is added to the starting materials. To form a sialon bond, alumina is added to the starting mixture. The various nitride phases all possess non-wetting properties, relatively low thermal expansion and high strength. The selection of an appropriate nitride bond is dependent upon the degree of oxidation in the service environment. Generally, the sialon bond possesses the highest degree of oxidation resistance. Nitride-bonded SiC-shapes are used in aluminum melting & refining applications, as well as in the bosh & lower in wall of blast furnaces.

Self-bonded SiC refractories are made by first forming a shape using conventional ceramic binders and then firing the shape at very high temperatures 2000 to 2250°C in an atmosphere of silicon vapors. Fine carbon placed in the starting materials reacts with the silicon vapors to form a silicon carbide secondary phase. Such shapes have a "recrystallized" appearance and are readily distinguishable. Self-bonded SiC is typically used in heating elements and structural supports.

Important properties of silicon carbide refractories (which contains > 85% SiC) which make them suitable for use in ceramic, metal working and process industries are their:

- high thermal conductivity and high refractoriness (decomposes at 2200°C and does not melt).
- high thermal spalling resistance and temperature load bearing capacity.
- excellent mechanical properties and chemical inertness particularly to acid slags.
- light weight (true sp.gr. = 3.1).

These refractories are used for nozzles for rocket motors, nose pieces for rockets, skids of reheating furnaces, electrical resistor bars etc.

ZIRCON

Zircon is a silicate of zirconium ($ZrSiO_4$) containing about 67% zirconia and 33% silica. Zircon refractories are made by blending beneficiated zircon sands, milled zircon sands & a plasticizer or temporary binder, forming a desired shape, and firing to an elevated temperature. The firing temperature of zircon is limited to temperatures beneath 1500 to 1650°C, because within this temperature range zircon dissociates.

The actual temperature at which zircon dissociates into zirconia and silica is influenced by mineralizers, such as alkalies & fluorides. Brick consisting essentially of zircon are typically made by a forming process called impact pressing. This process uses a rapid series of air hammer impacts to compress the mix into simple shapes, such as rectangles. Other, more intricate shapes are made by air ramming or vibration casting. The main advantages of zircon refractories are their relative chemical inertness against acidic slags and their good thermal shock resistance.

ZIRCONIA

Zirconia (ZrO) is usually obtained through a chemical process involving zircon or by fusing zircon with coke in an electric furnace. Zirconia has excellent chemical inertness. It has an extremely high melting point of 2680°C. Use of zirconia is limited, because of two major drawbacks, its high cost and its tendency to change crystal form upon heating.

Zirconia can occur in three polymorphs; monoclinic, tetragonal, and cubic. The typical room temperature phase is monoclinic which is stable to 1160°C to 1190°C upon heating. Heating through the monoclinic-tetragonal transition causes a volume contraction; cooling through the transition causes a volume expansion. The phase change transition through cooling is instantaneous and results in spontaneous failure of the zirconia crystal: This failure results in the cracking and/or disintegration of the refractory shape.

The cracking problem can be overcome using either of two contrasting approaches. One is to mill monoclinic zirconia to a fine size (less than one micron) and disperse it within a refractory body so that destructive micro cracking is avoided. The dispersed phase works as a stress absorber as energy is absorbed by zirconia to convert from one phase to another. In this way, the dispersed phase is said to produce a "toughening" effect.

The other approach is to stabilize the cubic structure with lime or magnesia by heating zirconia with one of these oxides within the temperature range 1500°C to 1700°C. The cubic form of zirconia has a uniform thermal expansion, whereas thermal expansion of the other polymorphs reflect volume changes which occur upon heating. A disadvantage of stabilized zirconia is its tendency to thermally age. That is, the stabilizer tends to migrate out of the structure when the material is exposed to long term temperatures within 800 to 1400°C range.

Due to its high cost, zirconia refractories are only used in critical applications, such as metering nozzles used in continuous casting and inserts in the bore area of slide gates. In these applications, control of the bore diameter during casting is vital. Some zirconia is used to make crucibles for refining special alloys where purity of the molten metal is of concern. However, the main use of zirconia in the refractories industry, is as an additive to increase the thermal shock or slag resistance of the refractory.

SIALON REFRACTORIES

These are ceramic products of oxide and monoxide compounds which combine the properties of high density and high strength. A solid solution of alumina in silicon nitride from which its name Si-Al-O-N (Sialon) has been derived, is a specific example of this group of ceramic material. Because of their (Sialons) poor sintering characteristics, hot pressing at a pressure of 18-30 MPa (180—300 kgf/cm^2) and 1700 - 1760°C in graphite moulds is done to get a low porosity dense product. Hot pressing of a mixture of silica and aluminium in a nitrogen atmosphere at 1400°C also produces 'Sialon' which is a solid solution of AlN and alpha- Al_2O_3 in beta- Si_3N_4. It can also be produced by reaction sintering at 1700-1800°C.

Sialons have good resistance to oxidation and action of molten metals like Al, Zn, Cu, Fe & steel. They are not affected by H_2SO_4, HCl, borax & alkalis and hence are suitable for storage & transportation of these materials.

CERMETS

Cermets are combination of a metal or an alloy and non metal e.g. an oxide, carbide, nitride or boride. Cermets are produced from the mixture of ceramics (*e.g.* oxides, carbide, borides etc.) and metals or alloys (ceramic + metal = cermet). Cermets combine the good properties of both the ceramics (which are brittle, have low thermal shock resistance and high strength at high temperatures) and the metals or alloys (which are ductile but loose strength when heated). Ceramic materials used in the production of cermets are generally high purity oxides (e.g. Al_2O_3, Cr_2O_3, SiO_2, MgO, BeO, ZrO, UO_2 etc.) or high purity carbides (e.g. that of Si, Ti, Tu etc.) or borides (*e.g.* that of Al, Ti, Zr etc.). Metal phase used in the production of cermets may be Al, Be, Cr, CO, Fe, Ni, Mn, Stainless steel etc.

Cermets are produced by processing of the powders of ceramic phase & metallic phase to compact shapes by pressing and sintering as is done in powder metallurgy. Metal phase acts as a bonding material for ceramic phase grains. High strength and resistance to high temperatures make them suitable for use in very special and more recent applications like those in space vehicles, missiles and nuclear power plants.

Cermets being very hard and strong are used for the fabrication of strong wear resistant tools like rock boring drill bits, cutting off wheels and parts of space vehicles & missiles which require high strength at high temperature and under high wear conditions. This group of materials are examplified by TiC-Ni & WC-Co compositions having high temperature strength, high hardness, toughness and also Al_2O_3-Cr compositions having high temperature properties & resistance to thermal fatigue. Oxide cermets are used in space vehicles, missiles etc. while carbide cermets are suitable for cutting and boring tools. Uranium oxide dispersed in a matrix of stainless steel can be easily moulded into various shapes making it suitable as a nuclear power reactor fuel. Moreover, it has good mechanical strength at high temperature, good stability under irradiation conditions and good corrosion resistance under pressurised water environments.

FUSED AND ELECTROCAST REFRACTORIES

Fused cast refractories is formed by melting refractory compounds in a fusion furnace and casting the liquid melt into a simple shape, such as a block. Special attention is needed to the cooling rate of the melt to prevent cracking of the shape and localized defects, such as shrinkage cavities.

The advantage of using fused cast refractories to contain molten metal or slag is their lack of interconnected porosity, a feature inherent to sintered refractories. The absence of open porosity enables this type of refractory to resist corrosion and infiltration of corrosive agents. The main disadvantage of fused cast refractories is their great sensitivity to thermal fluctuations. The sudden temperature changes which occur in many applications are simply too rapid to prevent cracking or shattering of fused cast refractories.

Fused cast refractories are normally available in the following compositions: alumina, alumina-zirconia-silica, alumina-silica, magnesite-chrome, zircon and spinel.

Fused cast alumina is primarily used is in refining and superstructure areas of glass tanks. The predominant type of fused cast is alumina-zirconia-silica. It is used in melting and superstructure areas. Fused cast magnesite -chrome refractories are generally composed of 50 to 60% dead burned magnesite and 40 to 50% chrome ore. They offer excellent resistance to fluid corrosive basic slags; however, their relatively high cost and the development of suitable alternative refractories at lower cost has limited their wide-spread usage.

Fusion cast refractories are produced by fusing the desired mixture of raw materials in an electric arc furnace and then moulding the required shape in graphite moulds by pouring the molten mass into the mould. Cooling should be slow and done in about a week's time. Annealing of the finished product is done to get better and uniform properties. Alumina-silica, alumina- zirconia, chrome-alumina and alumina are the common raw materials for producing fusion cast refractories.

Since, complete mullitisation does not occur during firing of alumino-silicate refractories at high temperature without actual fusion of the whole mass, the fused mullite refractories produced from pure silica and alumina are superior to them in many physical & chemical properties. Electrocast mullite refractories are neutral, retain their rigidity & strength at high temperatures upto 1800°C, have high thermal spalling resistance and withstand the attack of molten metals & slags. Fused mullite refractories are used for lining the ladles used in high temperature steel melting operations, cupola slag tap holes, reheating furnace hearths, open hearth slag pocket arches, upper section of regenerator checkers and those parts of reheating, annealing & heat treatment furnaces which are subjected to thermal shock. Electrocast mullite is also crushed/ground and mixture of its various sizes is rebonded to various standard shapes like saggars, burner blocks, grooved heating element carrier shapes, tiles, tubes etc. used for special purposes.

Electrocast alumina is produced by fusion of high quality alumina (containing > 99% Al_2O_3) or bauxite at 2000°C. It has high refractoriness, RUL, abrasion resistance, chemical stability and inertness to highly reducing & oxidising conditions. They are used in lining carbon black reactors, domes of chemical reactors in fertiliser plants, reactors in gasification of petroleum feedstock and other special units like combustion chambers in acetylene production, catalytic cracking & reforming units in refineries etc. where high pressure, temperature & highly reducing conditions (including the presence of hot hydrogen) are encountered. 90% alumina electrocast refractories are used in the rotary kilns, L.D. converters, blast furnace stacks, vacuum degassing chambers of special alloy steels and in the hearth construction of slab/billet reheating furnaces.

Fused corundum and mullite-corrundum are specially used in slide gate plates for steel teeming laddles. Plates impregnated with bakelite & heat treated,, ground and polished have open porosity of <7%. They are highly wear resistant and have high strength & heat resistance and low open porosity. Mullite-corundum refractories are also used in the equipment for refining and vacuum treatment of steel. Alumino-silicate castables are used for lining ascension pipe and doors of coke ovens.

Electrocast zirconia, magnesite and corundum are used for making crucibles of vacuum induction furnaces for melting precision alloys as they are highly resistant to molten metals. Electrocast zirconia in particular undergoes less reduction with the metals or less contamination by oxygen than refractories of Al_2O_3, MgO, BeO etc. when melting heat resistant steel/alloys in vacuum.

Basic fusion cast refractories of Al_2O_3, Cr_2O_3, MgO (with some Fe_2O_3) have low open porosity (0.7-1.8%), high C.C.S. (1050 kg/cm^2) and high RUL (1750°C under a load of 2 kg/cm^2). These are used for lining such parts which are liable to wear in absence of thermal shock viz. slag belt, hearth or bottom of continuous melting furnaces used for sulphide materials. Fused magnesia is now increasingly used in slide gates for steel teeming.

Fused/electrocast refractories are consumed mainly in modern glass tank furnaces, skid rails in steel slab reheating furnaces, slide gates for steel teeming laddles and in cyclones/ducts carrying dust laden particles at high velocity etc.

Hot face insulating linings of high purity alumina/fused mullite are used where very high temperature & highly reducing conditions are involved and where presence of iron or silica is harmful. Such light weight insulating refractories are used in high temperature laboratory furnaces for heating element carrier blocks and in kilns involving high firing temperature e.g. those used for firing spark plugs & other electrocast ceramics.

Fused silica refractory is produced by electrically heating quartzite (SiO_2) with a purity of at least 98% silica in a fusion furnace. In this process, the crystalline nature of the silica is transformed into an amorphous structure by rapid quenching from a molten condition.

Because of the extremely low thermal expansion of fused (amorphous) silica, this material has excellent thermal shock resistance. Fused silica also has excellent corrosion resistance in acidic media, such as strong acids. Fused silica powders are used in the electronics industry as resin extenders due to their excellent electrical insulating property. Fused silica filled resins are used to encapsulate electronic components to protect them from the environment.

Fused silica grains are classified into various sizes and formulated into a slip mix for casting into plaster molds. Using this technique, many intricate, complex shapes can be made. These shapes are used as coke oven doors, shroud tubes, glass-tank refractories nonferrous troughs & spouts, and as linings for chemical reactors. Fused silica shapes can be used in constant temperatures up to 1650°C and cyclical temperatures up to 1094°C. Beyond 1094°C, amorphous silica devitrifies into cristobalite which undergoes a volume expansion when heated, displaying a significantly higher thermal expansion than fused silica. Because of these properties, devitrified fused silica is not considered volume or thermally stable when heated to elevated temperatures.

INSULATING BRICK

Insulating brick are made from a variety of oxides, most commonly fireclay or silica. The desirable features of these brick are their light weight and low thermal conductivity, which usually results from a high degree of porosity. The high porosity of the brick is created during manufacturing by adding a fine organic material to the mix, such as sawdust. During firing, the organic addition burns out, creating internal porosity. Another way to accomplish high porosity involves the addition of a foaming agent to slip. Using this approach, insulating brick can be cast instead of dry pressed. Additions of lightweight aggregates like diatomite, haydite, etc., is another approach.

Because of their high porosity, insulating brick inherently have lower thermal conductivity and lower heat capacity than other refractory materials.

ASTM classifies fireclay and high- alumina insulating refractories in the following sequence: 16, 20, 23, 26, 28, 30 and 33. These numbers multiplied by 100 represent the nominal service temperature in degrees Fahrenheit to which the refractory can be exposed in service. Products numbered from 16 to 26 are made from a fireclay base and products numbered from 28 to 33 are made from a high-alumina base.

Typically, insulating refractories are used as backup materials, but they can also be used as working linings of furnaces where abrasion & wear by aggressive slag and molten metal are not a concern.

Where they can be used, insulating materials offer several distinct advantages:

- Savings in fuel cost due to decreased heat losses through the furnace lining and less heat loss to the refractory.
- Faster heat up of the lining due to the insulating effect and lower heat capacity of the insulating refractory.
- Thinner furnace wall construction to obtain a desired thermal profile.
- Less furnace weight due to the lower weight of the insulating refractory.

A variety of insulating brick provide a range of thermal efficiencies and strengths. By composition & property characteristics, lightweight insulating silica brick are similar to conventional silica brick with the exception of density and porosity. They have a maximum service limit of 1650°C and are used in the crowns of glass furnaces and tunnel kilns. Insulating brick based on fireclay aggregate are also available with a combination of high strength and low thermal conductivity. These brick offer a maximum service limit in the range of 1150 to 1260 °C. They are primarily used in rotary cement kilns and glass tanks.

For even higher temperature & corrosive applications, lightweight, insulating 90 & 99% alumina brick and alumina-chrome bricks are also made.

Refractory Cements/ Mortars, Castables and Ramming Masses

REFRACTORY CEMENTS/MORTARS

Masonry built of refractory brick consists of many relatively small units. The strength of the masonry depends upon the strength of the individual brick, the manner in which they are laid together and the nature of the mortar material used in the joints. The purpose of the mortar is to fill the joints and bond the individual brick together. It should protect the joints from attack by slag & other fluxes and provide resistance to infiltration by cold air & to the outward flow of gases.

Mortar material should be selected as carefully as the brick with which it is to be used. Poorly made joints, or joints filled with improper material, may greatly shorten the life of a refractory structure.

Refractory mortars/cements are refractory binders having high fusion temperatures (such as magnesite, zircon, silica, bauxite, fireclay or carbon base) and are used as joining materials for laying refractories. They should be able to develop strong bonds between the refractories and resist the action of prevailing furnace atmosphere similar to the main refractories used in the furnace. Depending on whether they are to be used for laying acidic, basic or neutral refractories, refractory cements/mortars are accordingly selected. Powdered quartzite mixed with oxalic acid usually produce acid proof mortars. Dry mortars are made into paste by adding water just before use. Mortar should not shrink much on heating, otherwise it will result in loosening of refractory joints.

Refractory mortars for basic, fireclay or silica refractory laying are made by mixing powdered used refractories (either basic, fireclay, quartzite base silica, sand or quartzite lumps), kyanite, plastic clay and additives. Similarly, bauxite based mortars (used for high alumina refractory laying) are made by mixing powdered bauxite and magnesite, firebrick grog, sodium silicate and other additives. Graphite or coke based mortars containing upto 25% fireclay are used for laying carbon refractories. *Silica cement* (used for laying silica refractories) is made by mixing powdered silica grog and plastic fireclay which on mixing with water produces a good slip. *Magnesium cement* (used for laying magnesite refractory) is made by mixing crystalline magnesia, magnesium chloride/sulphate and lightly burnt magnesia. Neutral chrome cement can be used for laying any type of refractory. It can also replace fireclay cement and eliminate shrinkage gaps where thick joints are unavoidable. Zircon based mortars are for acid and zircon base refractories linings. Aluminous cements (used for laying fireclay, super duty or high alumina refractories) contain mainly calcium aluminates with varying amounts of $Si0_2$ and Fe_20_3. It contains usually equal proportions of CaO and $A1_20_3$ (about 40%) with upto a maximum of 33% $Si0_2$ and ferrous & ferric oxide. High aluminous cement contains upto 70-90% alumina, 1% SiO_2 and iron oxides. Aluminous cements are also used for giving thin coating to the inner surface of the walls facing high working temperatures.

The *refractory binders* may be air, chemical or heat setting. Air setting mortars develop strength on drying due to the presence of sodium silicate which is added in minimum quantity just sufficient to develop air setting properties because it reduces the refractoriness of the mortars. *Heat setting mortars* develop strength only on heating to high temperature when vitrification and boiling occurs. Additives present in most of the mortars/cements result in cold setting strength in the refractory structure and a permanent strong bond is developed on heating the structure to high temperatures. *Air setting cements* are better than heat setting/ceramic cements because good bond strength is developed as the heating proceeds except that its strength drops near its fusion temperature.

Refractory mortar materials are divided into two classes: heat-setting mortars and air-setting mortars.

Most heat-setting mortars require relatively high temperatures to develop a ceramic set, in contrast with air-setting mortars which take a rigid set merely upon drying. Phosphate-bonded mortars develop a chemical bond at lower temperatures. Temperatures in excess of 370°C are necessary to permit formation of more stable phosphate bonds which are less susceptible to rehydration in high moisture conditions. Included in each of these groups are materials of various compositions for use in specific applications.

Among the factors included are workability, plasticity, water retention, fineness of grind, drying & firing shrinkages, chemical composition, refractoriness, cold & hot bonding strengths, vitrification temperature and resistance to chemical attack.

With excellent workability and water retention over a range of consistencies, a mortar can be used for dipped or trowelled joints, as a surface coating for walls, or for patching. The mortar should not shrink excessively upon drying or heating, nor should it overfire and become vesicular at the maximum service temperature. The thermal expansion of the mortar should be approximately the same as that of the brick with which it is used; otherwise, temperature changes will affect the bond between brick & mortar and cause surface coatings to crack or peel. If strong joints are needed, the mortar material must be affected sufficiently by the heat to develop a strong ceramic bond.

However, the refractoriness of the mortar must be high enough to resist melting or flowing from joints at high temperatures. In some cases, there must be adequate chemical reaction between brick and mortar to develop a strong bond between them, but in no case should the chemical reaction be sufficient to damage the brick. In many types of service, it is essential that joint material be highly resistant to chemical attack by the furnace charge, slag, dust, volatized fluxes or gases; and for certain uses the mortar material should not discolor nor otherwise contaminate the material being processed in the furnace. Mortars which do not develop a strong bond are often desirable for use in laying brick walls which are alternately subjected to soaking heat and cooling cycles.

Mortar types

Fireclay Mortars. Air-setting mortars containing a mixture of high fired, fireclay and high-alumina calcines & smooth working plastic clays are recommended for use in laying high-alumina brick in the 50 to 70 % range, as well as insulating brick. Mortars of this kind are available in a wet or dry form.Other air-setting mortars are available with high refractoriness, excellent intermediate temperature strength and smooth working properties.

High Alumina Mortars. Heat-setting mortars with very high refractoriness, volume stability,& resistance to attack by molten metal or slag are used in laying high-alumina and superduty fireclay brick in various applications, especially those where resistance to ferrous slags is required.

These mortars can be dipped or trowelled.

High-alumina air-setting mortars are used in applications up to 1760°C. They have high refractoriness and excellent resistance to attack by corrosive slags.

Phosphate-bonded mortars with high refractoriness & smooth working properties are used for laying high-alumina brick in a variety of applications.

Heat-setting mortars based on high purity tabular alumina calcines are available for use up to 1870°C. These mortars have exceptional stability & load bearing ability at high temperatures and are highly resistant to corrosion by volatile alkalies & slags in all types of furnaces. They are typically used for laying brick in the 90% alumina class.

Phosphate-bonded alumina-chrome mortars generate high bond strengths and show excellent resistance to corrosion by ferrous & nonferrous metals. Mortars of this kind are used when laying alumina-chrome brick, brush coating over refractory walls, or other applications where strong bonded joints and resistance to slag or metal penetration are desired.

Basic Mortars

Dry, air-setting mortars with a chrome ore base have excellent resistance to corrosive slags and fumes in chemical applications. They are used for laying basic brick of all types, but can also be used as a neutral layer between basic and acid brick.

Mortars which contain high quantities of penetration & corrosion inhibitors have high resistance to corrosive slags and are used for laying all types of basic brick, as well as some high-alumina compositions where slag attack & corrosion are especially damaging.

Mortars based on high purity magnesite are used for refractories with high MgO content. These are usually dry and may be used with other types of basic brick.

MONOLITHIC REFRACTORIES

Monolithic or monolith-forming refractories are special mixes or blends of dry granular or cohesive plastic materials used to form virtually joint free linings. They represent a wide range of mineral compositions and vary greatly in their physical and chemical properties. Some are relatively low in refractoriness, while others approach high purity brick compositions in their ability to withstand severe environments.

Monolithic refractories are rapidly replacing the conventional type fired refractories in many applications including industrial furnaces. The main advantages of monolithics are:

- Elimination of joints which is an inherent weakness.
- Faster application method.
- Special skill for installation not required.
- Ease of transportation and handling.
- Better scope to reduce downtime for repairs.
- Considerable scope to reduce inventory and eliminate special shapes.

- Heat saving ensured.
- Better spalling resistance.
- Greater volume stability.

Monolithic refractories are used to advantage over brick construction in various types of furnaces. Their use promotes quick installation, avoiding delays for the manufacture of special brick shapes. Using monolithics frequently eliminates difficult brick laying tasks, which may be accompanied by weakness in construction. They are of major importance in the maintenance of furnaces because substantial repairs can be made with a minimum loss of time and, in some cases, even during operations. Under certain conditions, monolithic linings of the same chemical composition as firebrick provide better insulation, lower permeability and improved resistance to the spalling effects of thermal shock.

Monolithic refractories are packaged in suitable containers for convenience in handling and shipping. With little or no preparation, they can be applied to form monolithic or joint-free furnace linings in new construction, or to repair existing refractory masonry.

Common usage divides monolithic refractories into the following groups:

- Castables
- Plastic Refractories
- Ramming Mixes
- Gunning Mixes

Castables consist of graded dry refractory aggregates combined with a suitable hydraulic activated bonding agent. Castables are furnished dry and form a strong cold set upon mixing with water. They are usually poured or cast in much the same manner as ordinary concrete, but are sometimes vibrated, trowelled, rammed or tamped into place, or applied with air placement guns. They form strong monolithic linings, possessing combinations of properties that make them ideal for many applications.

Plastic refractories are mixtures of refractory aggregates and cohesive clays, prepared in stiff plastic condition at the proper consistency for use without further preparation. They are generally rammed into place with pneumatic hammers, but may also be pounded into place with a mallet.

Ramming mixes consist essentially of ground refractory aggregates, with a semi-plastic bonding matrix which can be purchased ready-to-use or prepared by adding water in the mixer at the construction site. Ramming mixes are placed with pneumatic hammers in 2.5 - 3.75 cm layers. They supply a denser, stronger refractory body than plastics, but need some sort of form to restrain them when rammed.

Gunning mixes consist of graded refractory aggregate and a bonding compound, and may contain plasticizing agents to increase their stickiness when pneumatically placed onto a furnace wall. Typically, gunning mixes are supplied dry. To use, they are predamped in a batch mixer, then continuously fed into a gun. Water is added to the mix at the nozzle to reach the proper consistency.

Monolithics are put into place using various methods, such as ramming, casting, gunniting, spraying, and sand slinging. Ramming requires proper tools and is mostly used in cold applications where proper consolidation of the material is important. Ramming is also used for air setting and heat setting materials. Because calcium aluminate cement is the binder, it will have to be stored properly to prevent moisture absorption. Its strength starts deteriorating after 6 to 12 months.

In recent years, new installation technologies have been developed for refractories that parallel that of Portland cement. New equipment and improved refractory materials have led to new installation techniques like self leveling castables and pumping & wet spraying of refractory mixes. Pumping low cement and ultra low cement castables ina variety of applications is now an accepted installation practice. Wet spraying is different than gunning since this is a wet castable pneumatically applied with a set activator added at the nozzle. In gunning, dry material is conveyed to the nozzle and water added at the nozzle creating the wet mixture as it is applied to the wall. The great upsurge in the use of these installation techniques is sufficient evidence of its utility & effectiveness and its acceptance by all segments of the pyro- processing industries as an important methods for the placement of refractory concretes. The discussion above suggests the manner in which each class of monolithic refractory is most commonly installed. However, material of one group may be installed by a technique more common to another group. Specially designed plastics are sometimes gunned, as are many castables and ramming mixes. Gunning mixes are often cast or trowelled.

Monolithic refractories are single piece casts in the shape of equipment, such as a ladle as shown in Figure 8.1.

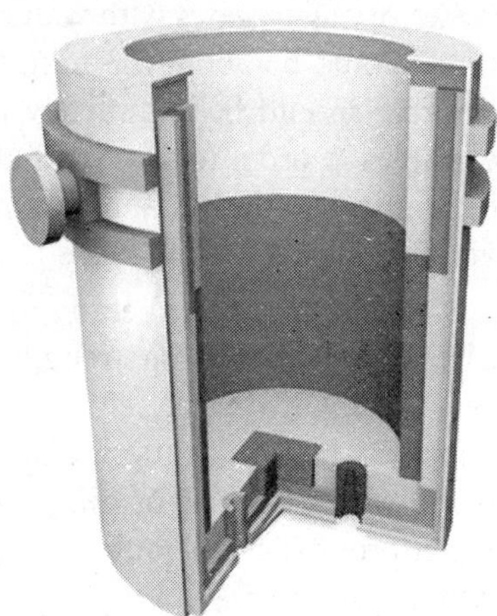

Fig.8.1: A Monolithic refractory Lining for Ladle.

When air-setting or hydraulic activated monolithic refractories are used, the entire thickness of a lining becomes hard and strong at atmospheric temperatures. The strength can be somewhat lower through the intermediate temperature range, but increases at higher temperatures with the development of a ceramic bond.

Heat-setting monolithic refractories have very low cold strength and depend on relatively high temperatures to develop a ceramic bond. In the case of a furnace wall having the usual temperature drop across its thickness, the temperature in the cooler part is usually not enough to develop a ceramic bond. However, with the use of a suitable insulating material as backup, the temperature of the lining can be high enough to develop a ceramic bond throughout its entire thickness.

When monolithic linings are used as the primary furnace lining, they are usually held in place with either ceramic or high temperature steel anchors. Each method of anchoring has advantages, depending upon furnace conditions and installation technique.

CASTABLES

Castable is an *unshaped / monolithic refractory* like ramming masses. It is dry mix hydraulic composition of refractory aggregates with a suitable binder which on addition of water produces a concrete like mass which forms useful castables. Castables have the highest demand out of all types of monolithics. Heating to high temperature develops ceramic bonding in cast shapes. Refractory aggregates (which do not undergo structural changes on heating) used for making castables comprise bauxite, grog of fireclay, sillimanite and other alumina materials or chromite etc. Refractory castables are used for producing monolithic linings beside patch work. The use of castables avoids storage & inventories of large number of complicated shapes and minimises the number of joints in the structure thereby increasing the lining life & reducing the maintenance jobs.

The *unshaped refractories / monolithics* like castables have following advantages over shaped refractories:

- It facilitates jointless construction with elimination of the weakest area of construction *viz.* brick joints.
- It requires reduced inventory and hence inventory control is easier.
- It is easier to install, leading to lower cost of construction.
- It facilitates easy repair of the areas of wear and tear for maintenance resulting in prolonged life of refractory lining/construction.

Castables should be mixed thoroughly in their dry form before use in order to avoid segregation of sized grains during transportation. There is a restricted shelf life for castables as the binder used in these products is of hydraulic setting type. Mechanical mixing (using a paddle mixer) of dry refractory aggregates with water produces castable with higher cold crushing strength. The mixing time should be about 2-3 minutes and right quantity of water should be used. The water mixed castable should be completely laid within 15-30 minutes after water has been added to the dry castables in order to avoid initial setting. The castables may be cast by *pouring, troweling* or *gunning* depending on the suitability of the purpose. It may also be applied by *tamping / ramming.* Use of vibrator during casting helps in removal of entrapped air.

There should not be any loss of water from the mixed castable during its installation till the casting sets because the strength of the castable is developed due to an exothermic chemical reaction between the hydraulic setting high alumina cement and water used during mixing causing a rise in temperature. The shuttering for the forms to be used for laying the castables is normally made of wood with a coating of oil, paraffin or metal sheet for making the forms impervious to water *i.e.* water proof. There should not be any foreign matter like paper, nail, wood etc. within the forms in which castables is to be laid. During repairs, the old, used and existing refractory surface should be made non absorbing by spraying sufficient water to saturate the refractory lining on which the fresh castable is to be applied.

Both metallic (*e.g.* alloy steel) and refractory anchors to withstand the temperature and other conditions at the place of application of castables are in practice. The surface of castables should not be allowed to dry during first 24 hours after installation (by regular water spray) and not fast during subsequent hours. The surface of the installed castable layer should never be finished smooth. Small holes (called weeping holes) are provided in the construction, in order to facilitate drying from the interior of the thickness of the castable layer. Although complete drying out is impossible at ambient temperature, the drying is completed during first heating up schedule at a very slow rate. The first heating of castable lining is done at a rate of 30°C/hour. The temperature of 110°C is maintained for 1/2 hour for each 25 mm thickness of castable layer. The temperature is then raised at the rate of 50-60°C/hour and again held at 220°C for about 1/2 hour. The temperature may even be maintained at 500 and 1000°C, if required. Finally the rate of heating is 100°C per hour beyond 1000°C upto the operating temperature. Castables undergo appreciable reduction in strength in the temperature range of 600-1100°C which can be checked by introducing stainless steel fibres (of

0.125-1.5 mm dia and 5-75 mm long) in the castable mix to the extent of 3% by volume. Drop in strength at around 700-800°C can also be eliminated by the use of powder aluminium phosphate upto 5% either separately or along with hydraulic setting high alumina cement. Depending upon the expansion and contraction characteristics of the castables at the working temperature, facilities for the expansion & contraction are provided. The expansion joints are provided by introducing combustible materials like wooden planks, paper cardboard, plastic sheets etc. at a different intervals. Similarly specific openings (called cut joints) are left out at definite intervals for adjusting the contraction of the castables.

Castables are generally referred to as refractory concretes. They are available in a wide variety of base materials and typically consist of a refractory aggregate, special purpose additives and a cement binder. The bonding systems used are often used to classify the types of castables into four categories: conventional, low cement, ultra-low cement and lime-free castables.

Conventional castables have a cement-bonded matrix where, typically, a calcium-aluminate type of cement fills in the spaces between aggregates. This kind of castable is the most versatile for placement purposes in that normally it can be poured, vibrated, rammed or gunned into place while maintaining its designed properties.

Low cement castables are materials with lime contents of roughly between 1 to 3%. High densities & strength are achieved by careful particle packing and the use of additives to reduce the water needed to cast.

Ultra-low cement castables contain from 0.2 to 0.8% lime. Like low cement castables, they consist of sized particles to achieve maximum particle packing. Because of the low cement content, these mixes are not usually as strong in the low to intermediate temperature range as other types of castables; but they tend to have higher hot strength and refractoriness compared to chemically similar mixes with conventional or low cement bonds.

Lime-free castables have been developed with bonding systems containing no cement. These castables have desirable properties for use in certain chemical applications and where the highest possible hot strength and high temperature load resistance is required, such as metallurgical operations and other high temperature furnace applications. Some of these materials can approach the properties of pressed and fired brick.

Fireclay Castables. Mixtures of high fired fireclay with a refractory cement binder are designed to impart high initial strength and maintain good intermediate temperature strength. At high temperatures, a strong ceramic bond forms, providing good strength throughout their working temperature range. These castables are used in many applications, including boiler furnace ash pits, piers, hoppers, annealing furnace tops, tunnel kiln car bottoms, flues, stacks, linings for chain sections in rotary kilns and sub bottoms of various types of furnaces.

Use of higher purity aggregates and higher purity cements can create castables with additional refractoriness. Primary uses for these castables include rotary cement and lime kilns in sections other than the burning zone.

Other types of fireclay castables include high purity conventional types developed for high strength and abrasion resistance. These are all purpose castables for applications up to 1538°C. In the petrochemical & mineral processing industries, fireclay castables designed for extreme abrasion applications in the intermediate temperature range, have been developed. They have outstanding intermediate strength and abrasion resistance.

Low cement fireclay castables are used where high strength and refractoriness are needed. The lower lime content and porosity compared to conventional fireclay castables make them ideal for many applications.

High Alumina Castables. It consists of accurately sized high alumina aggregates with low iron refractory cements. This mixture provides good all purpose castables with service temperature limits of 1650°C.

Upgraded high-alumina conventional castables based on low alkali, high purity alumina-silica aggregate & super purity cements offer better high temperature strengths and more total refractoriness than castables made from lower purity cements. These products have a wide range of uses up to 1705°C.

High purity alumina-bonded conventional castables with super high purity cement typically have extremely high refractoriness and chemical resistance up to 1816°C. These types are extremely low in silica content, making them quite effective where silica could react with furnace constituents. This form of castable is used in many severe abrasion and chemically corrosive applications.

Low cement high-alumina castables with excellent intermediate to high temperature properties are another alternative in this category. When properly vibrated into place, they provide high density, strength and excellent abrasion resistance. The lower lime content provides good chemical resistance to furnace atmospheres that can attack lime.

Alumina contents of 60 to 70% are typical of many low cement castables and are used for applications up 1705°C, such as kiln floors, doors, car tops, cement kiln coolers, and precast shapes.

High-alumina castables also include 85% alumina low cement castables for use up to 1760°C. They are chemically similar to 85% phos-bonded plastics, but can develop greater intermediate strengths. Their uses include rotary kiln lifters and cooler curbs.

Low cement castables based on high purity aluminas with alumina contents from 90 to 98% are also used. Because of their high purity and very low silica content, they have outstanding hot strength at elevated temperatures and are excellent for metal contact areas.

Ultra-low cement castables in the 70% alumina category exhibit excellent high temperature strength and thermal shock resistance. Bauxite-based,85%alumina ultra-low cement castables offer excellent hot strength and thermal shock resistance.

Silica Baesd Castables. It includes those made with vitreous silica as the raw material with extremely low thermal expansion, giving them excellent resistance to cracking under repeated thermal cycling to 1093°C. Their maximum service temperature is 1316°C under continuous service conditions. Primary applications are coke oven doors, zinc induction furnaces, glass forming dies and aluminum transfer ladles.

Other silica-based castables include high strength castables containing a fortified matrix and silica aggregate. This type of composition has substantially lower density& thermal conductivity than fireclay extra strength castables with comparable strengths and abrasion resistance. This combination of properties allows it to be used as a single component lining where a dense castable with a lightweight castable backup would otherwise be used.

Ultra-low cement silica based castables with chemistry, refractoriness, density and porosity equivalent to high quality fired silica brick are also produced.

Basic Refractory Castables. This class of castables includes chrome ore base products with hydraulic cement binders. These compositions have outstanding strength & abrasion resistance and resist chemical attack & thermal spalling. Typical uses include patching for rotary kiln linings.

Basic castables also come in the form of chrome-magnesite mixes with chemical air-setting bonds. They can be cast, rammed or gunned and have many uses. Other high strength, air- setting magnesite castables have bonding systems which can give them extremely high hot strength.

PLASTIC REFRACTORIES

Plastics refractories include all the fireclay, clay-graphite, high-alumina, high- alumina graphite and chrome types adapted for different operating conditions.

Heat-setting superduty fireclay plastics form a solid monolithic surface highly resistant to thermal shock and many acid slags. They have excellent workability and very low shrinkage, making them the ideal choice for rotary kiln hoods, incinerators and other general super duty plastic requirements. Cold setting versions of superduty fireclay plastics are available with many of the same features.

Superduty heat-setting plastics with graphite exhibit excellent resistance to wetting and corrosion by molten metal and slags. This type of composition is typically used in metal applications.

Plastics in the 50% alumina class typically serve as upgrades to superduty plastics. They are resistant to spalling due to thermal shock and many types of acid slags.

Heat-setting 60% alumina class plastics offer higher refractoriness over superduty plastics, with increased strength and volume stability throughout their temperature range. Application areas include cement kiln coolers and bull noses.

Air-setting high-alumina plastics in the 80% alumina class are primarily used where improved refractoriness over 60% alumina plastics is desired. They offer good resistance to fluxing oxides and slags up to their maximum service temperature.

Phosphate-bonded high-alumina plastics ranging in alumina contents from 70 to 90% are widely used in many applications as primary lining materials and for patching existing refractory linings. These products typically have high density and strengths, combined with excellent volume stability throughout their temperature range. Plastics ranging from 70 to 85% alumina content are often used in applications where resistance to slags and metal wash are required. The excellent abrasion resistance of 85% alumina plastics make them suitable for use in high abrasion conditions in petrochemical applications. Additional uses include both ferrous & nonferrous metal applications, where slag and metal penetration are wear mechanisms.

Plastics in the 90% alumina class are based on high purity aluminas. These products typically have high strengths at high temperatures and are often used in the foundry and steel making process.

Phosphate-bonded alumina-chrome plastics have very high strength at high temperatures. These compositions, based on high purity alumina aggregates & chromic oxide, form an alumina-chrome solid solution bond at high temperatures which has extremely good resistance to high iron oxide slags of an acid to neutral nature and to attack by coal slag.

Other phosphate bonded alumina- chrome plastics with lower alumina contents have been developed for specific service conditions. These include mullite based alumina-chrome plastics which have outstanding slag resistance to acid to neutral slags. The presence of mullite grain allows for earlier reaction, making this type of plastic refractory ideal for slagging applications between 1370 and 1593°C.

Silicon-carbide based phosphate- bonded plastics with an aluminum phosphate bond are also available. These have high conductivity and high abrasion resistance as well as non wetting properties to many acid slags & nonferrous metals.

Plastic refractories are used to form refractory monolithic linings in various kinds of furnaces, and are especially adaptable for making quick, economical emergency repairs. They are easily rammed to any shape or contour.

They are typically packaged in strong, easy to handle, moisture proof cartons. Special gunning versions are also available and are shipped in granulated form in moisture resistant pallet packages. These are prepared at the proper consistency, ready to use.

The high refractoriness, the range of compositions, and the ease with which they can be rammed into place make plastics suitable for many important applications. Plastic refractories are often highly resistant to destructive spalling

Plastic refractories are packed in moisture proof packings which should be unpacked just before installation. These are always available in moist clot form unlike dry powder form of castables. Usually, no further water addition is needed in their installation. The clot size is normally chosen so that a single clot layer is the full thickness of the lining to be installed. The plastic refractory clots are placed in position either by manual tamping or pneumatic ramming. Like castables, anchors (preferably refractory) are used in installing plastic refractories also. Plastic refractories have high erosion resistance and are used at places where tamping can be done conveniently and when pouripg or troweling is not possible. To facilitate drying, the exposed working surface should not be smoothened and measures similar to castables should be taken during the drying & heating up process.

RAMMING MASSES / MIXES

Refractory ramming masses/ mixes consist of refractory aggregates and a semi-plastic bonding phase. When properly installed, ramming mixes offer a way of placing a cementless monolithic lining at high density and relatively low porosity.

A well balanced selection of ramming materials includes compositions with base materials of silica, high-alumina, corundum, mullite, dead-burned magnesite, chrome ore, zircon and others. These materials are particularly suited for forming dense monolithic, lining construction and numerous other monolithic constructions. Ramming mixes are typically supplied in both wet and dry forms, depending on the binder system.

High-Alumina Ramming Mixes. Alumina-silicon carbide ramming mixes are designed for non-ferrous industries. High purity ramming mixes based on mullite grain are also used in steelmaking, burner blocks, ports and similar applications.

Ramming mixtures of 80% plus alumina content have excellent resistance to shrinkage and thermal spalling at high temperatures. Other air-setting high-alumina mixes employ stabilized, chemically refined high purity aluminas. These have excellent resistance to thermal spalling at high temperatures and remarkable volume stability up to their temperature limit.

Phosphate-bonded, alumina-chrome ramming mixes can offer exceptionally high purity. They typically feature very high strength at high temperatures and extremely good resistance to acid to neutral slags, including coal ash slags.

Alumina-graphite ramming mixes are designed for the steelmaking industry. Their combination of high-alumina aggregate and slag inhibitors gives them excellent slag resistance to acid to slightly basic slags.

Basic Refractory Ramming Mixes. Dry ramming mixes based on high purity magnesite and a sintering aid have been designed for steelmaking.

Magnesite ramming mixtures of exceptional purity & stability are used primarily as lining materials for coreless-type induction furnaces melting ferrous alloys. Magnesite-chrome fused grain ramming mixes provide exceptional density and strength. The bond of this composition provides adequate strength at low temperatures until direct-bonding occurs at higher temperatures.

Fireclay/alumina, silica, zirconia, magnesia, chromite or fireclay graphite base compositions with sufficient workability are prepared and used as ramming masses for obtaining monolithic working faces as acid, base or neutral linings depending on their characteristics. These refractory compounds *(i.e.* ramming masses) are used in repair and construction of various furnace parts where refractory linings should have high hot strength, high resistance to slag attack and abrasion and high volume stability.

Ramming masses like plastic refractories are not hydraulic setting. They are normally ceramic setting, however, some ramming masses are air or chemical setting also. The ramming masses are available in the form of either dry or moist powder. In dry powder, water or some other binder is added but moist powders can be used as such. Usually, the bottom or hearth of the furnace is built up with the ramming mass, but they can be used for repair jobs at any position of the refractory lining wherever ramming is possible. Ramming mass is applied by ramming (hence so named) and it does not require anchorage. Drying up and heating up schedules for ramming masses are faster than that for castables and plastic refractories.

REFRACTORY GUNNING MIXES

Refractory gunning mixes include siliceous, fireclay, high-alumina, dead-burned magnesite and chrome types. Many castables, ramming mixes and specially designed plastics can also be applied with pneumatic guns. Acid gun mixes are normally pre damped and fed through a continuous dual chamber or rotary gun. Magnesite and hot gun mixes are not pre damped and are placed in a batch pressure gun. Gun mixes should wet up well, have as wide a water range as possible, and provide excellent coverage in a variety of applications.

Fireclay Gunning Mixes. Fireclay gunning mixes include multi purpose hard-fired fireclay & standard calcium-aluminate cement compositions formulated for easy installation and low rebounds. These mixes are used in boilers, incinerators, process heaters, stacks, breechings and a variety of other medium service areas.

There are also fireclay gunning mixes with high purity calcium-aluminate bonding systems, designed for more severe service conditions experienced in incinerators. Other versions are for high alkali applications such as in cement kiln preheaters; or to provide excellent CO and load deformation resistance, for use in steel making.

High-Alumina Gunning Mixes. These are high purity alumina mixes which provide exceptional refractoriness, volume stability and a service temperature up to 1650°C. Gun mixeswhich combine high fired alumina aggregate with a high strength, high purity calcium-aluminate binder have excellent strength and chemical purity which allow them to withstand severe environments with a maximum service temperature of 1820°C. Primary application areas are primarily chemical.

There are also gunning mixes developed specifically for hot gunning maintenance. These are based on high- alumina content aggregates, providing a 1650°C service temperature limit. They have good slag and corrosion resistance.

Silica and Silicon Carbide Gunning Mixes. A gun mix based on vitreous silica and a special combination of calcined fireclay and high purity calcium-aluminate cement binder gives excellent strength & abrasion resistance coupled with outstanding thermal shock resistance & low thermal conductivity.

Silicon carbide-based gunning mixes are designed for preheater buildups and cooler bull noses. They are available in a variety of silicon carbide levels and have a high purity calcium-aluminate cement bond. The cement bond has the advantage of forming a room temperature set and a water insoluble bond at low temperatures.

Basic Refractory Gunning Mixes. A series of gun mixes are available for hot electric furnace maintenance. These range in magnesia content from 60 to 95% and are available with or without a phosphate bond. Gunning mixes of this kind are provide an even feed in a batch gun, wet up well and stick to a hot furnace wall.

In some industrial furnaces, turnaround time and installation costs are the major factors when choosing a refractory lining. In other cases, repairs need to be made with little or no downtime. In both circumstances, pneumatic conveying of material, or gunning, is the method of choice. Dense, homogeneous monolithic linings can be gunned without the use of forms and with a marked savings in time.

Characteristics of some typical mortar/cements, ramming masses and castables are tabulated in Table 8.1, Table 8.2 and Table 8.3 respectively.

Table 8.1: Characteristics of Some Typical Refractory Mortars/Cements

Name / quality	Composition, %	Fusion temperature, °C	Characteristics/applications
Orind MH Bond	Al_2O_3 = > 30 + 2 Fe_2O_3 = <2	1659	Used for laying fireclay refractories-air setting
KFS-ASC (series)	Al_2O_3 = 35 – 85	1400-1750 (service temp.)	A series of air setting cements for laying fireclay and high alumina refractories.
Bronex	SiO_2 = 80 – 85	—	Air setting silica mortar for providing gas tight joints in coke oven silica walls and other silica brick works.
Belset—MG	MgO = > 85	—	Magnesite mortar (air setting)
Zircon bond	ZrO_2 = 62 – 63 SiO_2 = 32 – 34	1850	Air setting cement for laying zircon refractories.

Velspray laddles	Al_2O_3 = 35 – 40	Not applicable	Chemically setting mortar used for hot and cold repair of laddies by spraying.
KFS silica	SiO_2 = > 90	1655	Ceramic setting cement for laying acidic refractories.
Calundum	Al_2O_3 = 46.9, SiO_2 = 4.4, Fe_2O_3 = 6.6, CaO = 35.7, TiO_2 = 4	1350-1380	Calcium aluminate binder.
Moriar chrome	Cr_2O_3 = 46 MgO = 12		General purpose chromite bricks mortar.
Mortar MAG—1	MgO = 84	—	General purpose magnesite bricks mortar.
Mortar mag-chrome	MgO = 45, Cr_2O_3 = 20	—	General purpose magnesite chrome brick mortar.
Mortar chrome-mag	Cr_2O_3 = 36 MgO = 25	—	General purpose chrome magnesite mortar.

Table 8.2: Characteristics of Typical Ramming Masses

Name 1 Quality	Composition, %	Temperature, °C		Characteristics / Applications
		Fusion	Service (Maximum)	
Z.I.T. Ram	Al_2O_3 = ≥ 40, Fe_2O_3 = ≤1.5	1683	1400	Chemically setting fireclay ramming mass used for zinc, lead and tin melting furnaces.
Valram-sill	$A1_2O_3$ = 8—60	> 1730	1600	Chemically setting, used in high temp, glass tank and other furnaces where high abrasion resistance, refractoriness and volume stability required.
Oriswan	SiO_z = ≥ 95	1683	1650	Ceramic setting, sufficient workability, used for induction furnace monolithic lining.
Cupolmix	SiO_2 = ≥ 92	1655	1600	Pure quartzite grains embedded in a suitable bond clay, ceramic setting, used in cupolas and locations where good molten iron resistance is needed.
Magchrome—I	MgO = ≥ 40 Cr_2O_3 = £ 32	—	1700	Chemically setting used for the repair/ construction of basic steel and other metallurgical furnaces.
Orimagfet/ Orimag—Ram	MgO = ≥ 80 Fe_2O_3 = £ 5.5	—	1750	Used for fettling/bottom making in electric arc and basic open hearth furnace.
Orichrome I,	Cr_2O_3 = ≥ 40	—	1400	Ceramic setting, used in high pressure boilers and between acid and basic refractories.

Chrome	Cr_2O_3 = ≥40	1743	1500	Air setting neutral ramming mass.
ZIR—RAM (ZS/ ZA)	ZrO_a = ≥60/52	—	1700/1500	Chemically setting zircon ramming mass.
Belspray—ZS	ZrO_a = ≥8	–		Chemically setting zircon spray compound used for hot repairs of silica walls of glass tank and other zircon lined furnaces.

Table 8.3: Characteristics of Some Typical Refractory Castables

Name	Composition, %	Cold Crushing Strength		Temperature, °C		Characteristics / Applications
		Temp., °C	kg/cm²	Fusion	Service (Max.)	
Orind Dense cast	Al_2O_s = 40 ± 2	110 1300	250–300 300–350	1564	1350	Used for rotary kilns, coolers, chain covers etc. It has high abrasion resistance.
Orind Super	Al_2O_3 = 75	110 1500 0	250 >500	1717	1500	Used for furnace arches, walls, car tops, lintels, and for guniting, it has excellent spalling resistance.
Oiicast SO	AljOj = 90 ± 2	110 1450	150 200–250	1743	1550	Non wetting to aluminium melt, used where dense/ impervious structure is required e.g. in burner blocks, ladle linings etc.
Supercast (1600-1825 series)	Al_2O_3 = 60–95 Fe_2O_3 = 1.5–3	110 1550	150–300 300–600	1717 1804	1600/1850	Used in B.F., soaking pit covers, carbon reactors, heat treatment furnaces, reverberatory furnaces etc. involving high working temperatures.
Velcast (120/140)	Al_2O_3 = 20–30/ 38–42	110 1400	150/175 upto 250	1450 1600	1200/1450	Used in annealing furnaces, iron blast furnaces, soaking pit covers, burner blocks, furnace side walls and arches, rotary kilns etc., it is hydraulic setting.
Belcrete (FH14/14HS)	Al_2O_3 = 38–45 CaO = 5–12 Fe_2O_3 = 2.5–3.5	NA	NA	1500–1600	1450/1350	It is fireclay castable and is hydraulic setting.
Belcrete (SH-165/5EH)	SiO_2 = 95–97/ 93–96 CaO = 3–4.5/ 2.5–3.5	NA	NA	> 1650	1650	Hydraulic setting silica castable.

Belcrete CH-15	Cr_2O_3 = 25–30 MgO = 2.5–3	NA	NA	> 1700	1500	Hydraulic setting chrome castable.
Belcrete MH-15	MgO = 75–80 CaO = 2.5–3	NA	NA	> 1780	1700	Hydraulic setting magnesite castable.
Chrome crete	Cr_2O_a = 40 CaO = 10	110 1500	100–150 150–200	1743	1500	Hydraulic setting neutral castable.
KFS insulating–96	Al_2O_3 = ≥ 35 Fe_2O_3 = £8 CaO = £ 20	110 1100	14–16 14–16	1264	1100	Hydraulic setting insulating castable.
KFS insulating–25	Al_2O_3 = ≥ 33 Fe_2O_3 = £5 CaO = ≤16	110 1300	50–75 45–60	1430	1300	Hydraulic setting insulating castable.
Belcast	Al_2O_3 = 93.5 Fe_2O_3 = 0.2	NA	NA	> 1800	1850	Air setting high alumina castable.

Glossary of Terms (Related to Refractories)

Abrasion. The wearing of a surface by the mechanical action between solids.

Abrasion of Refractories. Wearing away of the surfaces of refractory bodies in service by the scouring action of moving solids.

Abutment. In furnace construction, the structural member which withstands the thrust of an arch. An abutment consists of a brick skewback and a steel supporting member.

Acid-Proof Brick. Brick having low porosity & permeability, and high resistance to chemical attack or penetration by most commercial acids and some other corrosive chemicals.

Acid Refractories. Refractories such as silica brick which contain a substantial proportion of free silica and which when heated, can react chemically with basic refractories, slags & fluxes.

After-contraction/After-expansion. The percentage permanent contraction or expansion, measured after cooling that takes place when a refractory material is heated for a specified period at a specified temperature. Also known as the permanent linear change on reheating (PLC) when measured linearly.

Ageing or Souring. A process in which the plasticity of a body is improved by allowing the moist body to lie for a period, *e.g.* by storage in a heap or in a pit.

Aggregate. As applied to refractories, a ground mineral material, consisting of particles of various sizes, used with much finer sizes for making formed or monolithic bodies.

Air ramming. A method of forming refractory shapes, furnace hearths or other furnace parts by means of pneumatic hammers.

Air-setting mortar. A refractory composition containing chemical agents that ensure hardening at room temperature.

Air-Setting Refractories. Compositions of ground refractory materials which develop a strong bond upon drying. These refractories include mortars, plastic refractories, ramming mixes and gunning mixes. They are marketed in both wet and dry condition. The dry compositions require tempering with water to develop the necessary consistency.

Alumina. Al_2O_3, the oxide of aluminum; melting point 2,050°C ; in combination with H_2O (water), alumina forms the minerals diaspore, bauxite and gibbsite; in combination with SiO_2 and H_2O, alumina forms kaolinite and other clay minerals.

Alumina-Silica Refractories. Refractories consisting essentially of alumina & silica, and including high-alumina, fireclay & kaolin refractories.

Amorphous. Lacking crystalline structure or definite molecular arrangement; without definite external form.

Anchor. (a) A metal or refractory device for holding a refractory lining in place. (b) Refractory or metallic pieces fixed to the casing and designed to ensure the stability of the refractory lining.

Andalusite. A brown, yellow, green, red or gray orthorhombic mineral; Al_2SiO_5. Specific gravity 3.1 - 3.2. Decomposes on heating, beginning at about 1,350°C to form mullite ($Al_6Si_2O_3$) and free silica.

Arch or Roof. Part of a structure bridging a gap in a wall *(e.g.* a door arch) or spanning the space between two walls *(e.g.* a furnace roof).

Arch brick. A brick in which the two large faces are inclined toward each other in such a way that, with a certain number of these bricks, an arch can be constructed.

Arch, Flat. In furnace construction, a flat structure spanning an opening and supported by abutments at its extremities. The arch is formed by a number of special tapered brick, and the brick assembly is held in place by the keying action of the brick. Also called a jack arch.

Arch, Sprung. In furnace construction, a bowed or curved structure which is supported by abutments at the sides or ends only, and which usually spans an opening or space between two walls.

Arch, Suspended. A furnace roof consisting of brick shapes suspended from overhead supporting members.

Attrition. Wearing away by friction; abrasion.

Auger Machine. A machine for extruding ground clays in moist and stiffly plastic form, through a die by means of a revolving screw or auger.

Baddeleyite. A mineral composed of zirconia (ZrO_2). Specific gravity 5.8. Melting point 2,700°C.

Baffle. Refractory or metallic partition which directs the gases through the banks of tubes.

Bag wall. One of the protective walls within a kiln around individual fireboxes.

Ball Clay. A highly plastic refractory bond clay of very fine grain, which has a wide range of vitrification and which burns to a light color. Often high in carbonaceous matter.

Basic Refractories. Refractories which consist essentially of magnesia, lime, chrome ore or mixtures of two or more of these and, when heated, can react chemically with acid refractories, slags & fluxes.

Batt or Bat. (a) A fired slab of refractory material of thin section, used in the firing of ceramic products. (b) A flat refractory slab.

Batter. (a) Incline of a wall surface; also designates the angle made by a wall surface with the vertical.

Battery. A series of ovens built side by side in a continuous structure.

Bauxite. An off-white, grayish, brown, yellow, or reddish-brown rock composed of a mixture of various amorphous or crystalline hydrous aluminum oxides and aluminum hydroxides (principally gibbsite, some boehmite), and containing impurities in the form of free silica, silt, iron hydroxides, and especially clay minerals; a highly aluminous laterite.

Bauxitic Clay. A natural mixture of bauxite and clay containing not less than 47% nor more than 65% alumina on a calcined basis.

Bell, Funnel or Trumpet. The top unit in the trumpet assembly.

Belly or Waist. The part of the blast furnace, of maximum cross-section, between the stack & the bosh, and also the part of the converter in which steel is gathered at the moment of tapping.

Bentonite. A kind of clay derived from volcanic ash and characterized by extreme fineness of grain. Its main constituent is the clay mineral montmorillonite. It is somewhat variable in composition and usually contains 5 to 10% of alkalies or alkaline earth oxides. One type has the capacity for absorption of large amounts of water, with enormous increase in volume.

Bevel brick. A brick shape in which one edge, between two adjacent faces, is replaced by a bevel.

Binder. A substance added to a nonplastic granular material to give it workability and green and/or dry strength.

Black core or Black heart. Dark central part of a fired product resulting from inadequate oxidation

Blank or Clot. The rough shape of a product ready for final shaping, or of a raw material prior to calcination.

Bloating. The permanent expansion, accompanied by the formation of a vesicular texture, which occurs when some types of clay are fired. Swelling of a refractory when in the thermoplastic state, caused by temperatures in excess of that for which the material is intended.

Bond. A material that binds together the discrete grains of a mix.

Bonded roof. A roof in which the joints are crossed in one or both directions.

Bonder. A brick one-and-a-half times as wide as a standard square, and sometimes with the large faces inclined towards each other from end to end.

Bosh. The truncated central part of the blast furnace, narrowing from the lintel at the top to the tuyere belt at the bottom.

Breakout. An escape of metal or slag through the furnace lining, especially through the hearth.

Breast or Banks. Sloping part joining the central area of the hearth to the walls in an open hearth furnace. The breast is the sloping part at the end of the hearth. The banks are those parts at the sides of the hearth, beneath the back and front walls.

Breast wall, (a) Side wall of combustion space of a tank furnace, above and generally behind the tank blocks, (b) Brick wall between pillars of a pot furnace, and in front of or surrounding the front of a pot.

Bridge cover. Blocks spanning the space between the bridge walls.

Bridge wall, (a) Brickwork, in a furnace, separating the combustible or fuel from the working chamber. (b) The wall between the slag chamber and the regenerator of an open-hearth furnace (c) or *Firebridge.* Refractory wall towards the end of the furnace, designed to ensure completion of combustion or to limit the combustion chamber.

Brucite. A mineral having the composition $Mg(OH)_2$. Specific gravity 2.38 - 2.40. A soft, waxy, translucent mineral which dissociates at moderate temperatures with the formation of MgO.

Buckstay or Buckstave. A steel member used in furnace construction to take the thrust of the brickwork.

Bullhead, Key brick, Cupola brick or Crown brick. A brick in which the side faces are inclined towards each other in such a way that one of the end faces is shorter than the other, used particularly at the apex of an arch.

Bullnose or Jamb brick. A brick in which one end face is completely rounded off to join one side face.

Burn. The degree of heat treatment to which refractory brick are subjected in the firing process. Also, the degree to which desired physical and chemical changes have been developed in the firing of a refractory material.

Burning (Firing) of Refractories*. The final heat treatment in a kiln to which refractory brick are subjected in the process of manufacture, for the purpose of developing bond and other necessary physical and chemical properties.

Bursting. Disintegration following a permanent increase in volume, particularly of chrome-magnesite refractories which have absorbed iron oxide.

Buttress wall or Pinion wall. Non-refractory end wall of a battery.

Calcination. A heat treatment to which many ceramic raw materials are subjected, preparatory to further processing or use, for the purpose of driving off volatile chemically combined components and affecting physical changes.

Calcite. A mineral having the composition $CaCO_3$. Specific gravity 2.71 for pure calcite crystals. Calcite is the essential constituent of limestone, chalk & marble and a minor constituent of many other rocks.

Cap or Crown. The arched roof of a furnace, especially a glass tank furnace.

Carbon Deposition. The deposition of amorphous carbon, resulting from the decomposition of carbon monoxide gas into carbon dioxide & carbon within a critical temperature range. When deposited within the pores of refractory brick, carbon can build up sufficient pressure to destroy the bond and cause the brick to disintegrate.

Carbon Refractory. A manufactured refractory comprised substantially or entirely of carbon (including graphite).

Carbon-Ceramic Refractory. A manufactured refractory comprised of carbon (including graphite) and one or more ceramic materials such as fire clay and silicon carbide.

Car deck. Refractory top of a car for the support of the ware in a tunnel kiln or truck chamber-kiln.

Carburettor. The refractory-lined chamber of a water gas plant, often filled with checkers on which oil is sprayed for enrichment of the water gas.

Castable refractory. A mixture of refractory aggregates and heat-resisting hydraulic cement. These products are generally cast into place.

Casting slip. Generally, a stable suspension of clay in water with or without the inclusion of other materials. This term can also denote other stable suspension, *e.g.* of alumina.

Centre brick, Crown brick or Distributor brick. Hollow refractory block which distributes molten steel from the trumpet assembly to the runner bricks.

Ceramic bond. The vitreous or crystalline material, formed on firing, between the coarser constituents of a ceramic body and giving cohesion and mechanical strength to the fired product.

Ceramics. Originally, ware formed from clay and hardened by the action of heat; the art of making such ware. Current usage includes all refractory materials, cement, lime, plaster, pottery, glass, enamels, glazes, abrasives, electrical insulating products and thermal insulating products made from clay or from other inorganic, nonmetallic mineral substances.

Cermet. A composite material containing both ceramic and metallic constituents.

Chamotte. Refractory clay that has been specially fired for use as a nonplastic material.

Checker or Chequer. Bricks or shapes set in such a way that the hot gases can pass between them; the filling of a regenerator chamber.

Checker brick or Chequer brick. Brick used for a checkers in regenerator.

Chemically-bonded brick. A brick in which mechanical strength is imparted by chemical binding agents instead of by firing.

Chemically-bonded refractory cement, (a) Air-setting refractory cement; Air-setting mortar. A refractory composition containing chemical agents, that ensure hardening at room temperature.

Chrome Brick. A refractory brick manufactured substantially or entirely of chrome ore.

Chrome-Magnesite Brick. A refractory brick which can be either fired or chemically bonded, manufactured substantially of a mixture of chrome ore and dead-burned magnesite, in which the chrome ore predominates.

Chrome Ore. A rock having as its essential constituent the mineral chromite or chrome spinel, which is a combination of FeO & MgO with Cr_2O_3, Al_2O_3, and usually a small proportion of Fe_2O_3. The composition, which is represented by the formula (Fe, Mg) $(Cr, Al)_2 O_4$, is extremely variable. Refractory grade chrome ore has only minor amounts of accessory minerals and has physical properties that are suitable for the manufacture of refractory products.

Cinder notch or Slag notch. A hole in the hearth wall which is opened to allow molten slag to flow from the furnace.

Circle brick. A brick, the two side faces of which are parts of concentric cylinders.

Clay. A natural mineral aggregate, consisting essentially of hydrous aluminum silicates.

Clinker of Dross, (a) Fused or partially fused fuel ash which may contain some incompletely burned fuel (b) only *Clinker*— Hydraulic cement before being crushed or ground (c) only *Clinker*— A dead-burned product usually stabilised dolomite.

Closer, Pup or Soap. A brick of the same length as a standard square but of smaller thickness and/or breadth, with end faces square or nearly square.

Cold set. The hardening or setting of a mortar material which takes place at room temperature. Bonding mortars containing agents which give them greater strength after drying than is obtained with a clay or other refractory base alone, are said to be air-setting or cold-setting.

Collapse. The failure of a material under an applied load particularly in the refractoriness-under-load test.

Collar. A short fireclay section of a horizontal retort which is placed between the main silica portion and the mouth piece.

Compo. A refractory composition used in the casting pit for ramming purposes.

Cone. Hollow refractory shape used in bottom pouring,- through which the liquid steel passes from the runner brick into the ingot mould. When made in one piece with the runner brick, it is known as *riser brick* or an *end runner.*

Congruent melting. The change of a substance, when heated, from the solid form to a liquid of the same composition —*e.g.* melting of ice.

Conversion. A generally slow and not immediately reversible change in the crystalline structure of a silica material resulting from heat treatment.

Corbel. An arrangement of bricks in which each course projects beyond an adjacent course.

Corrosion of Refractories. Deterioration or wearing away of refractory bodies largely at their surface through chemical action of external agencies.

Corundum. A natural or synthetic mineral consisting solely of alumina (Al_2O_3). Specific gravity 4.00 - 4.02. Melting point 2,050°C. **Hardness - 9 on Mhos scale.**

Course. A horizontal layer or row of brick in a structure.

Crank. A composite refractory structure which supports and separate a number of pieces of fiat-ware during glost and decorating firing.

Crazing. A network of surface cracks.

Creep or Flow. Isothermal deformation of a stressed product as a function of time.

Cristobalite. A mineral form of silica (SiO_2); stable from 1,470°C to the melting point, 1,723°C. Specific gravity -2.32. Cristobalite is an important constituent of silica brick.

Cross wall. One of the brick partitions bracing horizontal retorts and arranged inside the setting.

Crown. A furnace roof, especially one which is dome-shaped; the highest point of an arch.

Crucible. A refractory container for molten metal.

Curtain arch. Refractory arch supporting the wall placed between a gas producer and the gas uptake.

Curtain wall. A suspended horizontal structure, fixed or movable, which separates the combustion space of a flat glass furnace into two zones.

Cryptocrystalline. A crystalline structure in which the individual crystals are so small that they cannot be made visible by means of the petrographic microscope, but can be seen with an electron microscope. Various so-called amorphous minerals are actually cryptocrystalline.

Crystal. A homogeneous, solid body of a chemical element, compound or isomorphous mixture having a regularly repeating atomic arrangement that can be outwardly expressed by plane faces.

Dead-burned. Term applied to a material that has been made as inert and stable as possible by appropriate heat treatment.

Dead-Burned Dolomite. A coarsely granular refractory material prepared by firing raw dolomite with or without additives, to a temperature above 1,538°C, so as to form primarily lime & magnesia in a matrix that provides resistance to hydration & carbonation.

Dead-Burned Magnesite. A coarsely granular dense refractory material composed essentially of periclase (crystalline magnesium oxide); prepared by firing raw magnesite (or other substances convertible to magnesia) at temperatures sufficiently high to drive off all of the volatile materials, and to affect complete shrinkage of the resultant magnesia, thereby producing hard dense grains which are entirely inert to atmospheric hydration and carbonation and free from excessive shrinkage when again subjected to a high temperature.

Devitrification. Formation of crystals in a vitreous material. The change from a glassy to a crystalline condition.

Diaspore. A mineral having the theoretical composition $Al_2O_3 \cdot H_2O$ (85% alumina; 15% water). Specific gravity 3.45.

Diaspore Clay. A rock consisting of diaspore bonded by flint clay. Commercial diaspore clay of the purest grade usually contains between 70 and 80% alumina after calcination.

Diatomaceous Earth. A hydrous form of silica which is soft, light in weight and consists mainly of microscopic shells of diatoms or other marine organisms. It is widely used for furnace insulation.

Direct Bonded Basic Brick. A fired refractory in which the grains are joined predominantly by a solid state of diffusion mechanism.

Direct Bonded Magnesite-Chrome Brick. A term applied to fired magnesite-chrome compositions when the amount of bonding mineral phase (silicates, forsterite, etc.) present in the matrix is sufficiently low that under microscopic examination the chrome ore grains appear to be bonded "directly" to the magnesite grains. The actual bonding mechanism in this instance is usually a combination of types, of which one may be direct (diffusion) bonding.

Dobie. A moulded block of ground clay or other refractory material, usually crudely formed and either raw or fired.

Doghouse. The small arched chamber in the end wall, through which is inserted a metallic fuel burner.

Dolomite. The mineral calcium- magnesium carbonate, CaMg $(CO_3)_2$. Specific gravity 2.85 - 2.95. The rock called dolomite consists mainly of the mineral of that name and can also contain a large amount of the mineral calcite ($CaCO_3$).

Dome. A roof in the shape of a segment of a sphere.

Dome brick. A brick in which the large and side faces are inclined towards each other in such a way that, with a number of these bricks, a dome can be constructed.

Dot. A small refractory support used for separating cranks and setters.

Dozzle. Refractory material or insulation at the top of an ingot mould, designed to maintain a reservior of molten metal until the main part of the ingot has solidified completely. When the refractory is made in one piece, it is called a *Core* or *Dozzle* and is usually preheated before use.

Duct or Smoke pipe. Brick or metal flue.

Dusting. Conversion of a refractory material, either wholly or in part, into fine powder or dust. Dusting usually results from (a) chemical reactions such as hydration; or (b) mineral inversion accompanied by large and abrupt change in volume, such as the inversion of beta to gamma dicalcium silicate upon cool- ing.

Dutch oven. A combustion chamber built outside and connected with a furnace.

End face. In a standard square, one of the faces bounded by the two shortest dimensions.

Erosion of Refractories. Mechanical wearing away of the surfaces of refractory bodies in service by the washing action of moving liquids, such as molten slags or metals.

Eutectic temperature. The lowest melting temperature in a series of mixtures of two or more components.

Exfoliation. The property possessed by certain materials, for example vermiculites and hydro-biotites, when submitted to sudden heating, to expand permanently in a direction parallel to the C-axis, forming a laminar texture.

Expansion joint. Space left in brickwork to accommodate the expansion which occurs on heating the furnace.

Fantail or Fantail arch. Opening between the slag pocket and the regenerator.

Fayalite. A mineral having the composition Fe_2SiO_4. Specific gravity 4.0 - 4.1. Melting point 1,205°C.

Feed-end block. A refractory shape for use at the feed end of a rotary cement kiln.

Feldspar. A group of aluminum silicate minerals with a general formula MAl $(Al,Si)_3$ O_8 where M=K, Na, Ca, Br, Rb, Sr and Fe. The most important feldspars are. (a) the potash group, of which orthoclase and microcline (K) are the most common, and (b) the soda-lime group, of which albite (Na) and anorthite (Ca) form the end members of a continuous series of solid solutions. Specific gravity 2.55 - 2.76. Melting points 1,120° to 1,550°C.

Fillet. The concave curved junction of two surfaces which would otherwise meet at an angle Fillets are used at recentrant angles in the design of brick shapes, to lessen the danger of cracking.

Firebox. One of the small chambers located wholly or partly in the wall of a kiln and in which the combustion of fuel takes place.

Fire Clay. An earthy or stony mineral aggregate which has as the essential constituent hydrous silicates of aluminum with or without free silica; plastic when sufficiently pulverized & wetted, rigid when subsequently dried, and of sufficient purity & refractoriness for use in commercial refractory products.

Fireclay Brick. A refractory brick manufactured substantially or entirely from fire clay.

Firestone. A siliceous rock used in the natural state without further processing, other than cutting, as a furnace-lining material.

Flaking, Peeling or Shelling. The loss of part of the surface of a refractory material as a consequence of cracking behind the hot face.

Flash wall. A continuous protective wall built within a kiln in front of the fireboxes.

Flat arch or Jack arch. A sprung arch in which the outer surface is horizontal and the inner surface is either horizontal or arched with a large radius.

Flaw. A defect in the texture of a product.

Flint. A hard, fine-grained cryto-crystalline rock, composed essentially of silica.

Flint Clay. A hard or flint-like fire clay which has very low natural plasticity and which usually breaks with a smooth or shell-like fracture. Its principal clay mineral is halloysite.

Floor or Bottom. The refractory lining forming the base of the combustion chamber.

Flue. Duct constructed of brick or metal for the passage of gases.

Flux. A substance or mixture which promotes fusion of a solid material by chemical action.

Fluxing. Fusion or melting of a substance as a result of chemical action.

Forsterite. A mineral having the composition Mg_2SiO_4. Specific gravity 3.21. Melting point approximately 1,900°C.

Freeze-Plane. An irregular plane lying between the hot face and cold face of a refractory lining — any point on which the temperature corresponds to the freezing point of a liquid phase present on the hot face side of the plane.

Front wall, (a) The wall of an open hearth furnace facing the charging platform. (b) The wall at the working end of glass tank furnace.

Furnace Chrome. A mortar material prepared from finely ground chrome ore, suitable for laying brick or for patching or daubing in furnaces.

Furnace Magnesite. A mortar material prepared from finely ground dead-burned magnesite, suitable for use as a joint material in laying magnesite brick, and for patching or daubing furnace masonry.

Fused-Cast Refractories. Refractories formed by electrical fusion followed by casting and annealing.

Fused Quartz. Silica in the glassy state produced by melting clear quartz crystalline feed. It is clear without entrapped gas bubbles or other impurities or diluents. Synonyms include quartz glass and vitreous quartz.

Fused Silica. Silica in the glassy or vitreous state produced by arc-melting sand. It always contains gas bubbles. Synonyms include vitreous silica and silica glass.

Fusion. A state of fluidity or flowing in consequence of heat; the softening of a solid body, either through heat alone or through heat and the action of a flux, to such a degree that it will no longer support its own weight, but will slump or flow. Also the union or blending of materials, such as metals, upon melting, with the formation of alloys.

Fusion point. The temperature at which fusion takes place. Most refractory materials have no definite fusion points, but soften gradually over a range of temperatures.

Gable wall, Back wall or End wall. Vertical wall at the charging end of a tank furnace.

Ganister. A dense, high-silica rock (quartzite), suitable for the manufacture of silica brick.

Gibbsite. A white or tinted monoclinic mineral; $Al(OH)_3$. Specific gravity 2.3 - 2.4.

Git or Sprue. The metal which solidifies in the hollow casting-pit refractories, at the end of teeming.

Glass. An inorganic product of fusion which has cooled to a rigid condition without crystallizing.

Grog, (a) Specifically, crushed firebrick, for use as a non-plastic material, (b) Crushed firebrick or other calcined or fused material (unused or used) for incorporation in a batch as a non-plastic material, (c) Burned refractory material, usually calcined clay or crushed brick, added to a brick or mortar mix to reduce shrinkage in drying and firing.

Ground Fire Clay. Fire clay or a mixture of fire clays that have been subjected to no mechanical treatment other than crushing and grinding.

Grout. A suspension of mortar material in water, of such consistency that when it is poured upon horizontal course of brick masonry, it will flow into vertical open joints.

Guard ring, (a) In various methods of measuring thermal conductivity, the test piece on which the measurement is made is placed centrally in the apparatus and is surrounded by a guard ring, (b) A device for maintaining unidirectional flow of heat through the test piece, in the measurement of thermal conductivity.

Gunning. The application of monolithic refractories by means of air-placement guns.

Halloysite. One of the clay minerals; a hydrated silicate of alumina similar in composition to kaolinite, but amorphous and containing more water; $Al_2Si_2O_5(OH)_4 \cdot 2H_2O$.

Header. A brick laid on flat with its longest dimension perpendicular to the face of a wall.

Heat-Setting Refractories. Compositions of ground refractory materials which require relatively high temperatures for the development of an adequate bond, commonly called the ceramic bond.

Hearth. The lower part of the blast furnace in which the molten iron and slag are retained.

Hearth bottom. The refractory lining upon which the molten iron is contained in the hearth.

Heat baffles. Refractory work protecting a metallic beam from the heat.

Heating flue. Vertical duct in a wall through which the heating gases pass.

Hematite. The mineral Fe_2O_3 (red iron ore). Specific gravity 4.9 - 5.3.

High-Alumina Refractories. Alumina-silica refractories containing 45 % or more alumina. The materials used in their production include diaspore, bauxite, gibbsite, kyanite, sillimanite, andalusite and fused alumina (artificial corundum).

High-Duty Fireclay Brick. Fireclay brick which have a PCE not lower than Cone 31½ or above 32½ - 33.

Hood, Boot or Potette. Refractory shape introduced into the furnace and partly immersed in glass to protect the gathering point from the furnace gases and the surface scum.

Hot blast main. A circular duct which conveys the hot blast from a stove to a blast furnace.

Hot blast stove or Cowper stove. A vertical cylindrical furnace having a combustion chamber and checkers, which alternatively receive heat from the combustion of blast furnace gas and deliver heat to the hot blast.

Hot-setting bonding mortar. A refractory mortar material, which requires relatively high temperature for the development of a bond.

Hot-tops. Combination of hot-top bricks and containers.

Hydraulic refractory cement, Hydraulic setting mortar. A refractory composition containing a heat-resisting hydraulic cement that ensures setting and hardening at room temperature.

Hydraulic Setting Refractories. Compositions of ground refractory materials in which some of the components react chemically with water to form a strong hydraulic bond. These refractories are commonly known as castables.

Illite. A group of three layer, mica like minerals of small particle size, intermediate in composition & structure between muscovite and montmorillonite.

Impact Pressing. A process for forming refractory shapes in which the ground particles of refractory material are packed closely together by rapid vibration.

Incongruent melting. Dissociation of a compound upon heating, with the formation of another compound and a liquid of different composition from the original compound. For example, mullite ($3Al_2 0_3.2Si0_2$) melts incongruently to form corundum ($A1_2 0_3$) and a siliceous liquid.

Insulating Refractories. Lightweight, porous refractories with much lower thermal conductivity and heat-storage capacity than other refractories. Used mostly as backing for brick of higher refractoriness and higher thermal conductivity. These materials provide fuel economy through lower heat losses, increased production due to shorter heat up time, economy of space (size & weight) because of thinner walls and improved working conditions. Insulating refractories are available as brick or monoliths.

Inversion. (a) A generally rapid and immediately reversible change in the crystalline form of a material, resulting from a change in temperature (b) A change in crystal form without change in chemical composition, as for example, the change from low quartz to high quartz or the change from quartz to cristobalite.

Inversion point. The temperature at which a crystalline inversion takes place.

Iron spot. A dark-coloured spot formed from particles of iron or its compounds present in a brick.

Isomorphous mixture. A type of solid solution in which mineral compounds of analogous chemical composition and closely related crystal habit crystallize together in various proportions. Each mixture is a homogeneous crystalline phase.

Jacket, Shell or Casing. The metal casing of the furnace enclosing the refractory brick work.

Jamb. (a) A vertical structural member forming the side of an opening in a furnace wall; (b) a type of brick shape intended for use in the sides of wall openings.

Joint. Space between bricks or refractory pieces, generally filled with cement.

Jumbo or Transfer ladle. A refractory-lined container for the transport of hot metal

Kaolin. A white burning clay having kaolinite as its chief constituent. Specific gravity 2.4 - 2.6. The PCE of most commercial kaolins ranges from Cone 33 to 35.

Kaolinite. A common white to grayish or yellowish clay mineral; $Al_2Si_2O_5\,(OH)_4$. Kaolinite is the principal constituent of most kaolins and fire clays. Specific gravity is 2.59. The PCE of pure kaolinite is Cone 35.

Key. In furnace construction, the uppermost or the closing brick of a curved arch.

Key Brick. A brick shape having six plane faces (two sides, two edges and two ends), in which two faces (the edges) are inclined toward each other and one of the end faces is narrower than the other.

K-factor. The thermal conductivity of a material, expressed in standard units.

Kyanite (Cyanite). A blue or light-green triclinic mineral; Al_2SiO_5. Specific gravity 3.56 - 3.67. Decomposition begins at about 1,325°C with the formation of mullite and free silica.

Ladle. A refractory-lined vessel used for the temporary storage or transfer of molten metals.

Launder. A refractory-lined open channel along which molten steel flows from the furnace.

L-D Process. A process for making steel by blowing oxygen on or through molten pig iron, whereby most of the carbon and impurities are removed by oxidation.

Limestone. A sedimentary rock composed essentially of the mineral calcite ($CaCO_3$) or of calcite mixed with dolomite, $CaMg(CO_3)_2$. Specific gravity 2.6 - 2.8.

Limonite. A mineral consisting of hydrous ferric oxides; the essential component of "brown ore." Specific gravity 3.6 - 4.0.

Lintel. A horizontal member spanning a wall opening.

Loss on Ignition. As applied to chemical analyses, the loss in weight which results from heating a sample of material to a high temperature, after preliminary drying at a temperature just above the boiling point of water. The loss in weight upon drying is called "free moisture"; that which occurs above the boiling point, "loss on ignition."

Low-Duty Fireclay Brick. Fireclay brick which have a PCE not lower than Cone 15 nor higher than 28 - 29.

Magnesioferrite. One of the spinel group of minerals; $(Mg,Fe)Fe_2O_4$. Rarely found in nature; usually constitutes the brown coloring material in magnesite brick. Specific gravity 4.57 - 4.65.

Magnesite. A mineral consisting of magnesium carbonate; $MgCO_3$. A rock containing the mineral magnesite as its essential constituent.

Magnesite Brick. A refractory brick manufactured substantially or entirely of dead-burned magnesite which consists essentially of magnesia in crystalline form (periclase).

Magnesite-Carbon Brick. A refractory brick manufactured of substantially magnesite (dead-burned, fused, or a combination thereof) and carbon, which may be in the form of various carbon bearing materials. Conventional tar-bonded and tar-impregnated brick do not fall into this class. Magnesite-carbon brick are distinct in that carbon is present in the composition to provide specific refractory properties beyond filling pores or acting as a bond.

Magnesite-Chrome Brick. A refractory brick which can be either fired or chemically bonded, manufactured substantially of a mixture of dead-burned magnesite (magnesia) and refractory chrome ore, in which the magnesite predominates by weight.

Magnesium Hydroxide. The compound of magnesium oxide and chemically combined water; $Mg(OH)_2$. Naturally occurring magnesium hydroxide is known as brucite.

Magnetite. A black, isometric, strongly magnetic, opaque mineral of the spinel group; (Fe, Mg) Fe_2O_4. Specific gravity 5.17 - 5.18. Melting point about 1,594°C.

Medium-Duty Fireclay Brick. A fireclay brick with a PCE value not lower than Cone 29 nor higher than 31 - 311/2.

Melting Point. The temperature at which crystalline and liquid phases having the same composition coexist in equilibrium. Metals and most pure crystalline materials have sharp melting points, i.e. they change abruptly from solid to liquid at definite temperatures. However, most refractory materials have no true melting points, but melt progressively over a relatively wide range of temperatures.

Metalcase Brick. Basic brick provided with thin steel casings.

Mica. A group of rock minerals having nearly perfect cleavage in one direction and consisting of thin elastic plates. The most common varieties are muscovite and biotite.

Micron. The one-thousandth part of a millimeter (0.001 mm); a unit of measurement used in microscopy.

Mineral. A mineral species is a natural inorganic substance which is either definite in chemical composition and physical characteristics or which varies in these respects within definite natural limits. Most minerals have a definite crystalline structure; a few are amorphous.

Modulus of Elasticity. A measure of the elasticity of a solid body; the ratio of stress (force) to strain (deformation) within the elastic limit.

Monolithic Lining. A furnace lining without joints, formed of material which is rammed, cast, gunned or sintered into place.

Monticellite. A colorless or gray mineral related to olivine; $CaMgSiO_4$. Specific gravity 3.1 - 3.25. Melts incongruently at 1,499°C to form MgO and a liquid.

Montmorillonite. A group of expanding lattice clay minerals containing variable percentages of one or more of the cations of magnesium, potassium, sodium and calcium. A common constituent of bentonites.

Mortar (Refractory). A finely ground refractory material which becomes plastic when mixed with water and is suitable for use in laying refractory brick.

Mullite. A rare orthorhombic mineral; $Al_6Si_2O_{13}$. Specific gravity 3.15. An important constituent of fireclay and high-alumina brick. Melting point under equilibrium conditions approximately 1,850°C.

Mullite Refractories. Refractory products consisting predominantly of mullite crystals formed either by conversion of one or more of the sillimanite group of minerals or by synthesis from appropriate materials employing either melting or sintering processes.

Muscovite. A mineral of the mica group; $KA1_2(AlSi_3)O_{10}(OH)_2$. It is usually colorless, whitish or pale brown and is a common mineral in metamorphic & igneous rocks and in some sedimentary rocks.

Nepheline (Nephelite). A hexagonal mineral of the feldspathoid group; $(Na,K)AlSiO_4$. A common reaction product in furnaces wherein slags or vapors of high soda content come into contact with fireclay or high-alumina brick. Stable at 1,248°C at which temperature it inverts to the artificial mineral carnegieite, which has the same composition, but a different crystalline form. Natural nepheline contains a small amount of potash. Specific gravity 2.67.

Neutral Refractory. A refractory material which is neither acid nor base, such as carbon, chrome or mullite.

Nose or Mouth. Upper opening in a converter through which the gases escape, charging is carried out and deslagging and pouring are effected.

Nostril or Port. The orifice through which heating gas or air is admitted to a combustion chamber.

Nozzle. A hollow refractory shape through which molten steel flows from a ladle or tundish.

Nozzle Brick. A tubular refractory shape used in a ladle; contains a hole through which steel is teemed at the bottom of the ladle, the upper end of the shape serving as a seat for the stopper.

Olivine. (a) An olive-green, grayish-green or brown orthorhombic mineral; (Mg,Fe)2SiO_4. It comprises the isomorphous solid-solution series forsterite- fayalite. (b) A name applied to a group of min- erals forming the isomorphous system (Mg,Fe,Mn,Ca) 2SiO_4, including forsterite, fayalite, tephroite and a hypothetical calcium orthosilicate. Specific gravity 3.27 - 3.37, increasing with the amount of iron present.

Orton cone. The standard pyrometric cone in the U.S.A.

Oven. Closed space into which a large quantity of coal is placed for carbonizing.

Oven sole. Brickwork on which the coal charge rests in coke oven.

Over firing. (a) Excessive heat treatment, (b) A heat treatment which causes deformation or bloating of clay or clayware.

Pallet. Small portable platform on which product may be handled, dried, stored or transported.

Periclase. An isometric mineral; MgO. Specific gravity 3.58. Melting point approximately 2,800°C.

Perish. To disintegrate as a result of exposure to the atmosphere, as does calcined dolomite.

Perlite. A siliceous glassy rock composed of small spheroids varying in size from small shot to peas; combined water content 3 to 4 %. When heated to a suitable temperature, perlite expands to form a lightweight glassy material with a cellular structure.

Permeability. The property of porous materials which permits the passage of gases and liquids under pressure. The permeability of a body is largely dependent upon the number, size & shape of the open connecting pores and is measured by the rate of flow of a standard fluid under definite pressure.

Picrochromite. Magnesium chrome spinel.

Pillar. One of the upright parts of a crank.

Pinion wall or Buttress wall. Non-refractory end wall of a battery.

Pip. Refractory bottom with a point on its upper surface.

Plastic Chrome Ore. An air-setting ramming material having a base of refractory chrome ore and shipped in plastic form ready for use.

Plastic Fire Clay. A fire clay which has sufficient natural plasticity to bond together other materials which have little or no plasticity.

Plastic refractory or Mouldable refractory. (a) A refractory composition ready for use, which may contain chemical agents to ensure hardening at a low temperature. These compositions are almost always put into place by ramming, (b) A blend of ground fireclay materials in plastic form, suitable for ramming into place to form monolithic linings.

Plasticity. The property of a material, and especially of clay, that allows it to be shaped without cracking and to retain its new form.

Pocket setting or Boxing-in. An arrangement within a setting to prevent excessive loading on special shapes.

Polymorphism. The property by virtue of which some elements and compounds from two or more minerals of different crystal structures and physical properties. For example,, silica crystallizes as quartz as cristobalite and as tridymite—three minerals having distinctive crystal forms and physical properties.

Pores. As applied to refractories, the small voids between solid particles. Pores are described as "open" if permeable to fluids; "sealed" if impermeable.

Porosity of Refractories. The ratio of the volume of the pores or voids in a body to the total volume, usually expressed as a percentage. The "true porosity" is based on the total pore volume; "apparent porosity" on the open pore volume only.

Port. Furnace operating through which liquid or gaseous fuel and/or air for combustion is introduced.

Post or Prop. One of the uprights supporting the batts on a tunnel-kiln car.

Pot. A vessel which may be either open or closed, used for melting and working glass.

Pot arch. A furnace for preheating pots.

Power Pressing. The forming of refractory brick shapes from ground refractory material containing an optimum amount of added water by means of high pressure applied vertically in a power driven press

Printer's bit. One of the small refractory pieces used to separate decorated ware before and during the decorating fire.

Pug Mill. A machine used for blending and tempering clays in a moist or stiffly plastic condition.

Pyrite. The most common sulfide mineral; FeS_2. Specific gravity 4.9 - 5.2. Color, brass yellow. Used mainly for making sulfuric acid and sulfates.

Pyrophyllite. A mineral consisting of hydrated silicate of aluminum; $AlSi_2O_5(OH)$.

Pyroplasticity. The physical state induced by soaking heat which permits a refractory body to be readily deformed under pressure or by its own weight.

Pyrometric Cone. One of a series of pyramidal shaped pieces consisting of mineral mixtures and used for measuring time-temperature effect. A standard pyrometric cone is a three sided truncated pyramid; and is approximately either 25/8 inches high by_5/8 inch wide at the base or 11/8 inches high by 3/8 inch wide at the base. Each cone is of a definite mineral composition and has a number stamped on one face; when heated under standard conditions it bends at a definite temperature.

Pyrometric Cone Equivalent (PCE). The number of that Standard Pyrometric Cone whose tip would touch the supporting plaque simultaneously with a cone of the refractory material being investigated when tested in accordance with the Method of Test for Pyrometric Cone Equivalent (PCE) of Refractory Materials.

Pyroscope. A ceramic device permitting the visual assessment of temperature or degree of firing.

Quarl or Throat. An opening in the wall of a boiler furnace into which a burner is set.

Quartz. A common mineral consisting of silica (SiO_2). Sandstones and quartzites are composed largely of quartz. Specific gravity 2.65.

Quartzite. A hard compact rock consisting predominantly of quartz. There are two types. (a) metaquartzite, a metamorphic rock usually derived from sandstone; and (b) orthoquartzite, a sedimentary rock consisting of grains of silica sand cemented together by at least 10 % of precipitated silica.

Radial brick. A brick, the two end faces of which are parts of concentric cylinders.

Ramming material. A granular refractory composition, hardening by ceramic bonding under the influence of heat. Ramming material are generally put into place by ramming after the addition of a suitable liquid.

Ramming mix. A ground refractory material which is mixed with water and rammed into place to form monolithic furnace linings. **Usually of a less plastic nature than plastic refractories.**

Ramp. The part of the roof of an open-hearth furnace that slopes upwards from the end of the main roof to the top of the end wall.

Recuperator. A continuous heat exchanger in which the sensible heat of combustion products is transferred to incoming air or gas through metal or refractory walls.

Refractories. Nonmetallic materials suitable for use at high temperatures in furnace construction. While their primary function is resistance to high temperature, they are usually called on to resist other destructive influences such as abrasion, pressure, chemical attack and rapid changes in temperature.

Refractory Clay. An earthy or stony mineral aggregate which has as the essential constituent hydrous silicates of aluminum with or without free silica; plastic when sufficiently pulverized and wetted, rigid when subsequently dried and of sufficient purity and refractoriness for use in commercial refractory products.

Regenerator. A periodic heat exchanger which alternatively receives heat from products of combustion and gives up heat in incoming air or gas.

Regenerator Checkers. Brick used in furnace regenerators to recover heat from hot outgoing gases and later to release this heat to cold air or gas entering the furnace; so-called because of the checkerboard pattern in which the brick are arranged.

Relieving arch, (a) An arch built into a wall to relieve that part of the wall below the arch from the weight of brick-work above the arch, (b) Brick arch placed so as to enable repairs to be made to the grate and the door without pulling down the whole gas producer.

Ribbed roof. A roof in which some of the courses have an increased thickness to give continuous protruding ribs.

Rider bricks or Sole-flue port bricks. Group of perforated or spaced bricks on which the checker brick of the regenerator rests.

Ring. A row of bricks in a roof, unbonded with the other rows.

Rise. The vertical distance between the lowest and highest part of the inner surface of an arch.

Rock. A naturally mineral aggregate consisting of one or more minerals. For example, quartzite rock is an aggregate consisting essentially of crystals of the mineral quartz; granite is an aggregate consisting essentially of crystals of feldspar and quartz.

Rotary Kiln. A cylindrical, refractory-lined, gas-fired kiln that rotates at an angle and in which the charge is introduced into the higher end and travels down the slope of the kiln to the discharge end.

Rowlock. A course of bricks consisting of headers laid on edge.

Runner. The channel along which molten iron or slag runs from the blast furnace.

Rutile. A mineral consisting of titanium dioxide (Ti_2O). Specific gravity 4.18 - 4.25.

Saddle. Firing support in the form of a rod of triangular section.

Saggar. A refractory container for protecting ware during firing.

Sand seal. A trough, filled with sand or other granular refractory material, parallel to the walls of a tunnel kiln and forming a seal with a metal plate (apron) on the side of each kiln car.

Scone brick or Split. A brick having the same length and breadth as a standard square but with a thickness less than that of the smallest square.

Scotch block. A monolithic gas port in an open hearth furnace, generally rammed *in situ* around a metal template either while the furnace is hot or when it is cooled.

Seat. A site prepared on the siege of a furnace to receive a pot.

Seating block, A shaped refractory material on which a boiler or part of a boiler rests.

Secondary expansion. The property, exhibited by some fireclay refractories, of developing permanent expansion at temperatures within their useful range; not the same as overfiring.

Segar cone. A type of pyrometric cone used in Europe.

Semi-Silica Fireclay Brick. A fireclay brick containing not less than 72% silica.

Serpentine. A group of rock forming minerals; (Mg, Fe) $3Si_2O_5(OH)_4$. Specific gravity 2.5 - 2.7. Also, a common rock consisting essentially of serpentine minerals.

Setter. A fired refractory having the contour of the article to be fired.

Setting. The process or the result of the arrangement of shaped goods for firing in a kiln.

Side arch. A brick in which the large faces are inclined towards each other in such a way that one of the side faces is narrower than the other.

Siege. The floor of a pot furnace.

Silcrete. A surface quartzite.

Silica. SiO_2, the oxide of silicon. Quartz and chalcedony are common silica materials; quartzite, sandstone and sand are composed largely of free silica in the form of quartz.

Silica Brick. A fired refractory consisting essentially of silica and usually made from quartzite bonded with about 1.8 to 3.5 % of added lime.

Silica Fire Clay. A refractory mortar consisting of a finely ground mixture of quartzite, silica brick and fire clay of various proportions; often called silica cement.

Silicon Carbide. A compound of silicon and carbon; SiC.

Silicon Carbide Refractories. Refractory products consisting predominantly of silicon carbide.

Sill. Horizontal structure member forming the bottom of a door frame in a furnace.

Sillimanite. A brown, grayish, pale green or white orthorhombic mineral; Al_2SiO_5. Specific gravity 3.24. At about 1,530°C it begins to dissociate into mullite and free silica.

Sintering. A heat treatment which causes adjacent particles of material to cohere at a temperature below that of complete melting.

Skewback or Springer. A brick shape having an inclined face from which an arch may be sprung.

Skimmer. A partly immersed block across the entry to a feeder channel.

Slag. Non-metallic material formed during the treatment or purification of a metal or substance resulting from the attack of a refractory product by materials in contact with it.

Slagging, A chemical reaction between a refractory material and an attacking agent, usually occurring at high temperature and with the formation of a liquid or viscous phase.

Slagging of refractories. Destructive chemical action upon refractories at high temperatures, resulting in the formation of a liquid.

Sleeper blocks or Throat cheeks. The side blocks of the throat passage.

Sleeves. Tubular refractory shapes used to protect the metal rod which holds the stopper head in the valve assembly of a bottom pouring ladle.

Slip-casting. Process of shaping by casting a slip in an absorbent mould.

Soaking pit. Reheating furnace for steel ingots which are changed and discharged from above.

Soapstone. A metamorphic rock consisting mainly of talc and derived from the alteration of ferromagnesian silicate minerals.

Soldier course. A course of bricks or blocks set on end.

Solid solution. A homogeneous crystalline phase with a variable composition. The term "solid solution" usually refers to isomorphous mixtures, but sometimes refers to the solution of small proportions of a material in a seemingly unrelated substance.

Spalling. The cracking or fracture of a refractory product caused by differential expansion due to thermal shock, the effect of a steep temperature gradient or a crystalline inversion.

Spalling of Refractories. The loss of fragments (spalls) from the face of a refractory structure, through cracking and rupture, with exposure of inner portions of the original refractory mass.

Span. Horizontal distance between the inner faces of the opposite supports of an arch.

Specific Gravity. The ratio between the weight of a unit volume of a substance and that of some other standard substance under standard conditions of temperature and pressure. For solids and liquids the specific gravity is based on water as the standard. The "true specific gravity" of a body is based on the volume of solid material excluding all pores. The bulk or volume specific gravity is based on the volume as a whole, i.e. the solid material with all included pores. The apparent specific gravity is based on the volume of the solid material plus the volume of the sealed pores.

Specific Heat. The quantity of heat required to raise the temperature of a unit mass of a substance one degree.

Spinel. (a) The mineral composed of magnesium aluminate; $MgAl_2O_4$. Specific gravity 3.6. Melting point 2,135°C. (b) A group of minerals of general formula; AB_2O_4 where A represents magnesium, ferrous iron, zinc or manganese or any combination of these minerals and B represents aluminum ferric iron or chromium.

Spout. A block by means of which glass is allowed to flow from a container.

Spring line. The line of contact between the under surface of an arch and the skewback.

Springer. Same as *Skewback.*

Sprung arch. (a) An arch supported by abutments, *i.e.* skewbacks only (*b*) An arch which is supported by abutments at the sides or ends only. A cross-section of a sprung arch taken at right angles to its axis, usually consists of a segment of a circular ring in which the inner and outer arch surfaces are represented by arcs of concentric circles.

Spur. Refractory tetrahedron with concave faces and very sharp points.

Standard pyrometric cone. A pyramid with triangular base, of specific shape and dimensions and of such composition that it bends (the tip bending to the level of the base) at a definite, temperature when it is heated under standardized conditions.

Stretcher. A brick laid flat, with its length parallel to the face of the wall.

Stud. A piece of metal welded into a water tube to anchor a monolithic refractory lining.

Superduty Fireclay Brick. Fireclay brick which have a PCE not lower than Cone 33 and which meet certain other requirements.

Suspended Arch. An arch in which the brick shapes are suspended from overhead supporting members.

Talc. A hydrous magnesium silicate mineral; $Mg_3Si_4O_{10}(OH)_2$. Specific gravity 2.7 - 2.8. Mho's Hardness - 1.

Tapping launder, Tapping launder or Tapping spout. Channel extending from the taphole.

Teeming, Pouring or Casting. The process of transferring molten steel from a ladle to mould or moulds.

Tempering. Mechanical treatment of a wet clay, with or without non-plastic to improve its workability.

Texture. Macroscopic relationship existing between the various shapes and sizes of pores and grains in a refractory product.

Thermal Conductivity. The property of matter by virtue of which heat energy is transmitted through particles in contact.

Thermal Expansion. The increase in linear dimensions and volume which occurs when materials are heated and which is counterbalanced by contraction of equal amount when the materials are cooled.

Thermal shock. A sudden change in temperature liable to cause spalling.

Thin section. A sample of a refractory material prepared for penological examination under the microscope by transmitted light.

Thixotropy. A reversible process by which certain materials become less fluid on standing and more fluid on stirring.

Throat. (a) The upper part of the blast furnace **at** the top of the stack, *(b)* The submerged passage in a bridge connecting into two compartments of a tank furnace, (c) The part of the furnace with the minimum cross-section separating the gas burner from the bath, in certain open-hearth furnaces.

Tridymite. A mineral form of silica; SiO_2. Stable from 870 to 1,470°C. Specific gravity 2.26. An important constituent of silica brick.

Trough. An open receptacle through which molten metal is conveyed from a holding device or furnace to a casting mold or another receptacle.

Trumpet assembly. The vertical refractory tube encased in a metal shell which is used in bottom pouring. It consists of a trumpet, a number of guide tubes and a bottom guide.

Tundish. A refractory-lined vessel or trough, having one or more nozzles in the bottom, which is interposed during teeming between a ladle and a mould or moulds.

Tuyere Brick. A refractory shape containing one or more holes through which air and other gases are introduced into a furnace.

Under-firing. Insufficient heat treatment.

Vacuum Pressing. A method of forming brick shapes by which they are subjected to a partial vacuum during pressing in a power press.

Vermiculite. A group of micaceous minerals, all hydrated silicates, varying widely in composition; $(Mg,Fe,Al)_3$ $(AlSi)_4O_{10}(OH)_2$. $4H_2O$. When heated above 150°C, vermiculite exfoliates and increases greatly in volume.

Vesicular. Having a cellular structure; applied to fire clays which have become bloated by overfiring.

Vibratory Pressing. A process for forming refractory shapes in which the ground particles of refractory material are packed closely together by rapid impact type vibrations of the top and bottom dies; also called impact pressing.

Vitrification. A process of permanent chemical and physical change at high temperatures in a ceramic body, such as fire clay, with the development of a substantial proportion of glass.

Wad. An extruded strip of clay used to seal the joints between two saggars or to level the saggars in the bung.

Warpage. The degree of bending, twisting or distortion which may occur during manufacture of a refractory brick, shape or block.

Water-smoking. The stage in the firing of clay products when water vapour is being evolved.

Weathering. The alteration of raw materials by exposure to the weather, resulting particularly in the oxidation and solution of certain impurities and their partial elimination.

Wedge. Same as *End Arch.*

Wedge Brick. A brick shape having six plane faces (two sides, two edges and two ends), in which two faces (the sides) are inclined toward each other and one end face is narrower than the other.

Whelp. A brick having the same thickness and breadth as a standard square but a greater length.

Wicket, (a) Wall built to close the entrance to the regenerator chamber of an open hearth furnace. (b) A wall built to close an entrance to a kiln.

Wollastonite. A triclinic mineral; $CaSiO_3$. Specific gravity 2.9. Inverts at 1,200°C to pseudo wollastonite. Melts incongruently at 1,544**°C.**

Workability. The ease with which a moistened material can be worked and shaped; this property depends on the plasticity and consistency of the body.

Young's Modulus. In mechanics, the ratio of tensile stress to elongation within the elastic limit; the modulus of elasticity.

Zircon. A mineral; ZrSiO4. Specific gravity 4.7. Begins to melt incongruently at 1,685°C forming ZrO_2 solid solution plus liquid; completely melted at about 2,650°C.

Zirconia. Zirconium oxide; ZrO_2. Specific gravity 5.8. Melting point 2,700°C. Its chief source is the mineral baddeleyite.

Short Questions & Answers on Refractory Technology

Q1. What is a refractory?

Ans. Refractories are heat resistant materials used in processes involving high temperatures and/ or corrosive environment. Refractories are inorganic, nonmetallic, porous and heterogeneous materials composed of thermally stable mineral aggregates, a binder phase and additives. These are typically used to insulate and protect industrial furnaces and vessels due to their excellent resistance to heat, chemical attack & mechanical damage.

Q2. What are the principal raw material used in the manufacture of refractories?

Ans. The principal raw materials used in the production of refractories are: the oxides of silicon, aluminum, magnesium, calcium & zirconium and some non-oxide refractories like carbides, nitrides, borides, silicates & graphite.

Q3. What are Refractories used for?

Ans. Refractories are used by the metallurgy industry in the internal linings of furnaces, kilns, reactors & other vessels for holding and transporting metal & slag. In non-metallurgical industries, the refractories are mostly installed on fired heaters, hydrogen reformers, ammonia primary & secondary reformers, cracking furnaces, incinerators, utility boilers, catalytic cracking units, coke calciner, sulfur furnaces, air heaters, ducting, stacks, etc. Majority of these equipments operate under high pressure, and operating temperature can vary from very low to very high (500 to 1600°C). The refractory materials are needed to withstand temperatures over and above these temperatures.

Q4. What are the basic classes of refractories?

Ans. Primarily, refractories are classified as basic, high alumina, silica, fireclay and insulating. There are also classes of "special refractories" which include silicon carbide, graphite, zircon, zirconia, fused cast and several others. Most refractory materials are supplied as preformed shapes. However, they also are manufactured in the form of special purpose clays, bonding mortars, and monolithics, such as hydraulic setting castables, plastic refractories, ramming mixes and gunning mixes. A variety of processed refractory grains & powders are also available for certain applications.

Q5. What are the Requirements of Right Refractory for a particular job?

Ans. The general requirements of a refractory material are as follows:

- Its ability to withstand high temperatures and trap heat within a limited area like a furnace.
- Its ability to withstand action of molten metal, hot gasses and slag erosion etc.
- Its ability to withstand load at service conditions.
- Its ability to resist contamination of the material with which it comes into contact.
- Its ability to maintain sufficient dimensional stability at high temperatures and after/during repeated thermal cycling.
- Its ability to conserve heat.

Q6. What are the various basis on which Refractories are Classified?

Ans. Refractories are classified on the basis of the:

- Chemical composition *eg.*; acid,basic & neutral refractory
- Method of Manufacture *eg.*; Dry Press Process, Fused Cast, Hand Molded etc.
- Physical Form *eg.*; *shaped* and *unshaped* refractories.
- Refractoriness (fusion point)
- Oxide content *eg.*; Single oxide refractories (Ex.: Alumina, Magnesia, Zirconia), Mixed oxide refractories (Ex.: Spinel, Mullite), Non-oxide refractories (Ex.: borides, Carbides, Silicates)

Q7. Describe refractories Classification Based on Chemical Composition with examples.

Ans. Refractories are classified on the basis of their chemical behaviour, *i.e.* their reaction to the type of slags. Accordingly the refractory materials are of three classes - Acid, Basic & Neutral.

Acid Refractories: Acid bricks contain at least 92% silicon oxide (SiO_2). Acid refractories are those which are attacked by alkalis (basic slags). These are used in areas where slag and atmosphere are acidic. Examples of acid refractories are: Silica (SiO_2), fireclay, Zirconia (ZrO_2),etc.

Basic Refractories: Basic bricks contain at least 60% magnesium oxide (MgO). Basic refractories are those which are attacked by acid slags but stable to alkaline slags, dusts and fumes at elevated temperatures. Since they do not react with alkaline slags, these refractories are important for furnace linings where the environment is alkaline; for example non-ferrous metallurgical operations. Important basic refractories are: Magnesia (MgO) - caustic, sintered and fused magnesia,Dolomite (CaO*MgO) - sintered and fused dolomite, Chromite -main part of chrome ore etc.

Semi-basic bricks contain at least 65% silicon oxide. but less than 30% alumina (Al_2O_3). Chemical characteristics of the furnace process usually determine the type of refractory required. Theoretically, acid refractories should not be used in contact with basic slags, gases and fumes whereas basic refractories can be best used in alkaline environment. Actually, for various reasons, these rules are often violated.

Neutral Refractories: Neutral bricks contain at least 30% alumina. Neutral Refractories are chemically stable to both acids & bases and are used in areas where slag & atmosphere are either acidic or basic. The examples of neutral refractories are: Carbon / graphite, Chromites (Cr_2O_3), Alumina etc. Out of these graphite is the least reactive and is extensively used in metallurgical furnaces where the process of oxidation can be controlled.

Q8. How are refractories Classified Based on Method of Manufacture?

Ans. The refractories can be manufactured in either of the following methods:

a. Dry Press Process
b. Fused Cast
c. Hand Molded
d. Formed (Normal, Fired or chemical bonded)
e. Unformed (Monolithic – Plastics, Ramming mass, Gunning, Castable, Spraying)

Q9. Describe refractory Classification Based on its Physical Form.

Ans. Refractories are classified according to their physical form. These are the shaped and unshaped refractories. The shaped refractory is commonly known as refractory bricks and the unshaped refractory as "monolithic" refractories.

Shaped Refractories: Shaped refractories are those which have fixed shaped when delivered to the user. These are what we call bricks.

Brick shapes maybe divided into two: standard shapes and special shapes. Standards shapes have dimension that are conformed to by most refractory manufacturers and are generally applicable to kilns & furnaces of the same type. Special shapes are specifically made for particular kilns and furnaces. This may not be applicable to another furnaces or kiln of the same type.

Shaped refractories are almost always machine-pressed, thus, high uniformity in properties are expected. Special shapes are most often hand-molded and exhibit slight variations in properties.

Unshaped (monolithic) Refractories: Unshaped refractories are without definite form and are only given shape upon application. It forms joint less lining and are better known as **monolithic** refractories. These are categorized as Plastic refractories, ramming mixes, castables, gunning mixes, fettling mixes and mortars.

Q10. How are refractory classified based on refractoriness (fusion point)?

Ans. Classification based on refractoriness (fusion points).

Refractories are also classified based on their refractoriness as given below:

Type of Refractory	PCE Value (Segar Cones)	Refractoriness/ fusion point, °C
Low heat duty refractories (LHD)	19—28	1520—1630
Medium/Intermediate heat duty refractories (MHD)	28—30	1630—1670
High heat duty refractories (HHD)	30—33	1670—1730
Super duty refractories (SD)	> 33	> 1730

Q11. What are different steps involved in the manufacture of refractories?

Ans. The following steps are followed in the manufacture of refractories.

Preparation of Mixture:

- The raw materials are crushed into a uniform size, about 25 mm, which are further grinded by grinding machine into the size of 200 mesh.
- The undesirable materials are removed from raw materials by either of the screening processes: (i) settling, (ii) magnetic separation, and (iii) chemical methods.
- The screened materials are mixed thoroughly to get proper distribution of plastic materials. Then it is subjected for moulding.

Moulding:

- Moulding is carried out mechanically or manually under high pressure.
- Mechanical moulding produces refractories of good strength and high density (low porosity) while the refractories of low strength and low density (high porosity) are obtained from hand moulding process.

Drying:

In this method, the moisture from the refractories is removed by using hot air slowly and for a longer period to get good quality desirable refractory.

Firing:

The dried refractories are heated under high temperature range of 1400 to 1900°C in tunnel kilns, or shaft kilns or rotary kilns to get high strength, high density and heat & chemical resistant materials.

General Method of Manufacture of Refractory Bricks

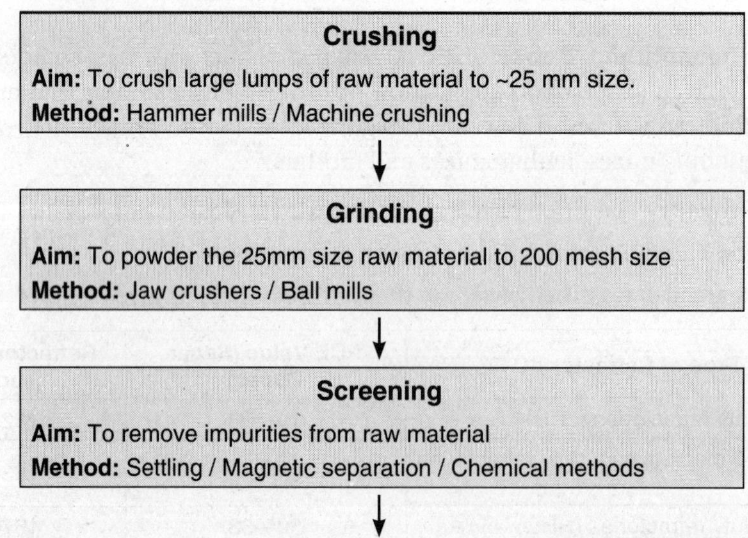

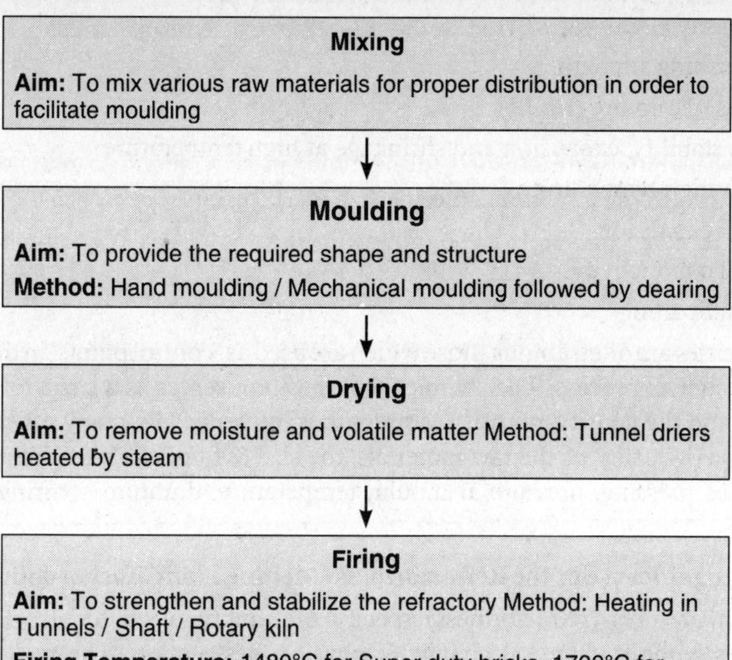

The raw materials are crushed, ground, screened and stored separately. Just before moulding they are mixed in the required ratio and processed. Moulding can be done by either one of the following ways:

- Hand moulding: For low density and low strength bricks.
- Mechanical Moulding: For high density and high strength bricks.

Dearing is essential for hand pressed refractory.

Dearing methods:

- Vacuum- Sucking air through pores
- Double pressing

Dearing of refractories increases their:

- Density
- strength

Q12. What are the important Properties of Refractories?

Ans. The quality of a refractory and its suitability for a particular application primarily depends on its physical, chemical & mineralogical properties. It may be possible to assess the quality of a refractory on the basis of a single property or a group of properties. The most common properties that are considered in selecting the optimum refractory are as follows:

- Refractoriness/melting point
- Refractoriness under load (RUL)
- Porosity

- Bulk density
- Cold crushing strength
- Modulus of rupture (MOR)
- Volume stability, expansion and shrinkage at high temperature
- Creep at high temperature
- Thermal expansion
- Thermal conductivity
- Thermal Spalling

These properties are often among those which are used as 'control points' in the manufacturing and quality control process. The chemical composition serves as a basic for classification of refractories and the density, porosity & strength is influenced by many other factors. Among these are type & quality of the raw materials, the size & fit of the particles, moisture content at the time of pressing, pressure at mould, temperature, duration of firing and the rate of cooling.

Q13. What is the significance of the Refractoriness/Melting Point /fusion point of refractories?

Ans. Melting temperatures (refractoriness) specify the ability of materials to withstand high temperatures without chemical change & physical destruction. The melting point of few elements that constitute refractory composition in the pure state varies from 1716 to 3482°C as indicated in the table below:

Refractory Element	Melting Temperatures (°C)
Graphite C Pure	3482
Thoria, ThO_2 Pure Sintered	3000
Magnesia, MgO, Pure Sintered	2800
Zirconia, ZrO_2, Pure Sintered	2700
Lime, CaO	2570
Beryllia, BeO, Pure Sintered	2550
Silicon Carbide, SiC, Pure	2250
Magnesia, 90-95%	2193
Chromite, FeO_2-Cr_2O_3	2182
Chromium Oxide	2138
Alumina, Al_2O_3, Pure Sintered	2050
Chromite, 38%, Cr_2O_3	1970
Fireclay	1870
Titania, TiO_2	1850
Kaolin	1816
Silica, SiO_2	1716

The melting point serves as a sufficient basis for considering the thermal stability of refractory mixtures and is an important characteristic indicating the maximum temperature of use.

Refractoriness of a refractory is the temperature at which is the material softens and hence is the measure of fusibility of the material. Addition of an impurity in the refractory material lowers its fusion point/refractoriness depending upon the amount of impurity present, the melting point of the lowest fusing constituent and the capacity of the lowest fusing constituent to dissolve the material of higher fusion point at comparatively low temperature.

Refractoriness is a property at which a refractory will deform under its own load. The refractoriness is indicated by PCE (Pyrometric cone equivalent). It should be higher than the application temperatures.

Refractoriness decreases when refractory is under load. Therefore more important is refractoriness under load (RUL) rather than refractoriness.

Q14. What do you mean by Pyrometric Cone Equivalent (PCE) of a refractory?

Ans. "Fusibility temperature range" is a more appropriate term for refractory materials as mixture of refractory materials do not have a sharp fusion/melting point. The heat resistance of refractory material is expressed in terms of Pyrometric Cone equivalents (PCE) which is determined by comparing the softening of a test pyramid cone with that of standard pyramid cones of known fusion temperature by heating them under similar conditions.

Temperature at which a refractory will deform under its own weight is known as its softening temperature which is indicated by PCE. Refractories due to their chemical complexity melt progressively over a range of temperature. Hence refractoriness or fusion point is ideally assessed by the cone fusion method. The equivalent standard cone which melts to the same extent as the test cone is known as the pyrometric cone equivalent (PCE).

The standard cones are called 'Segar cones' (as German standard) and 'Orton cones' (as British standard) and they are numbered according to their refractoriness/fusion points. PCE can be useful for quality control purposes to detect variations in batch chemistry that result from changes or errors in the raw material formulation.

End Point of Orton Pyrometric Cones			
Cone number	**End point (°C)**	**Cone number**	**End point (°C)**
12	1337	31	1679
13	1349	31½	1699
14	1398	32	1717
15	1430	32½	1730
16	1491	33	1741
17	1512	34	1759
18	1522	35	1784
19	1541	36	1796
20	1564	37	1830
23	1590	38	1850
26	1605	39	1865
27	1627	40	1885

28	1638	41	1970
29	1645	42	2015
30	1654		

Q15. How is refractoriness (fusion point) of a refractory measured in terms of Pyrometric Cone Equivalent (PCE)?

Ans. Refractoriness is defined as the ability of a material to withstand high temperature. It is measured in terms of Pyrometric cone equivalence.

This method was first adopted by Seger, hence also called Seger cone equivalence.

Pyrometric cones Test Method

- This method is used in ceramic industries to test 'refractoriness' of refractory bricks.
- The test material cone is mix of oxides that melt at specific temperatures.
- The cone has a equilateral triangular base (19mm).
- Height of the cone is 38 mm.
- The cone is heated at standard conditions (ie. at the rate of 10°C/min).
- Temperature at which the refractory brick or the cone bend (ie) the temperature at which apex touches the base is called refractoriness (units PCE)
- Refractory cannot be used above tnis temperature.

According to ASTM C24 - 01, PCE is measured by making a cone of the material and firing it until it bends to 3 O'clock.

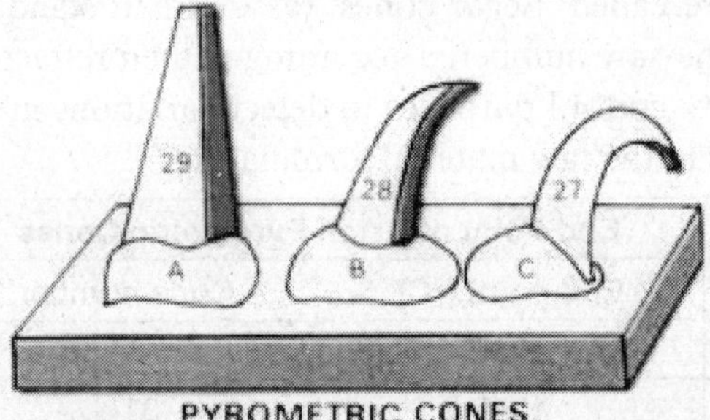

PYROMETRIC CONES

Cone # 29: Stiff at test temperature Cone # 28: Just started Softening

Cone # 27: Apex almost touching base

Thus in the figure above, refractoriness of Sample A is much higher than that of B and C. The pyrometric cone equivalent indicates only the softening temperature. But, in service the refractory is subjected to loads which would deform the refrectory at a much lower temperature than that indicated by PCE. With change in the environmental conditions, such as reducing atmosphere, the P.C.E. value changes drastically.

Representative PCE values for selected refractories include cones 33-34 for super duty fireclay, cones 29-31 for medium duty fire clay and cones 36-37 for a 60% alumina product. The cone values reported for refratories are based on a defined standard time – temperature relationship, so different heating rates will result in different PCE values.

PCE can be useful for quality control purposes to detect variations in batch chemistry that result from changes or errors in the raw material formulation.

Q16. What is Refractoriness under load (RUL)? Discuss its significance as an important property of the refractory.

Ans. Refractoriness refers to the resistance to extreme conditions of heat and corrosion when hot & molten materials are contained while being transported and/or processed.

The ability to withstand exposure to elevated temperatures without undergoing appreciable deformation is measured in terms of refractoriness. The refractoriness under load test (RUL test) gives an indication of the temperature at which the bricks will collapse, in service conditions with similar load. However, under actual service where the bricks are heated only on one face, most of the load is carried by the relatively cooler rigid portion of the bricks. Hence the RUL, test gives only an index of the refractory quality, rather than a figure which can be used in a refractory design. Under service conditions, where the refractory used is heating from all sides such as checkers, partition walls, etc. the RUL test data is quite significant.

In actual practice the refractoriness of the material is much less compared to that found by Seger cones method. This may be due to the load it is exposed during the process. In order to study the exact serviceable conditions, RUL is adopted which is a measure of deformation under load and temperature.

- **Test specimen:** rectangular in shape with a base area of 5cm^2 and height 75cm.
- **Test load:** 1.75 Kg/cm^2 (3.75 Kg/cm^2)
- Heating at a standard. rate of 10 °C/minute.
- Temperature at which 10% deformation occurs is RUL.

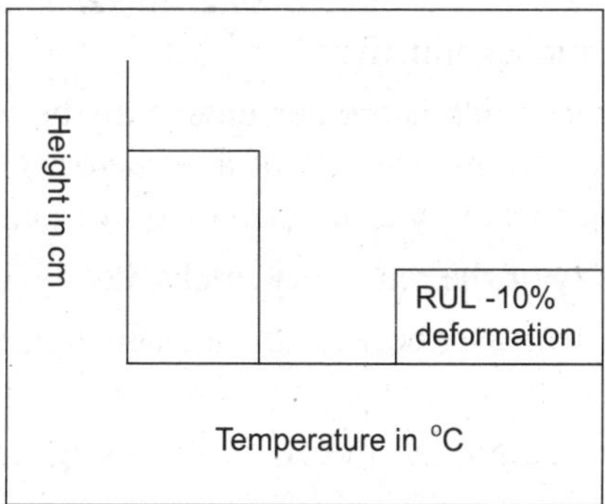

Refractoriness under load (RUL) of a refractory material is a measure of its failure resistance to the combined action of heat and load. RUL is the softening temperature of a refractory brick (indicated by breaking of the test specimen) when a load of 2 kgf/cm^2 is applied as per the standard test procedure.

Refractory bricks generally having more than two constituent compounds fail under load conditions at much lower temperature (*i.e.*, at RUL) than the fusion point (*i.e.*, refractoriness) of the highest melting and predominant constituent. That signifies that the temperature corresponding to the refractoriness under load of a refractory brick is much lower than its refractoriness (fusion/melting point). A refractory material composed of a single pure compound will have a closer value of its refractoriness and refractoriness under load.

Refractory failure is also influenced by its exposure time to load and heat. Longer exposure time results in failure of refractory under relatively low load and temperature conditions. Similarly refractories will fail at a lower temperature with increased load.

Q17. Describe the significance of the Porosity of refractories.

Ans. Porosity is a measure of the effective open pore space in the refractory into which the molten metal, slag, fluxes, vapors etc can penetrate and thereby contribute to degradation of the structure. The porosity of refractory is expressed as the average percentage of open pore space in the overall refractory volume.

High porosity materials tend to be highly insulating as a result of high volume of air they trap, because air is a very poor thermal conductor. As a result, low porosity materials are generally used in hotter zones, while the more porous materials are usually used for thermal backup.

Such materials, however, do not work with higher temperatures & direct flame impingement, and are likely to shrink when subjected to such conditions. Refractory materials with high porosity are usually not chosen when they will be in contact with molten slag because they cannot be penetrated as easily.

The apparent porosity is an important property in cases where the refractory is in contact with molten charge and slags. A low apparent porosity is desirable since it would prevent easy penetration of the refractory. Size and continuity of pores will have important influences on refractory behaviour. A large number of small pores is generally preferable to an equivalent number of large pores.

Porosity affects chemical attack by molten slag, metal and gases. Decrease in porosity increases strength and thermal conductivity.

True porosity of a refractory brick is the percentage of the volume of combined open and close pore space of its total volume. Density of a refractory brick is controlled by both open and close pores while its permeability is affected by only open (*i.e.*, interconnected) pores.

With increase in the porosity of the refractory bricks, its:

- Thermal conductivity decreases because the air which fills the pores acts as an insulator, which is why insulating bricks have higher porosity.
- Resistance to the penetration of metal, molten slag and flue gases reduces due to increased permeability.
- Temperature fluctuation sensitivity reduces as pores can accommodate the thermal expansion due to heat.
- Strength reduces due to decreased compactness obviating their usage in taller structure demanding higher strength.

Q18. How is porosity defined and determined?

Ans. In raw/green refractory materials the pores are all open fluids can pass through them. When these materials are fired some liquid formation due to fusion takes place and as a result some pores will be sealed. The apparent porosity as such takes the open pores and not the closed pores. True & apparent porosity is defined as:

$$T.P = (O+C)\times\frac{100}{E}$$

$$A.P = O\times\frac{100}{E}$$

Where,

T.P = True Porosity

A.P = Apparent Porosity

O = Volume of open pores

C = Volume of closed pores

E = External volume

Boiling Point Method for Porosity Determination:

The apparent porosity of a clean small refractory sample (measuring 6.5 x 6.5 x 4 cm) is measured by making three weighings. The first the dry specimen, the second the saturated specimen immersed in water and third the saturated specimen suspended in air. To ensure that the specimen is thoroughly saturated, it should be boiled in water for two hours. The apparent porosity is determined from these weighings by the equation:

$$A.P. = (W-D)\times\frac{100}{(W-A)}$$

where,

A.P. = % of apparent porosity

W = Weight of saturated specimen in air

D = Weight of dry specimen in air (dried at 110°C in an oven)

A = Weight of saturated specimen submerged in water.

The total porosity present can be found by relating the theoretical density (ρth), to the calculated density (ρ). The total porosity (T.P) of the sintered samples is also calculated from the following equation:

$$T.P.\ \% = (\rho th-\rho)\times\frac{100}{\rho th}$$

Q19. How the Bulk Density of a refractory is an useful property to indicate its quality?

Ans. A useful property of refractories is bulk density, which defines the material present in a given volume. The bulk density is generally considered in conjunction with apparent porosity. It is a measure of the weight of a given volume of the refractory. For many refractories, the bulk density provides a general indication of the product quality.

An increase in bulk density increases the volume stability, the heat capacity, as well as the resistance to abrasion and slag penetration.

While evaluating a refractory brand or comparing several products of equivalent type (except insulating types), it is considered that the refractory with higher bulk density (ie; lower porosity) will be better in quality. The structure of a refractory having higher bulk density will be denser, resulting in better resistance to chemical attack, decreased metal penetration, better abrasion resistance and other related benefits.

True density is the weight per unit volume of the refractory including the volume of its open and close pore space.

Q20. How the bulk density of refractories is determined?

Ans. The bulk density of any refractory is measured by the equation:

$$\text{Bulk Density} = \left(\frac{\text{Total weight}}{\text{Total volume}}\right) \text{of refractory}$$

Bulk density is measured using either direct measurement method or direct volume determination method.

Direct Measurement Method. The rectangular refractory test specimen brick is accurately weighed (to nearest 50 g) and its dimension is noted using a hook rule for calculating its volume. The bulk density is calculated using the formula:

B = W/V

where,

B = Bulk density of brick, g/cc,

W = Weight of the test specimen, gm

V = Volume of the test specimen, cc.

Direct Volume Measurement Method. The test specimen of refractory measuring (6.5 × 6.5 × 4) cm is cleaned of its adhering dust. Its dry weight (D) is recorded (after drying it at 110°C in an oven).

The specimen is then boiled in deionized water for more than 1 hour and cooled to room temperature. The immersed weight in deionized water (S) is measured at room temperature. The test specimen is then lifted up slowly from the immersion liquid by means of the sling thread. Saturated weight (W) of the test specimen is found by weighing the soaked specimen (after wiping off the liquid droplets from the specimen surface by a blotting paper) while suspended in air. The bulk density is then calculated by:

$$B = \frac{D}{(W - S)}$$

The direct volume determination method (generally used for deeply branded and irregular refractories) gives more accurate results than the direct measurement method (used for rectangular shapes of refractory only).

This is one of the main criterias for assessing the quality aspect of the particular brick. The true specific gravity of the material composing a brick as distinct from the bulk density of the brick as a whole (including the pore spaces) is of special significance only for silica bricks and magnesite bricks. This is due to almost constant composition of other bricks.

The specific gravity of firebrick is of little significance owing to the complex nature of the material.

Q21. What is the importance of strength of refractory in its utilization?

Ans. The physical strength of refractories in both cold & hot conditions, are characterized as measures of the use of a refractory. Cold strengths indicate the handling and installation of the refractory, whereas hot strengths indicate how the refractory will perform when used at elevated temperatures. Initial strength develops in refractory materials during the forming process. For shaped refractories, the strengths develop during the physical processing of the products and sometimes followed by higher temperatures where the refractory goes through a firing process. For monolithic refractories, the initial strength develops during the installation or forming process (for precast shapes), and the final strength develops while in application.

High-temperature strengths of refractories is more important rather than cold strengths, since refractories are used at elevated temperatures and not at room temperatures. Strengths of refractories are measured as cold compressive strength, cold modulus of rupture, or hot modulus of rupture. Hot modulus of rupture provides the best indication of the performance of a refractory material in use.

Strengths of refractories are usually reported in terms of the three-point bend strength or the flexural strength, frequently called the modulus of rupture (MOR). In the case of refractories, there exists a standard test for this simple strength measurement. It is based on the familiar formula:

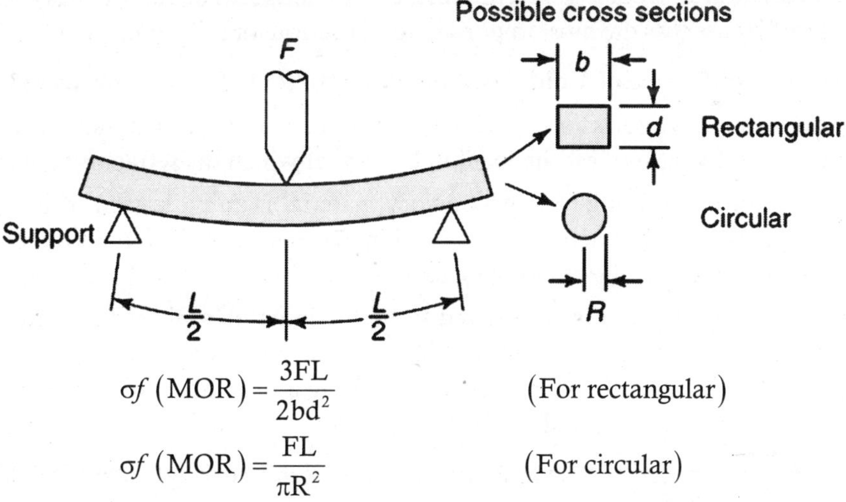

$$\sigma f\,(\text{MOR}) = \frac{3FL}{2bd^2} \qquad \text{(For rectangular)}$$

$$\sigma f\,(\text{MOR}) = \frac{FL}{\pi R^2} \qquad \text{(For circular)}$$

Where: σf is the strength in three-point bending, L is the length of the test span, b is the specimen width, and d is the specimen height. The fracture load is F. When the cross section is circular, R is the specimen radius.

Characteristic flexural strength values for several ceramic materials are given in Table below. The flexural strength will depend on sample size, since with increasing size there is an increase in the probability of the existence of a crack-producing flaw, along with a corresponding decrease in the flexural strength.

Q22. What is Modulus of rupture (MOR) of a refractory?

Ans. The modulus of rupture (MOR) is the flexural breaking strength of a refractory. MOR is measured at room temperature and expressed in kg/cm^2.

Q23. What is Cold Modulus of Rupture (CMOR) of a refractory?

Ans. The cold modulus of rupture of a refractory material indicates the flexural strength and its suitability for use in construction. It is indicative of the strength of the bonding system of the refractory product. Since the test is done at room temperature, it can only show its suitability and its use in construction. It provides no indication of how the refractory will behave at elevated temperatures.

Q24. What is the significance of Hot modulus of rupture (HMOR) of refractory in its utilization.?

Ans. The hot modulus of rupture (HMOR) is the flexural breaking strength of a refractory at a chosen elevated temperature or over a range of temperatures. The structural integrity and general abrasion characteristics of a refractory can be estimated from HMOR, making it an essential property to determine the suitability of a refractory in a certain temperature profile for a certain set of application conditions.

The hot modulus of rupture provides the indication of a refractory material about its flexural strength at elevated temperatures. Since refractories are used at elevated temperatures, the hot modulus of rupture is the true indicator of the suitability and performance of a refractory at high temperatures. Hence, in recent years, the hot modulus of rupture has been prescribed and required by users as the most important test criterion for selection and use of refractories.

Q25. What is the significance of Cold Crushing Strength (C.C.S) of a refractory?

Ans. The cold crushing strength is the capacity of a refractory to provide resistance to a compressive load at room temperature. It is the load, in kg/cm^2 at which the refractory breaks.

The cold compressive strength of a refractory material is an indication of its suitability for use of refractories in construction. It is a combined measure of the refractory for the strength of the grains and also of the bonding system.

The cold crushing strength is considered to be of doubtful relevance as a useful property, other than it reveals little more than the ability to withstand the rigorous of transport. It can be seen as a useful indicator to the adequacy of firing and abrasion resistance in consonance with other properties such as bulk density & porosity.

Q26. What is the importance of Size and Dimensional Stability of refractory bricks?

Ans. The size and shape of the refractories is an important feature in design since it affects the stability of any structure. Dimensional accuracy & size is extremely important to enable proper fitting of the refractory shape and to minimize the thickness & joints in construction.

Dimensional stability is the measure of ability of a refractory material to resist any change in the volume during working conditions. Dimensional changes could be reversible or irreversible. Irreversible or permanent changes may be due to contraction or expansion of brick material. Phase transformations may also occur during heating and cooling of refractories.

Reversible change occur in most bricks at temperature less than 750°C. Irreversible changes occur in the following cases:

i. Shrinkage

$$\underset{\text{(amorphous)}}{\textbf{Magnesite}} \xrightarrow{\text{Heat}} \underset{\text{(crystalline)}}{\textbf{Periclase}}$$

ii. Expansion

$$\textbf{Quartz} \xrightarrow[\text{(crystalline) 870°C \& load}]{\Delta} \textbf{Tridymite} \xrightarrow[\text{1470°C \& Load}]{\Delta} \textbf{Crystobalite}$$

The contraction or expansion of the refractories can take place during service. Such permanent changes in dimensions may be due to:

- The changes in the allotropic forms which cause a change in specific gravity.
- A chemical reaction which produces a new material of altered specific gravity.
- The formation of liquid phase.
- Sintering reactions.
- It may also happen on account of fluxing with dust and slag or by the action of alkalies on fireclay refractories, to form alkali-alumina silicates, causing expansion and disruption.

Q27. What is the significance of Volume stability, expansion and shrinkage at high temperature?

Ans. The contraction or expansion of the refractories can take place during service. Such permanent changes in dimensions may be due to:

- The changes in the allotropic forms which cause a change in specific gravity.
- A chemical reaction which produces a new material of altered specific gravity.
- The formation of liquid phase.
- Sintering reactions.
- It may also happen on account of fluxing with dust and slag or by the action of alkalies on fireclay refractories, to form alkali-alumina silicates, causing expansion & disruption. This is an example which is generally observed in blast furnaces.

While it is desirable that all these changes are effected during manufacturing, it is not possible due to economic reasons, as the process is time dependent. Permanent Linear Change (PLC) on reheating and cooling of the bricks give an indication on the volume stability of the product as well as the adequacy of the processing parameters during manufacture. It is particularly significant as a measure of the degree of conversion achieved in the manufacture of silica bricks.

Q28. How does refractory Creep at high temperature affects its performance?

Ans. Creep is a time dependent property which determines the deformation in a given time and at a given temperature by a material under stress. Refractory materials must maintain dimensional stability under extreme temperatures (including repeated thermal cycling) and constant corrosion from very hot liquids and gases. The criterion of acceptance usually adopted is; that compressive creep (deformation at a given time and temperature under stress) for normal working conditions of load and temperature shall not exceed 0.3% in the first 50 hours of the test. This figure has been found to indicate that the creep rate falls by a negligible amount at the end of the stipulated period, and therefore the refractory can be considered safe to use for a much longer time.

Dimensional changes take place in a refractory under a compressive load at elevated temperature. The dimensional change could be linear on increasing the temperature is known as thermal expansion under load (TEUL). The dimensional changes due to extended period of holding/soaking of a refractory at pre-selected temperature is nonlinear, and leads to plastic deformation known as creep. More specifically, creep is the heat-activated plastic deformation of a body under stress as a function of time. TEUL and creep are typically determined in sequence in the same test, using the same sample.

Q29. Describe the significance of thermal conductivity of refractory bricks.

Ans. Thermal conductivity depends upon the chemical & mineralogical compositions as well as the glassy phase contained in the refractory and the application temperature. The conductivity usually changes with rise in temperature. In cases where heat transfer is required though the brick work, for example in recuperators, regenerators, muffles, etc. the refractory should have high conductivity. Low thermal conductivity refractory brick is desirable for conservation of heat by providing adequate insulation.

The provisions for back-up insulation, conserves heat but at the same time it increases the hot face temperature and hence the demand on the refractory quality increases.

Accordingly, insulation on the roof in open hearth furnaces is normally not provided, otherwise it would cause failure due to severe dripping. Depending on the characteristic of the refractory used in the hot face, such as the high temperature load bearing capacity, it may be required that the quality of the brick be increased to match the rise in temperature caused by over insulation.

Light weight refractories of low thermal conductivity find wider applications in the moderately low temperature heat treatment furnaces, where its primary function is usually conservation of energy. It is more so in case of batch type furnaces where the low heat capacity of the refractory structure would minimize the heat storage during the intermittent heating and cooling cycles.

Q30. How Thermal Conductivity of refractory bricks play important role in its utilization?

Ans. Thermal conductivity is defined as the quantity of heat that will flow through a unit area in direction normal to the surface area in a defined time with a known temperature gradient under steady state conditions. It indicates general heat flow characteristics of the refractory and depends upon the chemical & mineralogical compositions as well as the application temperature.

High thermal conductivity refractories are required for some applications where good heat transfer is essential such as coke oven walls, regenerators, muffles and water cooled furnace walls. However, refractories with lower thermal conductivity are preferred in industrial applications, as they help in conserving heat energy.

Porosity is a significant factor in heat flow through refractories. The thermal conductivity of a refractory decreases on increasing its porosity. Although it is one of the least important properties as far as service performance is concerned, it evidently determines the thickness of brick work.

Q31. What is the effect of the structure and the microstructure of refractories on the thermal conductivity?

Ans. After firing process of refractories, the inner structure contains three phases (crystal phase, glass phase and gas phase (pores)). Effect of the structure and the microstructure of refractories on the thermal conductivity is as follows:

- The highest conductivity are achieved in the cluttered structure, ie, in structures consisting of single element, structure made up elements of similar atomic weight, and structure with no extraneous atoms in solid solution. Graphite is good example of single element refractory structure having high thermal conductivity.
- Refractories composed of elements of similar atomic weight and size, have high thermal conductivity. Good examples of this type of refractory are BeO, SiC and B_4C. The refractories having large difference in atomic weight and size have low thermal conductivity; for example UO_2 and ThO_2.
- Impurities, porosity and microcracks exert the main influence on the effective thermal conductivity. Presence of these defects can reduced the thermal conductivity of refractories.

A more porous brick has, in general, lower thermal conductivity. However, the size and number of the pores is also important as heat transfer can take place across the pores by radiation and is prominent at high temperature.

Refractories used in melting furnace should have low thermal conductivity to ensure least heat losses and maximum heat efficiencies whereas recuperators require refractory with high thermal conductivity to ensure maximum heat transfer.

Q32. What is the importance of Thermal Diffusivity of refractories?

Ans. Thermal diffusivity is the thermal conductivity divided by density and specific heat capacity at constant pressure. It measures the ability of a material to conduct thermal energy relative to its ability to store thermal energy. It has the SI unit of m^2/s or m^2/hr. The formula of thermal diffusivity (a) is:

a = Thcrmal conductivity /Heat storage capacity = = $k/\rho.Cp$

Where,

k : Thermal conductivity (W/(mK)) or (Kcal/m.hr.°C)

: Density (kg/m^3)

Cp : Specific heat capacity (J/(kgK)) or (Kcal/kg.°C)

Q33. What do you mean by Thermal Expansion of refractory?

Ans. This is a measure of the refractory about its linear stability when it is exposed to different ranges of high temperatures and then cooled to room temperatures. It is defined as a permanent linear change (PLC) and is measured by the changes in the longest linear dimensions. Most refractory materials expand when heated. Hence, when refractories are installed at room temperatures, the whole structure tightens up when heated.

Linear Thermal Expansion of refractory: When the temperature of a ceramic is changed ΔT, the change of its length ΔL is very nearly proportional to its initial length multiplied by ΔT. The Linear Expansion equation is:

$$\Delta L = \alpha L_0 \Delta T$$

Where:

α = the Coefficient of linear expansion.

L_0 = Initial length of the object.

ΔL = Length change of the object.

ΔT = Temperature change of the object.

Area Thermal Expansion of refractory: When the temperature of a surface is changed ΔT, the change of its area ΔA is very nearly proportional to its initial area multiplied by ΔT. The Area Expansion equation is:

$$\Delta A = \gamma A_0 \Delta T$$

Where:

γ = the Coefficient of Area expansion.

A_0 = Initial area of the object.

ΔA = Area change of the object.

ΔT = Temperature change of the object.

Volume Thermal Expansion of refractory: When the temperature of a refractory is changed ΔT, the change of its volume ΔV is very nearly proportional to its initial volume multiplied by ΔT. The Volume Expansion equation is:

$$\Delta V = \beta V_0 \Delta T$$

Where:

β = the Coefficient of volume expansion

V_0 = Initial volume of the object

ΔV = Volume change of the object

ΔT = Temperature change of the object.

Q34. Describe the factors on which thermal expansion of ceramic material depend?

Ans. Thermal expansion of ceramic materials depends on bonding strength between atoms. Ceramic with high degree of ionic bond have high thermal expansion (behavior similar to metals). As the bond strength increases or as the percent of covalent bond increases, the thermal expansion decreases. Both melting temperature and thermal expansion are controlled primarily by bond strength & the magnitude of thermal vibration. As the bond strength increases, the melting temperature increases and thermal expansion coefficient decreases.

The expansion characteristic correlates with the nature of the structure, and the atomic bonding. Al_2O_3, TiO_2 and mullite have moderately close packed structures and atomic bond strength. As a result, they have moderate thermal expansion. Within the graphite layers, the atomic bonding is very strong (C-C covalent bond) and the coefficient of expansion is low (1×10^{-6} /°C). The atomic bonding between layers is very week (Van der Waal's bonds) and the thermal expansion coefficient is high (27×10^{-6}/°C).

Q35. What is the significance of thermal expansion of refractories?

Ans. Many ceramic applications expose to a range of temperature. Mismatch in thermal expansion behavior between two adjacent materials, can result in high enough stresses to fracture. Moderate close thermal expansion match is necessary to minimize thermal stress. Some application requires very low thermal expansion. A well-known domestic application is the use of (LAS) based polycrystalline ceramic ($LiAlSi_2O_6$) for heat-resistant cooking ware.

Q36. What do you mean by Reversible Thermal Expansion of refractories? What is its significance in refractory utilization?

Ans. Any material when heated expands, and contracts on cooling. Thermal expansion is the intrinsic characteristic of refractory products to expand on heating and contract on cooling. The dimensional changes of a refractory due to thermal expansion are commonly expressed in permanent linear change (PLC,%) and the coefficient of thermal expansion (length per unit length).

The reversible themal expansion is a reflection on the phase transformations that occur during heating and cooling. The PLC and the reversible thermal expansion are followed in the design of refractory linings for provision of expansion joints. *Refractories with a lower thermal expansion co-efficient are less susceptible to thermal spalling.*

Q37. What do you mean by Thermal Shock of refractories? Discuss its effect on refractories.

Ans. Thermal shock is the fracture of a body resulting from thermal stresses induced by rapid temperature changes. Because ceramic materials are brittle, they are especially susceptible to this type of failure. The thermal shock resistance of many materials is proportional to the fracture strength & thermal conductivity, and inversely proportional to both the modulus of elasticity & the coefficient of thermal expansion.

Thermal shock is an important property for a refractory material. Most high-temperature processes experience heating and cooling. Thermal shock can be indicated by the number of cycles to withstand such temperature fluctuations. Thermal shock resistance can be defined as partial or complete fracture of material due to gradient temperature. The factor that affects the thermal shock resistance includes the shaping methods and the firing temperatures.

Thermal shock refers to the thermal stresses that occur in a component as a result of exposure to temperature difference between surface and interior of the component. Thermal stress occurs at the surface during cooling refers to the following equation:

$$\sigma th = E\alpha\Delta T/(1-\vartheta)$$

Where, σth : thermal stress, E : the elastic modulus, α : the coefficient of thermal expansion, ΔT : the temperature difference, and ϑ : Poisson's ratio

Above equation indicates that the thermal stress increases as the elastic modulus & thermal expansion coefficient of the refractories increases. The temperature difference ( ) can be decreased by increasing the thermal conductivity of the refractories.

Thermal shock resistance (R) of refractories can be calculated from the following equation:

$$R = \sigma \frac{(1-\vartheta)}{E.\alpha}$$

Where,

σ: The symbol stands for the strength of the ceramic material. As ceramic materials subjected to thermal shock generally fail in tension rather than in shear or compression, the tensile strength is generally used as the criterion for failure rather than the compressive or shear strength. As the thermal shock resistance of a ceramic material is governed mainly by its mechanical & thermal properties, as indicated above, an improvement of the thermal shock resistance is achieved by the high Young's modulus of ceramics.

Q38. What is Spalling of refractories?

Ans. Spalling is defined as the fracture of the refractory brick or block due to any of the following causes:

- A temperature gradient in the brick due to uneven heating or cooling that sets up stresses causing failure. (Thermal Spalling).
- Compression in a structure of refractories due to expansion of the whole from a rise of temperature causing shear failure.
- Variation in coefficient of thermal expansion between the surface layer and the body of the brick, due to the surface slag penetration or due to structural changes in service resulting in shearing off the surface layer. As a general rule, those with a lower thermal expansion co-efficient are less susceptible to thermal spalling.

Q39. What are the effects of Thermal Spalling of refractory? How it can be reduced?

Ans. Failure of refractory material under the influence of high temperature due to uneven expansion and contraction of refractory material is called thermal spalling.

Effects of thermal spalling include:

- Slag penetration
- Corrosion
- Abrasion

Thermal Spalling can be reduced by:

- Using materials with low coefficient of expansion and avoiding sudden temperature fluctuation.
- Using high porosity bricks.
- Using good thermal conductivity bricks.
- Overfiring refractory bricks at high temperature.
- By designing the furnace such that stress is alienated.

Q40. Write a short note on Fireclay refractory.

Ans. Fire-clay bricks are essentially hydrated aluminum silicates with minor proportions of other minerals. Typical composition consists of SiO_2 <78% and Al_2O_3 < 44%. They are sufficiently pure to serve as raw materials for refractories. The principal mineral in fireclays is kaolinite. While other clay minerals may be present, the formula $Al_2O_6.2SiO_2.2H_2O$ can usually represent the clay fraction

They are extremely versatile; least costly of all refractory bricks and are extensively used in the iron & steel industry, non ferrous metallurgy, glass industry, pottery kilns, cement industry and by many others. They constitute about 70 % of total refractory production.

Characteristically, fire-clay brick begin to soften far below their fusion temperature and under load actual deformation takes place. The amount of deformation depends upon the load, and, once started, this deformation is a slow but continuous process unless either the load or the temperature is reduced. It is for this reason that fire-clay brick are not well adapted for use in wide sprung arches in furnaces operating continuously at high temperatures.

Common fire clays contain 25 to 44 % Al_2O_3 and 50–78 % SiO_2. Clays with higher Al_2O_3 are higher alumina refractories.

Properties of fireclay bricks are:

- Porosity varies from 8 to 24% depending on the firing temperature.
- At high temperatures fire clay refractory combine with alkalis such as soda and potash.
- Cold crushing strength is 950 kg/cm^2 at 20 °C and decreases drastically at 800 °C to 555 kg/cm^2.

Refractoriness of various types of fireclay bricks are tabulated below.

Type of fireclay bricks	Al_2O_3 (%)	Refractoriness, °C
Medium duty	30-32	1665
High duty	38-40	1700-1717
Super duty	42 – 44	1743-1763

Used in: Furnaces, regenerators, ovens, and kilns.

Q41. What are different types of fireclay refractory?

Ans. There are five different types of fireclay brick: superduty, high-duty, medium-duty, low-duty and semi-silica fireclay brick. These classes cover the range from approximately 18 to 44% alumina, and from 50 to 80% silica.

A blend of clays is commonly used in the manufacture of high-duty and superduty fireclay brick. Flint clays and high-grade kaolin impart high refractoriness; calcined clays control the drying & firing shrinkages and plastic clays facilitate forming & impart bonding strength. The character and quality of the brick to be made determines the relative proportions of clays used in a blend.

Superduty fireclay brick have good strength & volume stability at high temperatures and an alumina content of 40 to 44%. Some superduty brick have superior resistance to cracking or spalling when subjected to rapid changes of temperature. There are several possible modifications in the superduty fireclay class, including brick fired at temperatures higher than the usual product. High temperature firing enhances the high temperature strength of the brick, stabilizes their volume & mineral composition, increases their resistance to fluxing, and renders them practically inert to disintegration by carbon deposition in atmospheres containing carbon monoxide gas.

High-duty fireclay bricks are used in large quantities and for a wide range of applications. Because of their greater resistance to thermal shock, high duty fireclay brick can often be used with better economy than medium-duty brick for the linings of furnaces operated at moderate temperatures over long periods of time but subject to frequent shutdowns.

Medium duty brick are appropriate in applications where they are exposed to conditions of moderate severity. Medium-duty brick, within their serviceable temperature ranges, can withstand abrasion better than many brick of the high duty class.

Low-duty fireclay brick find application as backing for brick with higher refractoriness, and for other service where relatively moderate temperatures prevail.

Semi-silica fireclay brick contain 18 to 25% alumina and 72 to 80% silica, with a low content of alkalies and other impurities. With notable resistance to shrinkage, they also have excellent load bearing strength and volume stability at relatively high temperatures.

Table below shows that as the quantity of impurities increases and the amount of Al_2O_3 decreases, the melting point of fireclay brick decreases.

Brick	% SiO_2	% Al_2O_3	Other Constituents	PCE, °F
Super Duty	49-53	40-44	5-7	3175-3200
High Duty	50-80	35-40	5-9	3075-3175
Medium Duty	60-70	26-36	5-9	2975-3075
Low Duty	60-70	23-33	6-10	2770-2900

Q42. What is high alumina refractory?

Ans. The term high-alumina brick refers to refractory brick having an alumina (Al_2O_3) content of 47.5% or higher. This descriptive title distinguishes them from brick made predominantly of clay or other alumino silicates which have an alumina content below 47.5%.

Alumina is one of the most chemically stable oxides known, which offers excellent hardness, strength and spalling resistance. It is insoluble in water and super heated steam, and in most inorganic acids and alkalis. Alumina refractories carry the all purpose characteristics of fire-clay brick into higher temperature ranges that makes it suitable for lining furnace operating up to 1800 °C. It has a high resistance in oxidizing and reducing atmosphere and is extensively used in heat processing industries.

The refractoriness of high alumina refractories increases with increase of alumina percentage. Manufacturing cost and, therefore, price of these brick increase more rapidly with % alumina content, so it is essential to determine experimentally or by test installations the most economical alumina content for each service.

Q43. How are high alumina brick classified?

Ans. High-alumina brick are classified by their alumina content. The 50%, 60%, 70% and 80% alumina classes contain their respective alumina contents with an allowable range of +/- 2.5% from the respective nominal values. The 85% and 90% alumina classes differ in that their allowable range is +/- 2% from nominal. The final class, 99% alumina, has a minimum alumina content rather than a range, and this value is 97%.

High-alumina brick are also classified by their alumina content according to the following ASTM convention. These are:

Mullite refractory: Mullite brick is about 72% alumina with 28% silica. It predominantly contains the mineral phase mullite ($3Al_2O_3 \cdot 2SiO_2$). These have excellent volume stability and strength at high temperatures. They are highly suitable for electric furnace roofs, blast furnaces & blast furnaces stoves, and the superstructure of glass tank furnaces.

Corundum refractories: The 99% alumina class of refractories is called corundum. These refractories comprise single phase, polycrystalline, and alpha-alumina.

There are several other special classes of high-alumina products:

- Chemically bonded brick - usually phosphate-bonded brick in the 75 to 85% Al_2O_3 range. An aluminum orthophosphate ($AlPO_4$) bond can be formed at relatively low temperatures.
- Alumina chrome brick - typically formed from very high purity, high-alumina materials and chromic oxide (Cr_2O_3). At high temperatures, alumina and chromia form a solid solution which is highly refractory.
- Alumina carbon brick - high-alumina brick (usually bonded by a resin) containing a carbonaceous ingredient such as graphite.

Q44. What are the properties and uses of high alumina refractory?

Ans. Some of the properties of high alumina refractory are:

- High refractoriness
- Better resistance to slag and spalling
- Higher load bearing capacity.
- Fusion point >1850 °C.
- Low coefficient of expansion hence high resistance to thermal spalling
- Appreciable RUL and Abrasion resistance

High-alumina bricks serve as a multi-purpose refractory material for severe environments. They are used extensively in the steel industry for such applications as hot metal cars, electric furnace roofs, piers & muffles for a variety of furnaces, and numerous applications where strength at high temperature is an essential requirement. The aluminium and glass industries use high-alumina refractories to keep the melt in thc moltcn state.

High alumina bricks are most commonly used in cement, lime & ceramic kilns, glass tanks, crucibles for melting a wide range of metals, hearth & shaft of blast furnaces and in lead drossing furnaces. They are very economical for the lower sections of soaking pits in the steel industry, because of their resistance to iron oxide slags.

Q45. Write a short note on mullite refractories.

Ans. Mullite is about 72% alumina with 28% silica. The manufacturing procedures are designed to maximize the formation of the compound mullite ($6Al_2O_6.2SiO_2$). A refractory with 71.8% alumina and 28.2% silica will be composed of only mullite, if fired at equilibrium conditions.

However, the extent to which well-developed mullite crystalline form occurs in a refractory depends on the purity of the raw materials used and the manufacturing processes, particularly firing. Therefore, all high-alumina bricks with around 70 % alumina may have a well-developed mullite phase.

Mullite refractories have excellent volume stability and strength at high temperatures. They are highly suitable material for electric furnace roofs, blast furnaces & blast furnace stoves, hot metal cars, and the superstructure of glass tank furnaces.

Q46. What is sillimanite refractory? Discuss its applications and properties.

Ans. Sillimanite refractory is a high alumina refractory with a higher percentage of mullite. High alumina refractories are particularly suitable for high-temperature applications (such as in the open-pot furnace for glass melting), with typical process parameters like thermal stress and chemical environment.

High-alumina refractories having the following characteristics are suitable for furnace floor construction:

- High temperature resistance (at least up to 1050 °C).
- High corrosion resistance (alkali resistant).
- Resistance to thermal fluctuation.

Table below shows physical & chemical properties of Sillimanite refractories.

Parameters	Value
$A1_2O_3$	60% minimum (preferably (around 70%)
Fe_2O_3	0.5% maximum
PLC (at 1500 °C)	± 0.2%
Apparent porosity	15 –17%
Cold crushing strength	400 kg/cm^2 minimum
Bulk density	2.50 g/cc minimum
RUL	1580 °C
HMOR (1400 °C)	60–65 kg/cm^2
Mullite	50% minimum (indicative)

Q47. Describe the features of various types / classes of high alumina bricks.

Ans. Features of various classes of high alumina bricks are described below.

50% Alumina Class

A brick classified as a 50% alumina product has an alumina content of 47.5 to 52.5%. Chemically, such brick are not greatly different from superduty fireclay brick which can contain up to 44% alumina. Brick within the 50% alumina class are often upgraded versions of fireclay brick with the addition of a high alumina aggregate. Compositions of this class are designed primarily for ladles. These 50% alumina class brick have low porosity and expand upon reheating to 1600°C - desirable features for ladle applications since they minimize joints between brick, giving a near monolithic lining at service temperature. These brick are also characterized by low thermal expansion and good resistance to spalling. Many high-temperature industries use them as backup brick.

50% alumina products based on high purity bauxitic kaolin, and other ingredients in the matrix, provide exceptional load-bearing ability, alkali resistance & low porosity. These qualities make such brick an excellent choice for preheater towers and calcining zones of rotary kilns.

60% Alumina Class

The 60% alumina class is a large, popular class of products. These bricks are used in the steel industry, as well as rotary kilns. Brick in this class are made from a variety of raw materials. Some are produced from calcined bauxitic kaolin and high purity clay to provide low levels of impurities. As a result of firing to high temperature, these brick have low porosity, excellent hot strength & creep resistance, and good volume stability at high temperatures.

70% Alumina Class

This is the most frequently used high alumina product class because of its excellent and cost-effective performance in multiple environments. They find applications in the steel industry and cement & lime rotary kilns.

Most brick in this class are based on calcined bauxite and fireclay. Brick are usually fired to fairly low temperatures to prevent excessive expansion in burning which causes problems in final brick sizing. Expansion is caused by reaction of the siliceous ingredients with bauxite to form mullite. The brick typically undergo large amounts of secondary expansion when heated. This is advantageous in reducing the size of joints between brick and providing a tight vessel structure, e.g., a rotary kiln.

A higher cost and higher quality alternative to producing a 70% alumina brick is represented by brands based on high purity calcined bauxitic kaolin. These brick have superior high temperature strength & refractoriness and significantly lower porosity than typical products based on calcined bauxite. Due to their more homogeneous structure, they show somewhat less expansion on reheating than bauxite based products.

80% Alumina Class

These products are based primarily on calcined bauxite with additions of various amounts of other fine aluminas and clay materials. They are usually fired at relatively low temperatures to maintain consistent brick sizing. Most brick in this class have about 20% porosity, good strength & thermal shock resistance. Because they are relatively inexpensive, perform well & are resistant to most slag conditions present in steel ladles, they are used extensively in steel ladle applications.

90% and 99% Alumina Classes

These brick contain tabular alumina as the base grain and may include various fine materials such as calcined alumina, clay & fine silica. As these brick generally have low impurity levels, alumina and silica typically make up 99% of the chemical composition.

Usually, the only mineral phases present are corundum and mullite. Properties such as high hot strength, creep & slag resistance benefit from this purity level.

Q48. Describe the manufacture,properties and applications of Silica Brick.

Ans. Silica brick is a refractory material containing at least 93% SiO_2. Silica bricks are made from raw materials that are essentially quartz. During the initial firing, alpha quartz is first converted to beta-quartz, accompanied by an abrupt expansion at 576 °C. Slow rates of heating are required through this temperature range to prevent cracking, as the volume change is about 0.9%. Since the final firing temperature is somewhat over 1026 °C, the brick as it is put into service consists of cristobalite particles (properties of cristobalite are not affected by temperature fluctuation, provided the temperature does not drop below 600 °C), with possibly some having residual unconverted quartz cores.

The outstanding property of silica brick is that it does not begin to soften under high loads until its fusion point is approached. This behaviour contrasts with that of many other refractories, for example alumino silicate materials, which begin to fuse and creep at temperatures considerably lower than their fusion points. Other advantages are flux & slag resistance, volume stability and high spalling resistance.

Silica refractories are well adapted to high temperature service because of their high refractoriness, high mechanical strength and rigidity at temperatures almost up to their melting points, as well as their ability to resist the action of dusts, fumes & acid slags.

In addition to high fusion point multi type refractories, the other important properties are their high resistance to thermal shock (spalling) and their high refractoriness.

They have one major drawback too ie; their susceptibility to spalling at temperatures below 600 °C. Temperature fluctuations above 600 °C do not affect silica brick adversely and in this range it is classed as a good spalling resistant brick. Many furnaces, however, must be cooled to nearly room temperature at frequent intervals (*eg.*; Beehive coke ovens) and for such processes silica brick are not practical. However, there is no substitute for silica bricks use in the walls of by-product coke ovens in steel industry..

Silica brick give good service in contact with high siliceous slags and successfully used in the wide sprung arches of open hearth steel furnaces, glass tanks, and reverberatory furnaces for the melting & refining of copper and many other metals.

Various grades of silica brick have found extensive use in the iron & steel making furnaces.It finds typical use in glass making and steel industry (in By-product coke ovens).

Q49. Describe the manufacture, properties and applications of magnesite refractory.

Ans. Magnesite refractories are chemically basic materials, containing at least 85% magnesium oxide. These are made from naturally occurring magnesite ($MgCO_3$) and Silica (SiO_2). Good quality magnesite bricks usually results from a CaO-SiO_2 ratio of less than 2 with a minimum ferrite concentration, particularly if the furnaces lined with the refractory operate in oxidizing and reducing conditions.

Raw Magnesite ($MgCO_3$) is generally used in its calcined or dead burned forms. *Dead burned magnesite is a raw material used widely by the refractory industry.* More recently fused magnesia grain; magnesia sinter with larger crystallite size and very high purity magnesia sinter were introduced to further improve the corrosion resistance. Fused magnesia is produced by melting refractory grade magnesia or other magnesia precursor in an electric arc furnace.

The molten mass is then removed from the furnace, cooled, and broken up. The material then is used in manufacture of refractories.

The physical properties of this class of brick are generally poor, and their great value is primarily in their resistance to basic slags. The slag resistance is very high particularly to lime and iron rich slags.

The properties of magnesite refractories depend on the concentration of silicate bond at the operating temperatures. Good quality magnesite usually results from a CaO/SiO_2 ratio of less than 2 with a minimum ferrite concentration, particularly if the furnaces lined with the refractory operate in oxidizing and reducing conditions.

Properties of magnesite brick:

- It is a Basic refractory and exhibit excellent resistance to basic slag
- PCE is 2000°C.
- RUL is 1500°C at 3.5 Kg/ cm^2
- It undergoes very little shrinkage during use but has poor abrasion resistance.
- It is extremely sensitive to temperature changes and hence possess poor spalling resistance.

Applications of magnesite bricks:

- Lining in open hearth furnace, copper reverbarators and reverbarators used for smelting lead & antimony.
- Lining in furnaces used for refining Au, Ag and Pt.
- Used in lining of hot zones of cement rotary kilns.

Magnesite bricks constitute the most important group of refractories for the basic steelmaking processes. In addition to metallurgical furnaces, basic brick are now being successfully used in glass tank checkers and in lime & cement kilns.

Caustic calcined magnesite is used in effluent treatment, blended fertilizers and as feed stock for the chemical industry.

Q50. Describe the manufacture, properties & applications of dolomite refractory.

Ans. The natural double carbonate dolomite ($CaCO_3$. $MgCO_3$) can be converted to refractory dolomite (CaO. MgO) by high temperature firing. High purity dolomite is greater than 97% CaO + MgO. Dolomite brick are available in burned and carbon-bonded compositions. The carbon-bonded varieties include both tar/pitch and resin-bonded versions. Some of the carbon-bonded products contain flake graphite and are somewhat analogous to magnesite-carbon brick.

Dolomite refractories are found to be the most compatible material with the cement kiln clinker as it provides excellent coating stability, has very good resistance to thermal shock & alkali attack in varying kiln operating conditions. These refractories with Zirconia enrichment are used for crack arresting.

Dolomite brick are widely applied in applications as diverse as argon-oxygen decarburization vessels (AOD's), rotary cement kilns steel ladles and wear out lining of L.D. converter.

Q51. Describe the features of chrome magnesite and magnesite chrome bricks.

Ans. Chrome-magnesite brick usually contains 15-35% Cr_2O_3 & 42-50% MgO whereas magnesite-chrome refractories contain at least 60% MgO & 8-18% Cr_2O_3.

Chrome-magnesite refractories are made in a wide range of qualities and are used for building the critical parts of high temperature furnaces. These materials can withstand corrosive slags & gases and have high refractoriness. The magnesite-chrome refractory products are suitable for service at the highest temperatures and in contact with the most basic slags used in steel melting. Magnesite-chrome usually has a better spalling resistance than chrome magnesite.

Q52. What are basic refractories? Describe briefly the features of various basic refractories of industrial importance.

Ans. Basic refractories are those that exhibit resistance to corrosive reactions with chemically basic slags, dusts & fumes at elevated temperatures

Basic refractories generally fall into one of five following compositional areas:

- Products based on dead burned magnesite or magnesia.
- Products based on dead burned magnesite or magnesia in combination with chrome containing materials such as chrome ore.
- Dead burned magnesite or magnesia in combination with spinel.
- Dead burned magnesite or magnesia in combination with carbon.
- Dolomitic products.

One of the more important types of magnesite brick are those that have low boron oxide contents and dicalcium silicate bonds. These chemical features give the brick excellent refractoriness, hot strength and resistance to load at elevated temperatures. Another category of magnesite brick contains a higher boron oxide content to improve hydration resistance.

Basic refractories containing chrome is an important group of materials due to their excellent slag resistance, superior spalling resistance, good hot strengths and other features. Historically, silicates in the groundmass or matrix formed the bond between the chrome ore and periclase in the brick. However, the advent of high purity raw materials in combination with high firing temperatures made it possible to produce "direct bonded" brick, where a ceramic bond between the chrome ore and periclase particles exists. These direct bonded brick exhibit superior slag resistance and strengths at elevated temperatures.

Magnesite spinel brick have increased in importance due to a desire to replace chrome-containing refractories because of environmental concerns. Brick made with spinel and magnesite have better spalling resistance and lower coefficients of thermal expansion than brick made solely with dead burned magnesite. These features minimize the chance of the brick cracking during service.

Basic brick containing carbon include pitch impregnated burned magnesite brick with carbon contents up to 2.5%, pitch bonded magnesite brick containing about 5% carbon and magnesite-carbon brick which contain up to 30% carbon. Development of the more corrosion resistant magnesite-carbon brick has resulted in decreased consumption of pitch impregnated and pitch bonded magnesite brick. In addition, in many instances the magnesite-carbon brick have replaced magnesite-chrome brick in applications such as electric arc furnaces.

Dolomitic products are an important class of refractories that are used in rotary cement kilns and AOD's. Dolomite brick offer a good balance between low cost and good refractoriness for certain uses. They also offer good metallurgical characteristics for certain "clean steel" applications.

Q53. Describe the manufacture, properties and applications of Zirconia Refractories.

Ans. Zirconium dioxide (ZrO_2) is a polymorphic material. There are certain difficulties in its usage and fabrication as a refractory material. It is essential to stabilize it before application as a refractory. This is achieved by incorporating small quantities of calcium, magnesium & cerium oxide. Its properties depend mainly on degree of stabilization and quantity of stabilizer as well as the quality of original raw material.

Manufacture of zirconia bricks involves the following steps:

Crushing. Zirconite, CaO or MgO (acts as stabilizer), and alumina (acts as binding material) are crushed separately to ~25 mm size by Hammer mills / Machine crushing.

Grinding. The crushed ores are powdered to 200 mesh size separately by Jaw crushers / Ball mills.

Screening. Impurities are removed from raw materials by adopting one of the Settling / Magnetic separation / Chemical methods.

Mixing. The above concentrated raw materials are mixed for proper distribution and made into slurry in order to facilitate moulding.

Moulding. The plastic (*i.e.* flexible) slurry is poured into moulds to get required shape and size by either Hand moulding or Mechanical moulding followed by deairing.

Drying. The moisture and volatile matter are removed by heating the moulded brick in Tunnel driers heated by steam.

Firing. The dried bricks are fired at 1700 °C in Tunnels kiln.

Zirconia refractories have a very high strength at room temperature which is maintained up to temperatures as high as 1500 °C. They are, therefore, useful as high temperature construction materials for furnaces and kilns. The thermal conductivity of zirconium dioxide is found to be much lower than that of most other refractories and the material is therefore used as a high temperature insulating refractory.

Properties of zirconia bricks:

- Zirconia is a Special Neutral refractory.
- PCE is 2000 – 2600°C.
- RUL is 1900°C at 3.5 Kg/cm^2.
- It exihibits good thermal shock resistance as well as good resistance to spalling.

Applications of zirconia bricks:

Being expensive, its application is limited. Used as lining in high frequency electric furnace.

Since Zirconia exhibits very low thermal losses and does not react readily with liquid metals, it is particularly useful for making refractory crucibles & other vessels for metallurgical purposes. Zirconia is useful refractory material for glass furnaces primarily since it is not easily wetted by molten glass and because of its low reaction with them.

Q54. What are monolithic refractories?

Ans. Monolithic refractories are all unshaped refractory products / materials installed as some form of suspension that ultimately harden to form a solid mass. Most monolithic formulations consist of large refractory particulates (an aggregate), fine filler materials (which fill the interparticle voids) and a binder phase (that gels the particulates together in the green state).

Monolithic refractories are replacing the conventional type fired refractories bricks in many applications including those of industrial furnaces. The main advantages of monolithic refractories are as follows:

Various means are employed in the placement of monolithic refractories like ramming, casting, gunniting, spraying, sand slinging etc. Ramming masses are used mostly in cold applications where proper consolidation of the material is important. The same practice can be adopted with both air setting and heat setting materials.

Q55. What are the advantages of Monolithic Refractories?

Ans. Monolithic refractory is unshaped refractory materials installed as some form of suspension that ultimately harden to form a solid mass. Being single piece casts in equipment shape, they are replacing conventional refractories.

Advantages of monolithic refractory include:

- Elimination of joints
- Faster application
- Heat savings
- Better spalling resistance
- Volume stability
- Easy to transport, handle & install
- Reduced downtime for repairs

Monolithic linings are installed by casting the refractory in a mould or by spraying the furnace shell.

Main problem with use of monolithic refractories are:

- Long drying time
- Steam explosion.

There are two basic types of monolithic lining, namely castable refractory and plastic refractory.

Installation of the monolithic refractory is important. Due to relatively poor strength, durability of monolithic lining depends on the design and installation of the anchors.

Q56. List out the various Types of Monolithic Refractories.

Ans. Various types of monolithic refractory are as follows:

- Castable Refractories
- Insulating Castables
- Plastic Refractories

- Ramming Mixes
- Patching Refractories
- Coating Refractories
- Mortars
- Gunning Mixes
- Fettling Mixes

Q57. Describe the various types of Monolithic Refractories.

Ans. Various monolithic (unshaped) refractories are described below:

Castable Refractories: 'Castable' is a material of hydraulic setting in nature. These are the materials that contain cement binder usually aluminate cement, which imparts hydraulic setting properties when mixed with water. Following the heat up of the material the binder either transforms or volatilises facilitating the formation of a ceramic bond. The most common binder used in castables is HAC (high alumina cement). Other binders that are often used include hydratable aluminas and colloidal silica. These materials are installed by casting and are also known as refractory concretes.

Insulating Castables: Insulating castables are specialised monolithic refractories that are used on the cold face of applications. They are made from lightweight aggregate materials such as vermiculite, perlite, bubble alumina and expanded clay. Their main function is to provide thermal insulation. They are typically of low density and low thermal conductivity. Insulating refractories have inferior mechanical strength to that of conventional castables.

Plastic Refractories: Plastic refractory is mixtures that is prepared in stiff plastic condition and are delivered in blocks wrapped in polyethylene. During application, the blocks are sliced into pieces and without further preparation, are rammed or poured into place with pneumatic rammer. Plastic refractories are easily rammed to any shape or contour.

Ramming Mixes: Ramming refractory materials are very similar to plastic refractories but are much stiffer. The particle sizes are carefully graded and the final product is usually delivered dry and then mixed with a little amount of water just before application. Other ramming mixes are delivered in wet form and are ready for use immediately upon opening. Application is done with pneumatic rammer.

Patching Refractories: These materials are similar to plastic refractories though they have a very soft plasticity allowing them to be pounded into place.

Coating Refractories: This type of product is used to protect refractory linings usually against chemical attack. Coating refractories are normally intended to cover just the working surface of a lining. They tend to be fairly thin layers.

Mortars: Mortars are generally neither classified under refractory brick nor monolithic refractories. These are finely ground refractory materials, which become plastic when mixed with water. These are used to bond the brickwork into solid unit, to provide cushion between the slightly irregular surfaces of the brick, to fill up spaces created by a deformed shell, and to make a wall gas tight to prevent penetration of slag into the joints.

Gunning Mixes: Gunning mixes are granular refractory materials sprayed on application area using air placement guns. These are heat setting and are used for patching and maintenance works for kilns & furnaces.

Fettling Mixes: Fettling mixes are also granular refractory materials, similar to gunning mixes, but are applied by shoveling into the furnace needing patching.

Q58. Write a short note on Insulating Refractories.

Ans. These are high porosity refractories with low thermal conductivity used in order to reduce the rate of heat flow (heat losses) and thus to maximize heat conservation within the furnace. The function of insulating refractory is to reduce the heat loss through the walls of furnaces. Insulation is effected by providing a layer of material having low heat conductivity, which means heat does not readily pass through them.

Insulating refractories have lower densities in range of 64 to 1120 kg/m^3. The three basic types of insulating materials for industrial use are:

- thin (less than 20 micrometers), low-density (less than 192 kg/m^3) fibers made from organic or inorganic materials
- cellular material in closed or open cell form made of organic or inorganic material.
- flaked or granular inorganic materials bonded in the desired form. In most cases, glass (silica), mineral wool, high alumina, mulite, or zirconia are the base materials and can be used to temperatures as high as 1600 °C. This class of materials has a lower density varying from 64 to 192 kg/m^3 and offers higher thermal resistance compared to firebricks.

In all cases, thermal conductivity of the insulation increases significantly as temperature increases.

Q59. What are special refractories?

Ans. These are very expensive refractory materials used for making crucibles and furnaces for special / experimental purposes where the cost of refractory is no consideration.

They are not very common due to their manufacturing limitation. Special refractory include pure oxides (*eg.*. Magnesia, Silica, Alumina, Thoria etc), borides, nitrides, silicides, carbides *etc.*

Other special refractories are Sialons (Silicon Alumina Oxy Nitride), Zircons, bricks of mullite, magnesia & mixtures of chromite, bauxite and magnesia.

Q60. What are the Characteristics of Good Insulating Refractory?

Ans. The desirable feature of insulating bricks is the low thermal conductivity, which usually results from a high degree of porosity. Structure of air insulating material consists of minute pores filled with air which have in them very low thermal conductivity. The air spaces inside the brick prevent the heat from being conducted but the solid particles of which the brick is made conduct the heat. So, in order to have required insulation property in a brick a balance has to be struck between the proportion of its solid particles and air spaces. The thermal conductivity is lower if the volume of air space is larger. The thermal conductivity of a brick does not so much depend on the size of pores as on the uniformity of size and even distribution of these pores. Hence, uniformly small sized pores distributed evenly in the whole body of the insulating brick are preferred.

The high porosity of the brick is created during manufacturing by adding a fine organic material to the mix, such as sawdust. During firing, the organic addition burns out, creating internal pores. Other ways to accomplish high porosity are:

- by using materials which expand and open up on heating.
- by using volatile compounds like naphthalene.
- by using aluminum powder in combination with NaOH solution (called chemical bloating).
- by using substances which by themselves have open texture e.g. insulating brick grog, vermiculite, ex-foliated mica, raw diatomite etc.
- by using foaming agents to slip.
- by aeration etc.

Because of their high porosity, insulating bricks inherently have lower thermal conductivity and lower heat capacity than other refractory materials.

Q61. How are various Types of Insulating Refractories classified?

Ans. Insulating materials are classified with respect to application temperature as follows:

- Heat resistant insulating materials for application temperatures up to 1100 °C: calcium silicate materials; products from siliceous earth, perlite or vermiculite; silica based micro porous heat insulators; alumino-silicate fibers.
- Refractory insulating materials for application temperatures up to 1350 °C: lightweight chamotte and kaolin bricks; lightweight castables; mixed fibers and aluminum oxide fibers.
- High refractory insulating materials for application temperatures up to 1700 °C: lightweight mullite and alumina bricks; lightweight hollow sphere corundum castables and bricks; special high refractory fibers.
- Ultra-high refractory insulating materials for application temperatures up to 2000 °C: zirconia lightweight bricks and fibers; non-oxide compounds.
- Several other types of insulating refractories include castables, granular insulation, and ceramic fiber insulation, which are light weight. Extremely lightweight materials have a porosity of 75 to 85% and ultra-lightweight, high-temperature insulating materials have a total porosity greater than 85%.

Q62. What are the areas of Applications & Advantages of insulating refractories?

Ans. Insulating refractories are widely used in the crowns of glass furnaces and tunnel kilns. They can also be used as linings of furnaces where abrasion and wear by aggressive slag & molten metal are not a concern.

Insulating refractories offer following advantages:

- Decreased heat losses through the furnace lining and less heat loss to the refractory leads to savings in fuel cost.

- The insulating effect causes a more rapid heat-up of the lining and lowers heat capacity of the insulating refractory.
- Thinner furnace wall construction are required to obtain a desired thermal profile.
- It results in less furnace mass due to the lower mass of the insulating refractory.

Q63. Describe Some Drawbacks of Insulating Bricks.

Ans. Some of the Drawbacks of Insulating Bricks are as follows:

- The porosity in insulating refractories creates a large amount of free surface area. Although porosity decreases thermal conductivity & density of the brick, it also degrades the mechanical strength of the brick as compared to a dense refractory brick. The poor strength of insulating bricks due to their high porosity can pose structural design problems.
- Porosity in insulating refractory leads to poor chemical resistance as compared to dense refractories of similar compositions. Gases, fumes, liquids such as slags, molten glass etc. at high temperatures can penetrate porous bricks easily, making insulating bricks unsuitable for direct contact with such liquids or gases.
- Insulating fire bricks often suffer from thermal spalling problems, particularly in an environment of rapidly changing temperature. Since these bricks are good insulators, a substantial temperature gradient will occur between the hot and the cold face of each brick. The hot face will expand more than the cold face. The thermal gradient thus, gives rise to mechanical & thermal stress in the body of the brick.

Q64. Describe the manufacture, properties and applications of Ceramic Fibers.

Ans. Ceramic fibers are a family of insulating refractory products that is fluffy, white cotton like fiber and can be spun and fabricated into textiles, blankets, felts, boards, blocks, etc.

Ceramic fibre is a low thermal mass insulation material, which has revolutionalised the furnace design lining systems.

Ceramic fibre is an alumino silicate material manufactured by blending & melting alumina & silica at temperature of 1800 – 2000°C and breaking the molten stream by blowing compressed air or dropping the melt on spinning disc to form loose or bulk ceramic fibre. The bulk fibre is converted to various products including blanket, strips, veneering & anchored modules, paper, vacuum formed boards and shapes, rope, wet felt, mastic cement etc. for insulation applications.

Ceramic fibres are usually produced in two temperature grades based on Al_2O_3 content. A recent addition is ZrO_2 added alumino silicate fibre, which helps to reduce shrinkage levels thereby rating the fibre for higher temperatures. Continuous recommended operating temperature for fibres are given in the following Table:

Table - Continuous Recommended Operating Temperature for ceramic fibres

	Al_2O_3, %	SiO_2, %	ZrO_2, %
1150°C	43-47	53 - 57	—
1250°C	52-56	44 - 48	—
1325°C	33-35	47 - 50	17 - 20

These fibres are generally produced in bulk wool form and needled into blanket mass of various densities ranging from 64 to 190 kg/m^3. Converted products and over 40 different forms are made from blankets to suit various requirements.

Ceramic fibres have low thermal conductivity, very low heat storage, extremely light weight, immunity to thermal shocks and are chemically stable. The light weight construction ensures that the required temperature in high-temperature furnaces is reached more rapidly, as only a small proportion of the temperature released into the processing furnace is used for the heating of the walls.

Ceramic fibers composed of 52 Al_2O_3 - 48 SiO_2 combinations can be applied as a hot face insulation material up to 1425 °C, whereas a 62 Al_2O_3 - 38 SiO_2 combinations impart greater refractoriness to fiber. 42% Al_2O_3, 52% SiO_2 and 6% ZrO_2 produces extra long staple fiber of 25 cms. or so and are used for manufacturing ceramic fiber textiles & ropes.

Ceramic fibres are used in metal treating furnaces, ceramic kilns, and numerous other periodic operations whose atmosphere do not negate their revolutionary thermal & lightweight qualities. Fibre mats also continue to be used in expansion joints & door seals, and in tunnel kilns and other exposed brick structures as either original or retrofit layers on the outside or cold-face surface.

Temperature-resistant ceramic materials for application temperatures up to 800°C are sometime regarded as thermal insulating materials and not refractory products

Q65. Describe some of the advantages and limitations of ceramic fibres.

Ans. The advantages of light weight ceramic fiber are:

- Better fuel economy (savings as high as 60% are feasible in the case of certain intermittent furnaces).
- Higher productivity & capacity of furnaces, due to reduced heat storage capacity.
- Higher service life of the furnace and reduced maintenance costs due to longer refractory life.
- Ease of installation.

Limitations of ceramic fibres are as follows:

- The chief limitation of ceramic fibre is its shrinkage at high temperatures. A high quality ceramic fibre blanket rated for continuous use at 1300 °C will have 5% shrinkage after 24 hr exposure at 1300 °C. Shrinkage will not continue past this level in normal operating conditions, but this shrinkage must be carefully considered in designing a furnace lining.
- Mechanical strength of ceramic fibers is poor. These are not really structural materials. Proper support must be given to all refractory fibre products.
- Tend to sag at high temperature due to softening of fibers, if improperly supported.
- Not suitable in severe hostile environments. These are handy repositories of dusts, fogs & combustible fumes; not to mention for process liquids like slags & metals.
- It is more expensive than conventional refractories however; installation labour savings and energy savings offset the high initial costs.

Q66. Describe the refractory applications in Iron and Steel industry.

Ans. Iron and Steel industry is the major consumer of refractories with about 70% of refractories being consumed by them.

The different areas of the steel manufacturing processes are exposed to different temperatures, slag & sulphur gases. The diversity in the operating conditions of different equipment's demands different grades of refractories for different areas of application.

Steel industry uses refractory for following applications:

- Coke Ovens
- Blast Furnace (B.F) & B.F. Stoves
- Steel & Hot metal ladles/Torpedo Ladles
- Basic Oxygen Furnace / L.D. converter Electric Arc Furnace (E.A.F)
- Soaking pits
- Reheating furnace
- Secondary Refining Ladles
- Stainless Steel Vessels
- Slide Gate Refractories
- Hot metal mixer
- Tundish Refractories
- Induction Furnace

Blast furnace, coke oven, L.D. converter & rolling mill furnaces (soaking pits, reheating furnace, annealing furnace etc.) are the focus of any integrated steel plant from refractory application point of view..

Coke oven walls are lined with silica bricks whereas oven top and regenerators are lined with fireclay bricks.

Blast furnace is used to reduce the iron ore to iron. The charge consisting of the sinter, lumps of iron ore, coke and limestone etc is fed from the top. Air heated in the blast furnace stoves, is supplied from the bottom of the furnace. The hot blast comes in contact with the descending charge in furnace and the iron ore gets reduced to iron due to reducing conditions on account of H_2 and CO in the furnace. CO provides further heat and a very high temperature is developed because of which the iron gets melted which, along with the slag is collected in the hearth from where these are tapped seperately from different tap holes.

B.F. hearth linings are normally constructed with high quality 42-62% Al_2O_3, Mullite, or conventional carbon block. Current furnace designs favor Carbon/Graphite block with super micro-pores because of its non-wetting characteristics. Monolithic linings based upon high MgO powders are rammed and contoured into the sidewalls. Some furnace operations do prefer brick hearths, generally based upon pitch impregnated fired magnesite.

B.F.Stoves use super duty fireclay brick in combustion chamber & high alumina bricks in the dome.

For L.D. Converters, basic refractory of magnesia constitute the most important group of refractories for the steelmaking processes. The conventional magnesite refractory may be pitch bonded; resin bonded, burned and impregnated magnesia brick.

Magnesia-carbon brick were originally designed for water cooled electric furnaces but later with improvement these are now widely used in basic oxygen converter, ladle slag lines, degassers, etc. Permanent lining of the L.D. converter is generally of chrome magnesite bricks whereas wear out lining is of magnesia bricks.Cheaper tar dolomite bricks are also used as wear out lining or as patching refractory.

Hot metal (Pig iron) mixers use chrome magnesite brick in the hearth and firebricks in the side walls.

Electric arc furnace roofs have a high degree of water cooling, and as such require little refractory material. The delta between the electrodes is invariably formed using a monolithic material or precast shape, but refractory bricks are often used to form the electrode ports.

Materials in this area require good thermal shock & abrasion resistance, and use of high alumina materials advantageous. Where the furnace roof is bricked, these are subjected to high thermal loading, and slag splash. Phosphate bonded heat treated materials offer good resistance to slag and spalling in the roof.

Soaking pits bottom is lined with chrome magnesite bricks whereas side walls are lined with fireclay bricks.

Reheating furnaces and annealing furnaces are lined with firebricks mostly.

Rotary kilns for calcination of limestone & dolomite are lined with chrome magnesite in the middle whereas with firebricks on both the charging & discharging sides.

Q67. Describe the refractory applications in Copper Industry.

Ans. In copper industry, "Direct Bonded Magnesia Chrome Brick" is the most popular choice worldwide.The bulk of copper smelting is produced using flash smelters followed by reverberatory furnaces. Flash furnaces have major benefits over reverberatory furnaces, particularly in fuel efficiency, and increased yield.

A comprehensive range of products for the traditional and advanced smelting processes is primarily based upon magnesite chrome products for the working lining with alumino - silicate materials for the intermediate & safety linings.

Primary smelting vessels, whether they are flash furnaces, reverberatory furnaces or electric arc furnaces; all subject the refractory lining to attack by slag and sulphur gases. The slag composition is normally high in Fe_2O_3 & SiO_2, both of which attack magnesia in the refractory to form magnesio-ferrite, and forsterite, whilst the sulphur present in the vapor phase also attacks magnesia to form magnesium sulphate. All reactions in the refractory system are accompanied by volume expansion, leading to crack formation behind the working face, which are then lost during temperature cycling.

Direct bonded magnesite chrome refractories are the preferred materials in contact with copper slags, owing to their greater tolerance to acidic slags. Improved products based upon pre-sintered clinker and/or fused grain clinkers also show further benefit.

The copper converter carries similar slags to those found in the smelting process, together with sulphurous gases, and consequently similar modes of refractory attack take place. Preferential wear frequently takes place at the tuyere line of the converter owing to the combination of slag attack & thermal shock. High hot strength chrome magnesite with good thermal shock resistance is a prerequisite.

There is little slag present in the Anode furnace, but the refractory lining is in contact with highly penetrative molten copper and cuprous oxide. Penetration of the refractory material with copper leads to disruptive failure owing to crack formation at the limit of penetration. Reaction between copper oxide and silica in the refractory leads to the formation of low melting point compound which exacerbate the formation of cracks, and loss of the working face.

The working environment of the Anode furnace demands chrome magnesite material with high hot strength to minimize copper penetration and disruptive loss of the working face.

The slag cleaning furnace has special requirements depending upon the slag practice. It is necessary to maintain a frozen slag layer on the refractory lining, therefore, a good slag and penetration resistance is required. Direct bonded Magnesite chrome materials are preferred.

Q68. Discuss the refractories used in Aluminum Industry.

Ans. Refractories in Aluminum industry are commonly utilized in several equipment's like:

- Anode Baking Furnaces
- Melting / Holding Furnaces
- Induction Furnaces
- Reduction Pots
- Ladles and Launders

Most common refractories used in Aluminum Industry include:

- Super duty Fireclay & High Alumina Bricks.
- Phosphate Bonded High Alumina Bricks – These are ideally suited to molten aluminum ladle/furnace applications due to their unique chemistry, including special aggregates and additives.

The special requirements for refractories in Aluminum industry are:

- Resistance to Carbon/Aluminum penetration.
- High Strength.
- High Corrosion resistance.

Range of products for Aluminum contact applications includes 'Insulating, dense, Low Cement and Ultra Low Cement Castables'. These products are not only reaction resistant to Aluminum contact but also totally non wetting.

Pot Shell Refractory Material. Pot shell is a refractory lined vessel composed of an outside steel shell and layers of refractory linings. The interior layer of the refractory lining is exposed to the process (molten metal, slag, corrosive gases). The linings between the shell and working lining are referred to as the "safety or insulating linings." Insulating linings are used to limit heat loss and to maintain the vessel shell temperatures at reasonable levels.

Typically, the pot shell lining consists of several layers of refractories blocks or bricks joined together using some monolithic material of varying qualities.

Refractories show significant expansion upon heating. In service, under given thermal & mechanical boundary conditions, the refractory linings are exposed primarily to through-thickness temperature gradients and to thermal expansion loading. The free thermal expansion of linings is controlled by the external structural steel shell. In most cases, the thermal stresses due to the restraint are considerably greater than the gravity weight stresses.

The cylindrical lining remains in full contact with the outside steel shell due to the thermal expansion and the hydrostatic pressure (due to molten metal). Therefore, both loads develop a uniform pressure loading around the cylindrical shell.

The behavior of the refractory flat lining differs considerably from that of the cylindrical lining. The refractory flat linings are subjected to unstable and undesirable inward & outward displacements (with a risk of buckling) when subjected to cyclic pressure and expansion loadings.

Q69. Describe uses of refractories in Cement Industry.

Ans. In cement industry, the kiln, the preheating system & the cooler operate in a vey high temperature range and therefore lined with refractories to protect steel structure against heat, abrasion & chemical attack.

Cement Rotary Kiln is the heart of any cement plant directly influencing the plant productivity. Nature of stresses is different in different areas, necessitating different quality of refractories for different zones.

Common refractory materials used in cement plant are:

- Dolomite
- Magnesia-Chrome Spinel
- Magnesia-Alumina Spinel
- Fireclay
- High Alumina

The cement kiln clinker area is usually provided with Dolomite refractories. Advantages of Dolomite refractories for cement are well established – it is the most compatible material with the cement kiln clinker as it provides excellent coating stability, has very good resistance to thermal shock and alkali attack in varying kiln operating conditions. The refractories with Zirconia enrichment are used for crack arresting.

The Transition zone / Preheat zone / Preheater /precalinator areas are usually provided with Magnesia-Alumina spinel variety with high thermo elastic and thermo mechanical property and various grades of high alumina bricks (45-80% Al_2O_3) with high strength.

Q70. Describe the refractory used in Glass Industry.

Ans. The glass industry furnaces are categorized by:

- High operating temperature.
- High Insulation of crown.

- High pull rates and low melting Rates.
- Corrosive vapors and Batch dust.

The bonded refractories are utilized in melter, refiner, regenerator, dog-house and ports of furnace.

Melter/Refiner: Refractories are needed for superstructure, crown and bottom. The important characteristics of the refractories in these areas are:

- High Alkali Vapor Resistance.
- High hot strength.
- High Corrosion Resistance.

Crown: The furnace crowns are heavily insulated to reduce the energy loss. The important characteristics of refractory to meet this requirement are:

- Very low flux factor <0.5; high hot strength.
- Very low residual quartz <0.5%; good volume stability at service temperature.
- High alkali vapor resistance.
- Close dimensional tolerances required for laying with zero joint concept.

Regenerators: Regenerator in single or multi pass design is a common energy saving measure adopted. The basic requirement of the regenerator is to maximize the heat recovery from outgoing gases apart from a long troubie free campaign life. This coupled with the increase in melting rates and use of alternative fuels has increased the severity of the working conditions in regenerator, subjecting the refractories to adverse chemical attack and high temperatures.

To achieve optimum energy saving and long campaign life, the refractories for such areas should be selected for:

- High thermal shock resistance.
- High hot strength.
- High creep resistance at service temperature.
- High heat transfer.

Regenerator checker system is the most vital component of a glass tank furnace for energy saving. To increase the efficiency of heat transfer and to withstand the onslaught of the batch dust, high temperature & thermal cycling, the products are characterized by:

- High thermal shock resistance.
- High resistance to the chemical attack of batch dust.
- High heat transfer.

The needs of the different regions of the regenerator are different with condensation zone being the most critical zone.

Zirconia is useful refractory material for glass furnaces primarily since it is not easily wetted by molten glass and because of its low reaction with them.

Q71. Describe various Furnace refractory maintenance practices.

Ans. The following refractory maintenance practices are commonly practiced:

Slag splashing

Slag splashing is done in steelmaking vessels *eg.*; L. D. converter. After steel tapping, some amount of slag is retained. Composition of slag with respect to FeO and MgO is adjusted. FeO makes the slag adhesive on the refractory lining and MgO makes the lining high temperature resistant. Nitrogen is blown from top to splash the slag. The splashed slag gets coated on the lining. To reduce excessive slag build up in the bottom, excess slag is then poured before charging hot metal.

In case of combined blowing practice formation of skull may result in a failure of the bottom stirring elements.

Slag coating and slag washing

The small amount of liquid slag is retained in the vessel after tapping. Slag is enriched with dolomite or raw dolomite to cool the slag and to increase its adhesive properties. Vessel is rocked several times to coat the bottom and bottom joint with a slag.

Hot patching

Self flowing refractory mixtures enable precise maintenance of the scrap impact zone, tapping pad and bottom joint.

Gunning

By gunning, *i.e.* maintenance of pre worn areas with special gunning mixtures, vessel lining life can be extended. Flame gunning involves simultaneous melting of a refractory powder and gunning at the hot surface.

Since the gunned repair material is dense and fused directly on the hot surface; excellent results on life of lining is obtained in LD converter and coke oven walls.

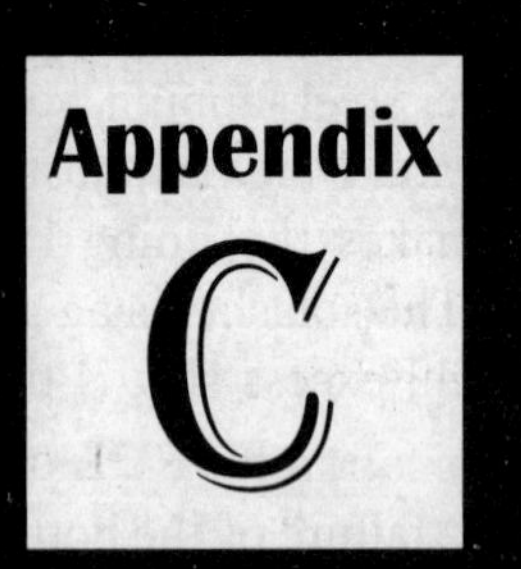

Multichoice Questions & Answer on Refractory Technology

1. High refractoriness of refractory bricks means, that it has a
   (a) High spalling resistance.
   (b) Low spalling resistance.
   (c) High resistance to fusion.
   (d) Low porosity.
2. Capacity of a refractory brick to withstand sudden changes in temperature is denoted by the property called
   (a) Spalling resistance.
   (b) Refractoriness.
   (c) Refractoriness under load.
   (d) None of these.
3. Permeability of bricks is a measure of the
   (a) Refractoriness.
   (b) Melting point.
   (c) Rate at which a fluid will pass through the pores.
   (d) Expansion during heating.
4. Pyrometric cone equivalent (PCE) of a refractory is the measure of its
   (a) Spalling resistance.
   (b) Fusion point.
   (c) Resistance to slag penetration
   (d) Resistance to carbon monoxide attack.
5. Fusion temperature of pure silica (SiO_2) is .... °C.
   (a) 1350 (c) 1715
   (c) 2570 (d) 2800
6. Fusion point of an acidic refractory material is
   (a) Increased by the addition of basic oxides.
   (b) Reduced by the addition of basic oxides.
   (c) Not affected by the addition of basic oxides.
   (d) Always more than 2500°C.
7. Fusion point of a basic refractory material is
   (a) Reduced by the addition of acid oxides.
   (b) Increased by the addition of acid oxides,
   (c) Not affected by the addition of acid oxides.
   (d) Always less than 1000°C.
8. Refractoriness under load (RUL) is the most important property for the refractory bricks used in the
   (a) Hearth bottom of the furnace on which stock is placed.
   (b) Furnace insulation.

Answers

1. (c) 2. (a) 3. (c) 4. (b) 5. (b) 6. (b) 7. (b) 8. (a)

(c) Roof of the furnace.
(d) None of these.

9. Refractories subjected to alternate cycles of heating & cooling are liable to loose their resistance to
(a) Thermal spalling.
(b) Slag attack.
(c) Fusion under load.
(d) CO attack.

10. Crushing strength of a refractory
(a) Increases with rise in service temperature.
(b) Decreases with rise in service temperature.
(c) Is unaffected with change in service temperature.
(d) Decreases with increase in porosity.

11. Thermal conductivity of refractory bricks
(a) increases with decrease in porosity.
(b) decreases with decrease in porosity.
(c) is independent of its porosity and is max imum for insulating bricks.
(d) increases with amount of air entrapped in pores.

12. Pick out the wrong statement.
(a) Electrical conductivity of refractory is not important, when these are to be used in electrical furnaces.
(b) Graphite & metals are the good electrical conductor among the refractories and others are all electrical insulators.
(c) Refractories used for lining electrical furnaces should ordinarily have very low electrical conductivity.
(d) Electrical conductivity of porous refractory materials is low.

13. Segar cones are used for the determination of the .... of refractories.
(a) softening temperature.
(b) spalling resistance.
(c) electrical conductivity.
(d) resistance to slag attack.

14. Test piece for determination for RUL of a refractory is heated in a/an
(a) oxidising atmosphere.
(b) reducing atmosphere.
(c) electric furnace.
(d) neutral atmosphere.

15. Panel test determines the .... of refractories.
(a) fusion point
(b) spalling resistance
(c) slag penetration resistance
(d) refractoriness under load

16. Spray test determines the.........of refractories.
(a) resistance to slag penetration
(b) resistance to CO attack
(c) RUL
(d) permanent linear change.

17. Which is an acidic refractory?
(a) Magnesite (b) Dolomite
(c) Fireclay (d) Chrome magnesite

18. Which is a basic refractory?
(a) Fireclay
(b) Silica
(c) Chrome magnesite
(d) None of these

19. Which is a neutral refractory?
(a) Graphite (b) Magnesite chrome
(c) Silica (d) Magnesia

Answers

9. (a) 10. (b) 11. (a) 12. (a) 13. (a) 14. (c) 15. (b) 16. (a) 17. (c) 18. (c) 19. (a)

20. Which is not an acidic refractory?
    (a) Silica
    (b) Fireclay
    (c) High alumina refractory
    (d) Carbon block.

21. Which is not a basic refractory?
    (a) Chrome magnesite
    (b) Magnesite
    (c) Dolomite
    (d) Silicon carbide

22. Insulating refractories should have
    (a) high porosity
    (b) low thermal conductivity
    (c) both (a) & (b)
    (d) neither (a) nor (b).

23. Which is not an alumino-silicate refractory?
    (a) Fireclay bricks
    (b) Mullite bricks
    (c) Tar dolomite bricks
    (d) High alumina bricks.

24. Water content is ground refractory material to be shaped into bricks by hand moulding is about .... percent.
    (a) 5 (b) 20
    (c) 40 (d) 55.

25. Hollow refractory bricks are made by
    (a) slip casting
    (b) hand moulding
    (c) pressing/machine moulding
    (d) extrusion

26. Refractories are dried in the
    (a) rotary kilns (b) tunnel kilns
    (c) sun (d) none of these

27. Which refractory must have controlled atmosphere (temperature & humidity) for its safe storage?
    (a) Tar bonded dolomite bricks.
    (b) Fireclay bricks.
    (c) Mullite bricks.
    (d) Magnesite bricks.

28. Which brick undergoes maximum shrinkage on drying?
    (a) Tar bonded dolomite bricks.
    (b) Fireclay bricks.
    (c) Magnesite bricks.
    (d) Chromite bricks.

29. An indication of degree of firing in silica brick is its
    (a) specific gravity (b) fusion point
    (c) RUL (d) none of these.

30. Mixing of ground refractory material and water is done in a.......................mill.
    (a) pug (b) ball
    (c) tube (d) rod

31. Colour of fireclay bricks is
    (a) light buff to reddish buff
    (b) yellow
    (c) black
    (d) none of these.

32. Porosity of fireclay refractories is .... percent.
    (a) 5-10 (b) 10-25
    (c) 25-35 (d) 35-50

33. Fireclay refractories have
    (a) low co-efficient of thermal expansion.
    (b) poor thermal spalling resistance.
    (c) tendency to expand unduly high during firing.
    (d) very high cost.

Answers

20. (d) 21. (d) 22. (c) 23. (c) 24. (b) 25. (a) 26. (b) 27. (a) 28. (b) 29. (d) 30. (a) 31. (a) 32. (a) 33. (b)

34. With increase in porosity, thermal spalling resistance of fireclay brick
    (*a*) increases
    (*b*) decreases
    (*c*) remains same
    (*d*) may increase or decrease

35. Refractoriness under load (RUL) of fireclay bricks (under a load of 2 kg/cm^2) is ....°C.
    (*a*) 500 (*b*) 100
    (*c*) >1350 (*d*) >2000

36. Cold crushing strength (C.C.S.) of fireclay bricks is about..............kg/cm^2.
    (*a*) 50—100 (*b*) 100—150
    (*c*) 200—400 (*d*) 500—1000.

37. Fireclay bricks are not used in the
    (*a*) blast furnace
    (*b*) hot blast stove
    (*c*) cupola
    (*d*) walls of coke oven.

38. Fireclay bricks are used in the lining of the
    (*a*) coke ovens regenerator.
    (*b*) L.D. converter.
    (*c*) hearth of blast furnace.
    (*d*) none of these.

39. Maximum alumina content in high alumina refractory can be as high as .... percent.
    (*a*) 30 (*b*) 50
    (*c*) 70 (*d*) 90

40. High alumina refractory compared to fireclay bricks have
    (*a*) less load bearing capacity.
    (*b*) less resistance to slag attack.
    (*c*) low refractoriness.
    (*d*) high resistance to thermal shock & creep.

41. Sillimanite is a......................refractory.
    (*a*) basic (*b*) neutral
    (*c*) high alumina (*d*) insulating

42. With increase of alumina content, the refractoriness of high alumina refractories
    (*a*) increases
    (*b*) decreases
    (*c*) remains same
    (*d*) may increase or decrease.

43. Refractoriness under high loads (RUL.) is quite close to the fusion temperature (PCE) for.....................bricks.
    (*a*) fireclay (*b*) silica
    (*c*) dolomite (*d*) very low alumina.

44. Si percentage in silica refractories used in the walls of coke oven is about
    (*a*) 45 (*b*) 60
    (*c*) 80 (*d*) 95

45. Silica refractories are not used in the
    (*a*) coke oven walls.
    (*b*) beehive coke ovens.
    (*c*) dome and upper portion of stoves.
    (*d*) open hearth furnace roof.

46. Dolomite bricks have good resistance to attack by
    (*a*) molten steel (*b*) iron oxide
    (*c*) lime slag (*d*) none of these

47. Periclase refractory contains mainly
    (*a*) CaO (*b*) MgO
    (*c*) Al_2O_3 (*d*) SiO_2

48. Firing temperature of magnesite bricks is about.....................°C.
    (*a*) 800-1000 (*b*) 1000-1200
    (*c*) 1600-1800 (*d*) 2400-2600.

49. Maximum apparent porosity of magnesite bricks is about....................percent.

Answers

34. (a) 35. (c) 36. (c) 37. (d) 38. (a) 39. (c) 40. (d) 41. (c) 42. (a) 42. (d) 43. (b) 44. (d) 45. (b) 46. (a) 47. (b) 48. (c) 49. (b)

(a) 8 (b) 24
(c) 44 (d) 58

50. Which of the following impurities reduces the refractoriness of magnesite bricks?
(a) Al_2O_3 (b) CaO
(c) SiO_2 (d) all (a), (b) & (c).

51. Magnesite bricks have poor resistance to attack by.........................slag.
(a) lime (b) basic
(c) acid (d) none of these

52. Slide gate in teeming ladle used for steel pouring in ingot mould is lined with the
(a) bakelite impregnated or fused periclase.
(b) silica bricks.
(c) semi-silica bricks.
(d) none of these.

53. Chrome magnesite bricks are
(a) acidic in nature.
(b) neutral in nature.
(c) having higher RUL than silica bricks.
(d) made by mixing 30% chromite & 70% periclase.

54. Chrome magnesite bricks are used in the
(a) side walls of soaking pits & arc furnaces.
(b) copper smelting furnaces & converters.
(c) burner block of glass tanks.
(d) all (a), (b) & (c).

55. Chrome magnesite is not used in the
(a) inner lining of L. D. converter.
(b) aluminium melting furnaces.
(c) wear out lining of steel melting furnaces.
(d) all (a), (b) & (c).

56. Magnesite chrome refractories .
(a) have better spalling resistance than chrome magnesite refractories.
(b) have very low thermal co-efficient of expansion.
(c) are not at all resistant to the corrosive action of iron oxide.
(d) have very low (50 kg/cm^2) cold crushing strength (C.C.S.) and cannot be used in 'metalcase' form.

57. Magnesite chrome bricks are used for the
(a) roof lining of basic open hearth and other basic furnaces.
(b) reheating furnaces.
(c) soaking pits.
(d) all (a), (b) & (c).

58. The most common bonding material for carbon refractory made from graphite is
(a) tar/pitch (b) clay
(c) lime (d) sodium silicate

59. Carbon refractories
(a) do not burn/oxidise when exposed to air on heating.
(b) are not attacked by slags as they are not wetted by melts.
(c) do not resist temperature fluctuations.
(d) have extremely low thermal & electrical conductivities.

60. Carbon refractories are exclusively used in the
(a) hearth of blast furnace.
(b) walls of coke oven.
(c) regenerators of coke oven.
(d) side wall of soaking pits.

61. Skull is not formed on the carbon blocks in the hearth of a blast furnace, when it becomes cold because of its
(a) non-wetting characteristic.
(b) high thermal conductivity.
(c) high crushing strength.
(d) none of these.

Answers

50. (d) 51. (c) 52. (a) 53. (c) 54. (d) 55. (c) 56. (a) 57. (d) 58. (a) 59. (b) 60. (a) 61. (a)

62. Which furnace consumes maximum refractory annually in an integrated steel plant?
   (a) Soaking pit (b) Blast furnace
   (c) L.D. converter (d) Coke ovens

63. The main raw material for the manufacture of silicon carbide refractories is
   (a) corrundum (b) carborundum
   (c) bauxite (d) periclase

64. Silicon carbide refractories have very low
   (a) refractoriness (<1700°C).
   (b) thermal conductivity.
   (c) thermal shock resistance.
   (d) none of these.

65. Zirconia refractory
   (a) does not react with basic slags.
   (b) is produced from baddeleyite.
   (c) cannot be used as an insulator.
   (d) has poor electrical conductivity at high temperature.

66. Zirconia refractories are not used in
   (a) making sheaths for thermocouple.
   (b) lining high temperature ceramic kilns.
   (c) furnaces subjected to fluctuating temperature.
   (d) high frequency induction furnaces in the form of inductors.

67. Dry mix hydraulic compositions of refractory aggregates with suitable bonding materials are called refractory
   (a) mortars (b) cements
   (c) castables (d) none of these.

68. Dynamometer is used for the determination of................................ of refractories.
   (a) Modulus of rupture.
   (b) Permanent linear change.
   (c) Resistance to CO attack.
   (d) RUL

69. A steel member used in furnace construction to take the thrust of the brick work is called the
   (a) buckstay (b) breast wall
   (c) armouring (d) baffle

70. Canister is a source of
   (a) silica (b) periclase
   (c) lime (d) none of these.

71. Thermal diffusivity of a refractory brick is high, when its.........................is high.
   (a) density
   (c) specific heat
   (c) thermal conductivity
   (d) none of these.

72. SiO_2 in firebrick is about .... percent.
   (a) 35-40 (b) 55-60
   (c) 80-85 (d) > 94

73. Which one contains minimum SiO_2?
   (a) Firebrick
   (b) Sillimanite
   (c) Semi-silica
   (d) Aluminous firebrick.

74. Which one contains maximum Al_2O_3?
   (a) Firebrick
   (b) Sillimanite
   (c) Magnesite
   (d) Aluminous firebrick

75. Which is the most stable crystalline form of silica at room temperature?
   (a) Quartz (b) Cristobalite
   (c) Tridymite (d) None of these

76. Rate of slag attack on refractories increases with rise in temperature due to

Answers

62. (b) 63. (b) 64. (d) 65. (b) 66. (c) 67. (c) 68. (b) 69. (a) 70. (a) 71. (c) 72. (b) 73. (b) 74. (b) 75. (a) 76. (a)

(a) decreased viscosity of slag.
(b) increased thermal conductivity of brick.
(c) oxidizing condition in the furnace.
(d) none of these.

77. Hot metal runner in the blast furnace are lined with the......................bricks.
(a) silica (b) carborundum
(c) fireclay (d) magnesite

78. Hot blast main (carrying air at 1000°C) in blast furnace are lined with the............. bricks.
(a) silica (b) fireclay
(c) magnesite (d) zirconia

79. Mullite is chemically represented as
(a) $Al_2O_3.2SiO_2$ (b) $3Al_2O_3.2SiO_2$
(c) $Al_2O_3.SiO_2$ (d) $2Al_2O_3.3SiO_3$.

80. Which is not a natural insulating material?
(a) Diatomaceous earth/kieselgur.
(b) Asbestos.
(c) Vermiculite.
(d) None of these.

81. Which property of refractories is the most important for top section of the blast furnace?
(a) Resistance to abrasion.
(b) Resistance to slag penetration.
(c) Stability of volume at high temperature.
(d) Resistance to corrosion by slag.

82. Lower part of hot metal mixer are lined with the..........................bricks.
(a) superduty fireclay
(b) high alumina
(c) silica
(d) carborundum.

83. Upper portion of hot metal mixer are lined with mullite bricks which helps in resisting the
(a) washing action of metal.
(b) corrosion by layer of slag.
(c) CO attack on bricks.
(d) none of these.

84. Bottom of basic open hearth furnace are constructed of the
(a) dead burnt magnesite ramming mass.
(b) porous fireclay bricks.
(c) semi-silica bricks.
(d) silicon carbide bricks.

85. Roof of a basic electric furnace is made of the.............bricks.
(a) superduty fireclay
(b) silica
(c) chromite
(d) none of these.

86. Lower courses of soaking pits are made of.................bricks to avoid the action of molten slag & scale.
(a) chrome or magnesite
(b) silicon carbide
(c) silica
(d) low duty fireclay

87. Walls, roofs & combustion chambers of annealing furnaces are made of the............................bricks.
(a) high duty fireclay
(b) silica
(c) mullite
(d) carborundum

88. Which is the stable form of silica below 870°C?
(a) Tridymite (b) Cristobalite
(c) Quartz (d) None of these

Answers

77. (c) 78. (b) 79. (b) 80. (d) 81. (a) 82. (d) 83. (a) 84. (a) 85. (b) 86. (a) 87. (a) 88. (c)

89. Which is the stable form of silica upto 1470°C?
    (a) Quartz *(b)* Cristobalite
    (c) Tridymite *(d)* None of these
90. Which is the stable form of silica between 1470°C and the melting point 1713°C?
    (a) Cristobalite *(b)* Tridymite
    (c) Quartz *(d)* None of these
91. Grog addition in fireclay during brick manufacture is done to
    (a) reduce its shrinkage on heating.
    (b) impart greater spalling resistance.
    (c) enhance the strength of fired refractories.
    (d) all *(a)*, *(b)* & *(c)*.
92. With increasing alumina content, the fusion point of high alumina refractories
    (a) increases
    (b) decreases
    (c) remains constant
    (d) may increase or decrease
93. Chromite refractories are
    (a) acidic refractory.
    (b) neutral refractory.
    (c) basic refractory.
    (d) fired at 600°C only.
94. Carbon refractory blocks
    (a) are wetted by molten iron.
    (b) are used in the hearth of blast furnace.
    (c) are acidic in nature.
    (d) all *(a)*, *(b)* & *(c)*.
95. Periclase is
    (a) crystalline form of MgO.
    (b) $MgCO_3$.
    (c) $ZrSO_2$.
    (a) amorphous magnesite.
96. Magnesite bricks are used in those parts of furnaces, which are
    (a) subjected to temperature fluctuation.
    (b) required to resist corrosive basic slag.
    (c) high load.
    (d) none of these.
97. Highest melting (m.p. = 3070°C) oxide refractory is
    (a) alumina *(b)* thoria
    (c) zirconia *(d)* magnesia
98. 'Super refractories' are made from pure
    (a) carbides *(b)* oxides
    (c) borides *(d)* nitrides
99. Carborundum used for making crucibles for melting non-ferrous metals is chemically
    (a) silicon carbide
    (b) silicon nitride
    (c) crystalline magnesia
    (d) zirconium sulphate
100. Chemically, mullite refractories is
    (a) Al_2O_3. $2SiO_2$ *(b)* Al_2O_3
    (c) $ZrSO_4$ *(d)* ThO_2.
101. ....bricks are used in the burning zone of a cement rotary kiln.
    (a) High alumina *(b)* Fireclay
    (c) Thoria *(d)* Silicon carbide
102. Roof of basic open hearth furnace is lined with the...............bricks.
    (a) silica *(b)* fireclay
    (c) dolomite *(d)* magnesite.
103. Fireclay bricks is not used for lining the
    (a) cupola
    (b) gas producer
    (c) bottom of hot metal mixer
    (d) roof of open hearth furnace

Answers

89. (c) 90. (a) 91. (d) 92. (a) 93. (b) 94. (b) 95. (b) 96. (b) 97. (b) 98. (b) 99. (a) 100. (a) 101. (a) 102. (a) 103. (c)

104. Chrome magnesite brick is not used for lining the
(a) hearth of soaking pits.
(b) bottom hearth of reheating furnace.
(c) coke oven regenerator.
(d) burning zone of limestone rotary kilns.

105. Porosity of silica bricks varies from .... percent.
(a) 5 to 10 (b) 20 to 30
(c) 45 to 60 (d) 60 to 75

106. Which of the following bricks should not be used, if the furnace is to be used intermit- tantly?
(a) Firebricks
(b) Silica bricks
(c) Silicon carbide bricks.
(d) Sillimanite

107. Which of the following has the lowest electrical resistivity?
(a) Graphite (b) Fireclay
(c) Alumina (d) Zircon

108. Refractoriness/fusion points of'superduty' refractories is................°C.
(a) 1520- 1630 (b) 1630- 1670
(c) 1730 (d) 2000

109. With decrease in porosity, the............of the refractories decreases.
(a) strength.
(b) thermal conductivity.
(c) spalling resistance.
(d) none of these.

110. ....is the measure of the strength of refractory under the combined effect of temperature & load.
(a) Porosity.
(b) RUL.
(c) Specific gravity.
(d) Thermal conductivity.

111. Spalling tendency of refractories is reduced by increasing its
(a) porosity.
(b) specific gravity.
(c) thermal conductivity.
(d) strength.

112. Refractories used in/for................should have low thermal conductivity.
(a) coke ovens (b) insulation
(b) regenerators (d) muffle furnaces

113. Except..............., all other refractories are bad conductors of electricity (*i.e.*, have low electrical conductivity).
(a) fireclay (b) carborundum
(c) graphite (d) chromite

114. Maximum water percentage in refractory mix meant for hand moulding may be as high as................precent.
(a) 2-3 (b) 5-7
(c) 7-12 (d) 14-20

115. Machine moulding of dry mixture of refractories requires a pressure of the order of..........................kg/cm^2.
(a) 10 (b) 500
(c) 100 (d) 1000

116. Maximum shrinkage in volume occuring during burning/firing of dried refractories may be as high as.....................percent.
(a) 10 (b) 15
(c) 20 (d) 30

117. Which of the following does not occur during firing/burning of refractories?
(a) Removal of water of hydration.
(b) Vitrification.
(c) Decrease in crushing strength.
(d) Development of stable mineral form.

Answers

104. (c) 105. (b) 106. (b) 107. (a) 108. (c) 109. (c) 110. (b) 111. (a) 112. (b)
113. (c) 114. (d) 115. (d) 116. (c) 117. (a)

118. Firing temperature is minimum (1250 - 1400°C) for the..................bricks.
(a) fireclay
(b) direct bonded basic
(c) silica
(d) magnesite

119. Cold crushing strength of ordinary fireclay brick is about 950 Kg/cm^2 . On exposure to a temperature of about 1500°C, its crushing strength may come down to as low as..............Kg/cm^2.
(a) 450 (b) 250
(b) 150 *(d)* 65

120. Maximum safe working temperature for fireclay bricks is about.....................°C.
(a) 1150 *(b)* 1300
(c) 1450 *(d)* 1550

121. The maximum linear expansion of silica bricks during firing is about.............. percent.
(a) 0.5 *(b)* 1
(c) 2 *(d)* 3.5

122. 10 to 30% magnesite is added to chromite to produce chrome-magnesite refractories. Magnesite addition is mainly done to improve the.....................of chromite.
(a) spalling resistance
(b) refractoriness
(c) crushing strength
(d) resistance to slag attack

123. The highest melting pure oxide (mp > 3000°C) is
(a) thoria *(b)* alumina
(c) beryllia *(d)* zirconia

124. Which is not a high alumina refractory?
(a) Mullite *(b)* Corrundum
(c) Bauxite *(d)* Dolomite

125. 'Spinel', a refractory mineral is chemically represented as
(a) $MgAl_2O_4$ *(b)* Al_2O_3
(c) $MgSO_4$ *(d)* $MgAl_2O_3 . 2H_2O$

126. In panel test for spalling resistance, the average face temperature of panel assembly is maintained at.................°C for 24 hours.
(a) 700 *(b)* 1000
(c) 1600 *(d)* 2000

127. With increase in the alumina content in fire bricks, its fusion point (refractoriness)
(a) decreases linearly
(b) remains unchanged
(c) increases
(d) decreases

128. Refractoriness of a typical silica brick corresponds to Segar cone number 32, which is equivalent to a temperature of..................°C.
(a) 1380 *(b)* 1520
(c) 1710 *(d)* 1915

129. Which form of silica has the highest specific gravity?
(a) Quartz
(b) Cristobalite
(c) Tridymite
(d) All have same specific gravity

130. The linear thermal expansion of .... bricks upto 1000°C is very low of the order of < 0.5%.
(a) fireclay *(b)* silica
(c) magnesite *(d)* corundum

131. Silica bricks have low spalling resistance below 600°C, due to their
(a) very high thermal conductivity.

Answers

118. (d)	119. (d)	120. (d)	121. (a)	122. (a)	123. (d)	124. (a)	125. (c)	126. (c)
127. (c)	128. (a)	129. (a)	130. (b)	131. (a)				

(b) high co-efficient of expansion upto this temperature.
(c) high thermal diffusivity.
(d) low refractoriness.

132. Electrical resistor bars are made of
(a) silicon carbide (b) alumina
(c) zirconia (d) graphite

133. Silica bricks are never used for lining the
(a) beehive coke ovens.
(b) by-product coke ovens.
(c) dome of blast furnace stoves.
(d) roof of open hearth furnace.

134. Rotary kilns meant for calcination of limestone are lined with chrome magnesite bricks in the.....................zone.
(a) preheating (b) cooling
(c) burning (d) all

135. Bauxite calcining rotary kins are lined with the....................bricks.
(a) fireclay (b) carbon
(c) 85% alumina (d) corundum

136. Vacuum steel degassing units are lined with the......................bricks.
(a) silica (b) low duty fireclay
(c) high alumina (d) graphite

137........................ nozzles are used in continuous casting of steel.
(a) Zircon (b) Thoria
(c) Carborundum (d) Beryllia

138. Cold crushing strength of a refractory does not depend upon its
(a) shape
(b) composition
(c) firing temperature
(d) texture

139. High thermal conductivity of a refractory material is not important, when it is to be used in the
(a) coke oven regenerators
(b) muffle furnace
(c) blast furnace
(d) recuperators

140. Basic bricks are not made of
(a) fireclay (b) magnesite
(c) foresterite (d) chromite

141......................of carbon blocks in the hearth of blast furnace helps in avoiding skull formation, when it becomes cold.
(a) High thermal conductivity
(b) Low porosity
(c) Non-wetting characteristic
(d) High density

142. Magnesite refractories are generally not used in the lining of
(a) electric furnace walls.
(b) steel melting furnace.
(c) open hearth furnace.
(d) burning zone of cement kilns.

143. Pick out the wrong statement.
(a) Insulating refractories used in place of regular refractory bricks are usually called light weight refractories and they have similar composition as heavy bricks.
(b) Graphite refractories are also called plumbago refractories.
(c) Superduty fireclay bricks correspond to a pyrometric cone equivalent of 26- 28.
(d) Calcined magnesite is also called dead burnt magnesite.

144. Pure bauxite is the best raw material for the manufacture of high alumina refractories in which maximum alumina content can be as high as...............percent.

Answers

132. (a) 133. (c) 134. (c) 135. (c) 136. (a) 137. (a) 138. (c) 139. (a) 140. (c)
141. (b) 142. (c) 143. (d) 144. (d)

(a) 55 (b) 70
(c) 80 (d) 90

145. Silliminite ($Al_2O_3.SiO_2$) refractory, which is a neutral refractory, is not used in the
(a) pottery furnace
(b) glass melting furnace
(c) crucibles
(d) gas retorts

146. Which of the following is not a neutral refractory?
(a) Silicon carbide (b) Magnesite
(c) Chromite (d) Graphite

147. Graphite or carbon refractories
(a) exhibit wetting characteristics.
(b) should be used in neutral or reducing atmosphere.
(c) exhibit high shrinkage on thermal treatment.
(d) are not resistant to corrosion by slag.

148. ....bricks should not be used in the oxidising atmosphere.
(a) Tar dolomite (b) Carbon
(c) Silica (d) Fireclay

149. Beryllia (which is used in making crucibles for melting uranium & thorium) is superior to alumina in all respects for high temperature (1900°C) use, except
(a) cost.
(b) electrical conductivity,
(c) thermal conductivity.
(d) fusion point.

150. Most Cermets, which normally have high thermal conductivity & high thermal shock resistance, comprises of ceramic & metallic components of.............percent respectively.
(a) 80 & 20 (b) 20 & 80
(c) 50 & 50 (d) 60 & 40

151. Pick out the wrong statement.
(a) Fluxing material like lime is added in clay to reduce the vitrification temperature.
(b) Main constituents of clay are alumina & silica.
(c) Addition of sand in ceramic materials makes it non-plastic, increases its fusion point & reduces its shrinkage on burning.
(d) Vitrification of fireclay material is done to increase its porosity.

152. Because of its very high refractoriness of the order of....................°C, silicon carbide refractories are used in zinc smelting furnace, muffle furnace & for supporting the wares in tunnel kilns.
(a) 1800 (b) 2200
(c) 2400 (d) 2700

153. Match the refractories used in various furnaces.

	List I	List II
(a)	Roof of basic open hearth furnace.	I. Chrome magnesite brick.
(b)	Burning zone of lime kiln.	II. High alumina brick,
(c)	Dome of blast furnace stove.	III. Carbon block.
(d)	Blast furnace hearth.	IV. Silica brick.

154. Match the values of refractoriness under load (RUL) at 2 Kg/cm^2 for various refractory bricks.

List I	List II
(a) MHD'firebrick	I. 1500°C
(b) Zirconia brick	II. 1300°C

Answers

145. (b) 146. (b) 147. (b) 148. (b) 149. (a) 150. (a) 151. (d) 152. (d) 153. (iv) (i) (ii) (iii) 154. (ii) (i) (iv) (iii)

(c) Carbon block	III. 1670°C
(d) Silica brick	IV. 1750°C

155. Match the examples of various classified type of bricks.

List I	List II
(a) Acidific refractory	I. Dolomite
(b) Basic refractory	II. Carbon
(c) Neutral refractory oxide + metallic	III. Ceramic alloys
(d) Cermet refractory	IV. Fireclay

156. Match the bulk density (gm./c.c) of various refractory bricks.

List I	List II
(a) Chrome brick	I. 1.7 - 1.9
(b) Zirconia brick	II. 2.7 - 3.0
(c) Magnesite brick	III. 4 - 4.6
(d) Silica brick	IV. 2.8 - 3.5

157. Match the refractory bricks used in various applications.

List I	List II
(a) Wear out lining of L.D	I. Fireclay bricks, converter.
(b) Regenerator of coke ovens.	II. Tar dolomite brick.
(c) Walls of coke ovens.	III. High alumina brick.
(d) Combustion chamber of	IV. Silica brick. B.F.stoves.

158. Spalling of a refractory means its
(a) softening.
(b) fracture due to uneven expansion at high temperature.
(c) resistance to compressive loads.
(d) resistance to chemical action of gases & molten fluxes.

159. An ideal refractory should have
(a) high spalling rate.
(b) high fusion point.
(c) shrinkage ability.
(d) none of these.

160. Refractory bricks having lower porosity have
(a) high insulating properties.
(b) low heat capacity.
(c) low thermal conductivity.
(d) greater strength.

161. The largest annual consumer of refractories is the..................industry.
(a) cement *(b)* metallurgical
(c) fertiliser *(d)* power

162. Which of the following consumes the maximum tonnage of refractories annually in an integrated steel plant?
(a) Soaking pits *(b)* Reheating furnace
(c) L.D. converter *(d)* Rotary lime kiln

163. Ceramic recuperators used for the waste heat recovery from high temperature flue gas going out of the furnace, is made of
(a) fireclay (b) silicon carbide
(c) corrundum *(d)* siliceous fireclay

164. PCE value (Segar cone) of superduty refractories is more than 33, which is equivalent to the temperature of................°C.
(a) 1520 *(b)* 1630
(c) 1670 *(d)* 1730

165. Cermets are combination of ceramic and metallic materials due to which they have high strength & resistance to high temperature. Cermets are used in the
(a) hearth of blast furnace.
(b) nuclear reactors, missiles & space crafts.
(c) insulation of high temperature furnaces.
(d) roof of electric furnaces.

Answers

155. (iv) (i) (ii) (iii) 156. (iv) (iii) (ii) (i) 157. (ii) (i) (iv) (iii) 158. (b) 159. (b) 160. (d)
161. (b) 162. (c) 163. (b) 164. (d) 165. (b)

166. Pick out the wrong statement.
   (a) Refractories used in muffle furnace should have low thermal conductivity.
   (b) The electrical resistivity of refractories drops rapidly with rise in temperature,
   (c) For reducing spalling tendency, the refractory should be well fired and its porosity should be more.
   (d) Refractoriness under load (RUL) of a refractory is always less than its refractoriness.

167. Pick out the wrong statement.
   (a) A ceramic material which becomes fluid upon heating and can be moulded in liquid/viscous state is termed as glass.
   (b) Ceramic materials do not undergo vitrification on heating.
   (c) Ceramic materials are brittle in nature.
   (d) Non-oxide ceramic materials generally act as a semi-conductor.

168. Refractory materials are never used in the construction of
   (a) segar cones   *(b)* orton cones
   (c) pressure vessels   *(d)* ovens & retorts

169. Spalling resistance of a refractory can not be increased by
   (a) increasing its porosity.
   (b) using a coarser grog during its manufacture.
   (c) decreasing its thermal co-efficient of expansion.
   (d) making it denser.

170. High density refractory bricks have lower
   (a) spalling resistance.
   (b) thermal conductivity.
   (c) fusion point.
   (d) slag penetration resistance.

171. Silica bricks are attacked by basic slags at high temperature. Which of the following is not used solely as a binding material?
   (a) aluminium phosphate
   (b) water
   (c) lime
   (d) plaster of paris

172. Carbon bricks are not used in the lining of the
   (a) combustion chamber of blast furnace stoves.
   (b) electric furnaces.
   (c) highly chemical resistant equipments.
   (d) blast furnace hearth.

173. ....is not a single oxide refractory.
   (a) Zirconia   *(b)* Silicon carbide
   (c) Magnesia   *(d)* none of these

174. Pure oxide refractories are generally monocrystalline in nature and are self bonded.................bricks are generally used as moderator in nuclear reactors.
   (a) Beryllia   *(b)* Carborundum
   (b) Corundum   *(d)* Thoria

175. Thoria is an expensive refractory material and is radioactive in nature. Thorium oxide is used in the manufacture of
   (a) segar cones.
   (b) muffles for muffle furnaces.
   (c) insulating bricks.
   (d) crucibles used for melting of high purity metals.

176. An insulating refractory brick should have high porosity & low thermal conductivity. Which of the following is not used for inducing porosity in the insulating refractory bricks during its manufacture?

Answers

166. (a) 167. (b) 168. (c) 169. (d) 170. (a) 171. (b) 172. (a) 173. (b) 174. (a)
175. (d) 176. (c)

(a) Cork
(b) Saw dust
(c) Sand
(d) Chemically prepared foam

177. Ferromagnetic ceramic material is not used in the
(a) thermal insulation.
(b) transformers.
(c) magnetic switches.
(d) television sets.

178. Spalling of silica bricks occurs due to abrupt volume changes, when it is cooled below a temperature of.......................... °C.
(a) 770 (b) 570
(b) 270 (d) 70

179. Pick out the wrong statement.
(a) Sand & coke is the main raw material for the manufacture of silicon carbide.
(b) Carbon refractories can not be used in the furnaces operating under reducing conditions.
(c) Mullite can be obtained by the heating of alusite, kyanite or silimanite.
(d) Silica occurs in nature in all cellular, amorphous or crystalline form.

180. Match the percentage of various constituents in different typical refractory bricks.

	List I	List II
(a)	Percentage of SiO_2 in fireclay bricks	I. 38-42
(b)	Percentage of Al_2O_3 in silica bricks	II. 20-35
(c)	Percentage of Cr_2O_3 in magnesite bricks	III. 0.5-2.0
(d)	Percentage of CaO in tar dolomite bricks	IV. 50-70

181. Match the fusion point (°C) of various pure refractory oxides.

List I	List II
(a) SiO_2	I. 2050
(b) Al_2O_3	II. 2570
(c) ThO_2	III. 1715
(d) CaO	IV. 3070

182. Match the specific gravities of various pure refractory oxides.

List I	List II
(a) SiO_2	I. 9.69
(b) Al_2O_3	II. 5.56
(c) ThO_2	III. 2.32
(d) ZrO_2	IV, 3.97

183. Match the refractory used in various equipments used in non-ferrous metal industries.

	List I	List II
(a)	Roaster used in the extrac-	I. Magnesite extraction of zinc
(b)	Blast, furnace (at tuyere	II. Dense fireclay level) used in the extraction of lead.
(c)	Flash smelter used in the bonded	III. Phosphate extraction of nickel bricks
(d)	Melting & holding mag- chrome	IV. Fusion cast furnaces in aluminium industry, bricks

184. The co-efficient of thermal expansion of refractory material should be
(a) low (b) high
(b) Medium (d) Very high

185. Alumina is a .... type of refractory.
(a) Neutral (b) Basic
(c) Acidic (d) None of these

Answers

177. (a) 178. (c) 179. (b) 180. (iv) (iii) (ii) (i) 181. (iii) (i) (iv) (ii) 182. (iii) (iv) (i) (ii)
183. (ii) (i) (iv) (iii) 184. (a) 185. (b)

186. The emissivity of a refractory material .... with increasing temperature.
(a) increases
(b) decreases
(c) remains unchanged
(d) increases exponentially

187. An increase in the bulk density of a given refractory increases its
(a) slag penetration resistance.
(b) heat capacity.
(c) volume stability.
(d) all 'a', 'b' & 'c'.

188. Cold crushing strength (CCS) of refractory is most useful for assessing its
(a) ability to withstand the rigors of transport.
(b) adequacy of firing,
(c) abrasion resistance.
(d) bulk density and porosity.

189. Refractory of low thermal conductivity is required for use in
(a) muffles (b) regenerators
(c) recuperators (d) back up insulation

190. Which of the following does not fall in the category of monolithic refractories?
(a) Ramming (b) Mouldables
(c) Castables (d) Fused cast.

191. Out of the following, which is the most important property for an insulating brick?
(a) Hot crushing strength.
(b) Cold crushing strength.
(c) Chemical property.
(d) Porosity.

192. The most common refractory used in general purpose furnace is...............brick.
(a) Silica (b) Fireclay
(c) Zirconia (d) Magnesite

193. The term blanket refers to
(a) castables
(b) ramming mass
(c) ceramic fibre
(d) insulating bricks.

194. Which of the following is added to aluminosilicate to help reduce the shrinkage level of ceramic fibre?
(a) Al_2O_3 (b) ZrO_2
(c) SiO_2 (d) $CaSO_4$.

195. Zirconia is a useful refractory for glass melting furnaces, primarily because it is
(a) having high thermal conductivity
(b) not easily wetted by molten glass
(c) very cheap
(d) not attacked by basic slag.

196. The use of ceramic fibre in a batch type furnace results in fuel economy is due to
(a) thermal shock resistance.
(b) lower heat storage.
(c) chemical resistance.
(d) lower maintenance.

197. Which property is the most important for an insulating brick?
(a) Porosity
(b) Chemical resistance
(c) Compact strength
(d) Mechanical strength

198. Waste heat recovery device in which high thermal conductivity refractory bricks are used for storing heat is
(a) thermo-compressor
(b) heat pump
(c) regenerator
(d) heat pipe

Answers

186. (d) 187. (a) 188. (d) 189. (d) 190. (d) 191. (b) 192. (c) 193. (b) 194. (b)
195. (b) 196. (a) 197. (c) 198. (d)

199. Which is not a property of ceramic fibre insulation?

(a) Thermal shock resistant.

(b) Low thermal conductivity.

(c) Light weight.

(d) High heat storage.

200. Ceramic fibre gives maximum savings when used in......................furnace.

(a) batch (b) continuous

(c) arc (d) induction

201. The insulation material suitable for low temperature application to prevent heat gain is

(a) fibre glass (b) polyurathane

(c) silica (d) mineral fibre.

202. The co-efficient of thermal expansion of refractory material should be

(a) low (b) medium

(c) high (d) very high.

203. Ceramic fibre is manufactured by blending and melting

(a) silica and chromite.

(b) alumina and silica.

(c) alumina and magnesia.

(d) silica and magnesia.

204. The heat stored in a ceramic fibre lining system is generally in the range of.................. Kcal/m^2.

(a) 270-400 (b) 2700-4000

(c) 27000-40000 (c) 27-40

205. A refractory property 'PCE' stand for

(a) preheated cone equivalant.

(b) pyrometric cone equivalant.

(c) pressurise cone equilibrium.

(d) pressurised cone equivalant.

206. Out of the following, cold crushing strength is highest for .... bricks.

(a) fireclay (b) silica

(b) high alumina (d) dolomite.

207. Which of the following is an organic insulating material?

(a) Mineral wool

(b) Expanded polystyrene

(c) Calcium silicate

(d) Ceramic fibre.

Answers

199. (a) 200. (b) 201. (a) 202. (b) 203. (b) 204. (b) 205. (b) 206. (b) 207. (b)

Appendix D

Technical Data on Refractories

Table 1. Composition and Properties of Some Typical Commonly Used Refractory Materials

Parameters	Units	Ordinary Firebricks	Aluminous Firebricks	Semi-silica	Silica	Stabilised Dolomite	Magnesite	Chrome-Magnesite	Silicon Carbide	Sillimanite	Ceramic Fibre
		(1)	(2)	(3)	(4)	(5)	(6)	(7)	(8)	(9)	(10)
Composition (%)											
Al_2O_3	%	36.5	43.0	12.0	1.0	2.0	2.0	10.5	1.5	61.5	43-47
SiO_2	%	57.5	51.0	85.0	96.0	14.5	3.0	4.5	4.5	34.0	50-54
Cr_2O_3	%	—	—	—	—	—	—	30.5	—	—	—
Fe_2O_3	%	2.5	2.0	1.0	0.5	4.0	4.5	12.5	1.0	0.5	1-1.5
CaO	%	1.0	0.5	0.5	2.0	39.0	2.5	1.0	—	1.0	0.1-1.0
MgO	%	0.5	0.5	0.5	0.2	40.5	88.0	41.0	—	0.5	Traces
TiO_2	%	1.0	2.0	0.5	0.1	—	—	—	—	2.0	1.5-3.5
Alkalies	%	1.0	1.0	0.5	0.2	—	—	—	—	0.5	0.5-2.0
B_2O_3	%	—	—	—	—	—	—		—	—	0.5-1.0
SiC	%	—	—	—	—	—	—	—	93.0	—	—
Properties:											
Bulk density	kg/m^3	1920-2080	1920-2080	1840-2000	1680-1840	2480-2640	2640-2880	2800-3360	2400-2640	2000-2160	60-130
Cold crushing strength (CCS)	kgf/cm^2	170-380	170-380	170-380	140-380	410-550	340-550	210-340	> 550	340-480	—
Refractoriness	Segar cone no.	28-31	32-34	29-31	31-33	38	38	38	38	35-36	Max. temp., 1260°C
Refractoriness	°C	1630-1700	1710-1760	1650-1690	1690-1730	1850	1850	1850	1850	1770-1790	
Refractoriness under load (RUL), at 2 kg/cm^2	°C	1440-1520	1550-1620	1550-1600	1660-1700	1550-1650	1500-1630	1500-1700	1700	1700	—

Thermal conductivity at 1000°C	$W.m^{-1}.°C^{-1}$	1.1	1.2	1.4	1.7	2.2	3.6	1.6	11.7	1.6	0.07 -0.22
		1	2	3	4	5	6	7	8	9	10
Maximum temperature of use	°C	1300-1450	—	1300-1400	1550-1700	—	1700-1800	1700-1750	1500-1600	1600-1800	1260
Resistance to basic slag		low		low	alkaline vapors low		very high	very high	very low	high especially to alkaline vapors	
Resistance to acid slag	—	moderate	—	moderate	high	—	low	moderate	low	high	—
Resistance to molten metal		high upto 1300°C		high upto 1300°C	high	—	high in absence of carbides and oxides	high to Fe, Cu, Sn, Pb, Bi, Sb	low to vapors, high to Zn, Cu	high	
Liability to spall	—	small or nil	—	small	high<600°C low >600°C	—	high	small	very low	very small	—
Resistance to oxidising atmosphere		very high		very high	very high		very high	very high	decomposes at 1100°C but protective film forms, can be used upto 1500°C	very high	
Resistance to reducing atmosphere	—	high upto 1350°C	—	high upto 1400°C	high upto 1700°C	—	high	moderate	high	high	—

Table 2. Chemical composition (%) of common refractory bricks

Refractory Bricks of the Alumina-Silica Type				
Types of brick	**Alumina (Al_2O_3)**	**Silica (SiO_2)**	**Alkalies (Na_2O, K_2O, Li_2O)**	**Others (TiO_2, Fe_2O_3, CaO, MgO)**
Fireclay Bricks				
Superduty	41.9	53.2	1.2	3.7
High-duty	41.0	51.6	2.0	5.3
Low-duty	30.0	62.0	3.7	4.3
High-Alumina Bricks				
60% class	58.0	38.0	0.1	3.9
70% class	69.2	26.2	0.2	4.4
85% class	86.5	8.9	0.1	4.5
90% class	88.5	11.0	0.2	0.3
Corundum class	99.2	0.4	0.2	0.2
Silica Bricks				
Conventional	0.5	95.9	0.1	3.4
Superduty	0.2	96.5	0.04	3.4
Basic Bricks				
Types of brick	Magnesia (MgO)	Chromia (Cr_2O_3)	Alumina (Al_2O_3)	Others (SiO_2, Fe_2O_3)
Magnesite, fired	98.2	–	0.2	1.6
Magnesite-chrome, fired	55.7	4.1	3.3	36.9
Magnesite-chrome, unfired	72.1	8.8	9.7	9.5
Chrome, fired	18.4	30.4	32.1	19.1
Chrome-magnesite, fired	36.1	24.8	22.3	16.8
Chrome-magnesite, unburned	31.5	26.6	23.4	18.5
Special Refractory Bricks				
Types of brick	**Silica (SiO_2)**	**Zirconia (ZrO_2)**	**Carbide (SiC)**	**Others (Al_2O_3, TiO_2, Fe_2O_3)**
Zircon	32.3	66.0	–	1.7
Silicon carbide	5.6	–	90.8%	3.6

Table 3. Typical Percentage Composition of Refractory Materials

Constituent of Brick	Type of Brick								
	Fire Brick	Aluminous Fire Brick	Silicious or Semi silica brick	Silica Brick	Sillimanite Brick	Magnesite Brick	Chrome Brick	Chrome Magnesite Brick	Dolomite Brick (Stabilised)
Silica-SiO_2	55-70	50-55	84-86	94-96	25-35	2-5	3-10	3-10	12-15
Alumuna – Al_2O_3	25-38	38-45	11-13	0.5-1.5	55-65	14	10-25	5-15	2-3
Titania – TiO_2	1-1.5	1-2	0.5-1	0.2-0.5	0.5-1	traces	traces	traces	traces
Iron-Oxide – Fe_3O_2	2-5	24	1-1.5	0.5-1.5	0.5-1.5	2-8	12-20	8-15	24
Chromic Oxide-Cr_2O_3	—	—	—	—	—	—	3545	20-35	—
Magnesia – MgO	0.2-0.8	0.2-0.6	0.2-0.5	0.1-0.5	0.5-0.8	84-92	15-35	40-60	3842
Lime – CaO	0.5-1	0.4-0.8	0.2-0.5	2-2.5	0.5-1	24	2-3	1-3	3842
Alkalies – K_2O + Na_2O	1-2	0.5-1	0.5-1	0.2-1	0.5-1	0.1-0.5	traces	traces	traces

References/ Bibliography/Further Reading

1. J.D. Gilchrist, 'Fuels, Furnaces and Refractories', Pergamon Press, London
2. J.H.Chesters, 'Refractories – Production & Properties', Iron and Steel Institute, London
3. F.H.Norton, 'Refractories', Mcgraw Hill Book Co. New York
4. A. B . Searle, 'Refractory Materials-Their Manufacture & Uses'
5. K.Shaw, 'Refractories and Their Uses', Applied Science Publisher Limited, London
6. G.R.Bashforth, 'The Manufacture of Iron & Steel (Volume III- Steelworks Fuels, Furnaces. Refractories & Instrumentation)', B.I. Publication, New Delhi
7. U.S. Steel Corporation, USA, 'Making, Shaping and Treating of Steel'
8. W.Trinks and M.H. Mawhinney, 'Industrial Furnaces-volume-I&II', John Wiley, New York
9. Harbinson Walker Refractories Company, Pensylvania, 'Modern Refractory Practice'
10. Relavant Internet Resources